Complete Guide To

PEST CONTROL

— WITH AND WITHOUT CHEMICALS

GEORGE W. WARE
UNIVERSITY OF ARIZONA

THOMSON PUBLICATIONS
P.O. Box 9335
Fresno, CA 93791

Printed in United States of America
Library of Congress Number 80-52306
ISBN 0-913702-09-9

Complete Guide to **PEST CONTROL** —WITH AND WITHOUT CHEMICALS

To my wife and children,
Doris, Cindy, Sam and Julie

DISCLAIMER CLAUSE

The author has made every attempt to provide up-to-date, scientific information for the novice to solve every variety of home pest control problems. Many sources of information have been gleaned to acquire these suggestions for both chemical and non-chemical controls. The pesticides recommended herein were registered for the uses described at the time of this writing and are known to be safe and effective if used according to the directions on the label. However, due to constantly changing laws and regulations, the author can assume no liability for these suggestions. The user assumes all such responsibility in following these as well as all other pest control suggestions involving pesticides. Those pesticides recommended in this book are registered with the Office of Pesticides Programs of the Environmental Protection Agency (EPA), and are safe for the uses indicated when following the directions for their use printed on the label or printed material accompanying the product. Suggestions for insect control are limited to uses in and around the home and do not apply to food-handling establishments.

Contents

Preface

This book was written to meet the needs of urban and suburban residents. It describes the pest problems that the average family or apartment dwellers might encounter anywhere in these United States, with accompanying safe and simple methods for their control in and around the home. Every effort has been made to include non-chemical control methods, along with chemical controls, when they were known.

Though practically everyone seems to know something about pests around the home and garden and their control, the layman's familiarity is generally fragmentary, and thus his understanding is frequently less than adequate. Ofttimes his concepts are even wrong, and he may pay tremendous fees to professionals for pest control services he could easily do himself. My experiences in talking to garden clubs, civic groups and college seminars convinced me that everyone enjoyed learning even the most intricate details of pest biology and control if the information were presented along old familiar paths. It seemed then that the subject matter on pests and pest control could be prepared in a digestible handbook form for persons from varied backgrounds. This book is the fruit of my efforts to accomplish that.

Its purpose is to present how-to-do-it information to the layman in a simple, understandable way — to give him an appreciation for the very important chemical and non-chemical tools and a knowledge of the biology and ecology of the pests they are intended to control. Discussions of individual pests or groups of pests are not restricted to bare definitions, but rather adequate information is given to convey something of their importance to the home or garden.

This book is a catalog of selected pests and their intelligent manipulation with and without the use of pesticides. The individual pests used for illustrating certain points were not selected to the exclusion of others, but were chosen based on their frequency of appearance as pests in and around homes in North America. Such memorable characteristics as a notorious history, catchy name, or unique biology make them even more interesting. The selection of individual pesticides for control purposes indicates that they will probably do the best job according to the authorities whose reports and research I have used.

While writing this fundamental book on pests and pest control and their important position in today's home life, I have attempted to adapt an everyday, factual, yet unemotional approach to a subject that is more often than not presented highly emotionally charged. Also an effort has been made to avoid the use of technical and scientific terms prior to their introduction and discussion. In trying to avoid the natural tendency of oversimplification, however, it was necessary to compromise between the too-simple and the incomprehensible.

No book of this subject could be complete because of the tremendous amount of material to be covered. It is intended to present a comprehensive picture of major pests and their management, and is not intended as an exhaustive study of every possible pest found in the many regions of this country. It will be of interest, not only to the homeowner, but also to owners of small farms, professional gardeners, grounds keepers, landscape maintenance persons, structural pest control servicemen and to students enrolled in various pest control, ecological, and environmental courses.

A great many ideas and some data are presented in various sections of the book without direct citation of sources. At the end of the book is a listing which acknowledges the sources of information and contributors whose work or writings were used.

During the three years of research and writing of this book I received valuable advice, constructive criticism, and assistance from many people. I want to express my appreciation in particular to Dr. Walter Ebeling of the University of California, Dr. Daniel A. Roberts of the University of Florida, Dr. W. E. Splittstoesser of the University of Illinois, and Dr. A. V. Barker of the University of Massachusetts for reviewing the manuscript and giving it the border-to-border and coast-to-coast application and focus where they were lacking. Special recognition is given to Mrs. Hazel Tinsley who meticulously typed a perfect manuscript from an incredibly difficult draft.

And, of course, to Doris, Cindy, Sam and Julie, my wife and children, who now know that butterflies come from caterpillars and writing isn't all fun, I owe a great debt of appreciation for permitting me the time to see this dream materialized.

Tucson, Arizona
June, 1980

George W. Ware

CHAPTER 1

. . . there came great swarms of flies into the house of Pharaoh and into his servants' houses, and in all the land of Egypt the land was ruined by reason of the flies.

Exodus, 8:24

THE PESTS – An Overview

WHAT ARE PESTS?

Pests are any unwanted plants, animals or microorganisms. Every homeowner and apartment dweller has pest problems. In fact, we spend more than $1 billion annually for professional pest control services in our homes and apartments. These pests may include, among other things, filthy, annoying and disease-transmitting flies, mosquitoes, and cockroaches; moths that eat woolens; beetles that feed on leather goods and infest package foods; slugs, snails, aphids, mites, beetles, caterpillars and bugs feeding on his lawn, garden, trees and ornamentals; termites that nibble away at his wooden buildings, books, and other cellulose products; diseases that mar and destroy his fruits, vegetables and plants; algae growing on the walls or greening the water of his swimming pool; slimes and mildews that grow on shower curtains and stalls and under the rims of sinks; rats and mice that leave their fecal pellets scattered around in exchange for the food they eat; dogs that designate their territories by urinating on automobile wheels, shrubs and favorite flowers, or defecate on lawns and driveways; alley cats that yowl and screech at night or catch song birds; and annoying birds that leave their feces on window ledges, sidewalks and statues of yesterday's forgotten heroes, or flock noisily together in the autumn by the thousands, leaving their tree roosts literally white-washed with their feces.

To the commercial grower pests could include insects and mites that damage crops; weeds that compete with field crops for nutrients and moisture; aquatic plants that clog irrigation and drainage ditches; diseases of plants caused by fungi, bacteria and viruses; nematodes, snails and slugs; rodents that feed on grain, young plants, and the bark of fruit trees; and birds of every imaginable species that eat their weight every day in young plant seedlings and grain from fields and feedlots as well as from storage.

IMPORTANCE OF PESTS AND THEIR DAMAGE

That sum of $1 billion plus spent every year for pest control services from professional operators does not include an equal amount of do-it-yourself applications, the subject of this book. The homeowner no longer must tolerate carpet beetles in his wall-to-wall floor covering, moths in his closets, or cockroaches in his kitchen. He now considers the presence of these pests in his castle a matter of social indignation as well as a health hazard and general nuisance.

It is estimated that in the United States alone, insects, weeds, plant diseases and nematodes account for losses up to $15 billion annually. This is for the agriculture sector. There is no way of estimating how much more than the $1 billion urban pest control costs the homeowner, for in addition to his own efforts and expenditures, many tax dollars are involved in municipal, county, and state pest control projects for flies, mosquitoes, cockroaches, and rats. It is because of the economic implications of such losses and savings that pests and their control have assumed their importance, both to the homeowner and the apartment dweller, whose only claim to a green thumb may be a window-box 32 floors above the concrete jungle.

Plants are the world's main source of food. They compete with about 80,000 to 100,000 plant diseases caused by viruses, bacteria, microplasma-like organisms, rickettsias, fungi, algae, and parasitic seed plants; 30,000 species of weeds the world over, with approximately 1,800 species causing serious economic losses; 3,000 species of nematodes that attack crop plants with more than 1,000 that cause damage; and over 800,000 species of insects of which 10,000 species add to the devastating loss of crops throughout the world.

An astounding one-third of the world's food crops are destroyed by these pests during growth, harvesting and storage. Losses are even higher in emerging countries: Latin America loses to pests approximately 40% of everything produced. The Food and Agriculture Organization (FAO) has estimated that 50% of cotton production in developing countries would be destroyed without proper insect control. The home gardener will lose approximately 25% of his products in the same fashion if pests are left uncontrolled.

Here are a few good examples of specific increases in yields resulting from the chemical control of insects in the U.S.: cotton, 100%; corn, 25%; potatoes, 35%; onions, 140%; tobacco, 125%; beet seed, 180%; alfalfa seed, 160%; and milk production, 15%.

Equally important are the agricultural losses from weeds. They deprive crop plants of moisture and nutritive substances in the soil. They shade crop plants and hinder their normal growth. They contaminate harvested grain with seeds that may be poisonous to man and animals. In some instances, complete loss of the crop results from disastrous competitive effects of weeds. And for the home lawn, if not properly managed, weeds can become the dominant species instead of the golf-green dream of its owner.

PESTS IN HISTORY

American history contains innumerable influences resulting from the mass destruction of crops by diseases and insects. In 1845-1851 the potato famine in Ireland occurred as a result of a massive infection of potatoes by a fungus, *Phytophthora infestans*, now commonly referred to as late blight. (Maneb or anilazine would control that handily today, with two or three applications.) This resulted in the loss of about one million lives and the cultural invasion of America by Irish refugees. The sad epilog, however, is that the infected potatoes were edible and nutritious, but a superstitious population refused to use diseased tubers. Even now, late blight still causes the annual loss of over 22 million tons of potatoes worldwide. In 1930, 30% of the U.S. wheat crop was lost to stem rust, the same disease that destroyed three million tons of wheat in Western Canada in 1954.

Aside from the historical plant epidemics, let's examine some of our more recent problems, animal and human disease, whose causal organisms are carried by insects. In 1971 Venezuelan equine encephalitis appeared in southern Texas, moving in from Mexico. However, through a very concerted suppression effort, involving horse vaccination, a quarantine on horse movement, and extensive mosquito control measures, the reported cases were limited to 88 humans and 192 horses. With the other arthropod-borne encephalitides, there were an average of 205 human cases in the U.S. annually from 1964 through 1973.

As late as 1955, malaria infected more than 100 million persons throughout the world. The annual death rate from this debilitating disease has been reduced from 6 million in 1939 to 2.5 million today. Similar progress has been made in controlling other important tropical diseases such as yellow fever, sleeping sickness, and Chagas' disease through the use of insecticides.

The Panama Canal was abandoned in the 19th century by the French because more than 30,000 — think of it! — 30,000, of their laborers died from yellow fever and other diseases. Since the first recorded epidemic of the Black Death or Bubonic Plague, it is estimated that more than 65 million persons have died from this disease transmitted by the rat flea. The number of deaths resulting from all wars appears paltry beside the morbid toll taken by insect-borne diseases.

There is currently the ever-lurking danger to man from such diseases as plague, encephalitis, typhus, relapsing fever, sleeping sickness, elephantiasis, and many others, which are transmitted by insects or mites.

Controlling our pests is neither a luxury nor mark of affluence — it is essential.

CHAPTER 2

You pays your money,
and you takes your choice.

Caption to cartoon by John Leech
in *PUNCH*, Jan. 3, 1846.

THE CONTROLS – A Choice

A CHOICE

When early man moved out of his tree into a cave, and later into a shelter of his own creation, he brought with him old and unwanted guests and encountered new ones. And today mankind has a magnificent variety of these pesky insect pests with which to compete. Even the Bible mentions ants, fleas, and moths as early pests of man. James wrote, "Your riches have rotted and your garments are moth-eaten". Similarly, Matthew recorded for us, "Do not lay up for yourselves treasures on earth, where moth and rust consume . . ." (It is highly probable that even then they practiced a form of moth control for their woolens found in the chapter on household insects, by exposing their garments to the noonday sun.)

Man hasn't taken all these pest problems sitting down. Undoubtedly in his early encounters with pests he learned that the easiest method of dealing with his competitors was to avoid them, live and move when and where the pests were not, by carefully selecting his habitat. When he was compelled to deal with the pests, regardless of the habitat, he may have covered himself with dust or mud, and eventually some forms of clothing, as barriers. Other methods included removing them by hand or killing them outright, forms of physical and mechanical control.

These were non-chemical methods of control. Man had only the sum of his skills to fight pests.

Chemical controls were slow in coming, and often as not, accidental more than intentional, and were based more on superstition than fact. The Greeks burned brimstone (sulfur) as an insecticide or purifier. The Bible refers to the use of ashes and salt as herbicides. That's about the way it remained until after the Civil War in the United States.

Historically most chemical control efforts were aimed at insects, because these were the most visible and despised. Pliny the Elder (A.D. 23-79) recorded most of the early insecticide uses in his Natural History, collected largely from the folklore and Greek writings of the previous two or three centuries. Among these were the use of gall from green lizards to protect apples from worms and rot. Since then a variety of materials have been used with dubious results: extracts of pepper, whitewash, vinegar, turpentine, fish oil, brine, lye and many others.

And really, nothing much happened in the way of chemical control, at least in this country, until 1868, when kerosene emulsions were developed for the control of various scale insects on citrus. Much of the rest you may know.

Today, unlike our ancestors, we have a choice of chemical or non-chemical control methods. It is the purpose of this book to present the best of both.

NON-CHEMICAL CONTROL — NATURAL ALTERNATIVES

The enthusiasm for natural or organic gardening has grown by leaps and bounds over the last decade. Included in this pattern of gardening is the enrichment of soil with naturally occurring products, including mulches, composts and animal manure, and controlling insects, diseases and weeds with other than chemical methods. These methods stress cultural, mechanical and biological controls along with strict avoidance or limited use of insecticides, fungicides and herbicides, unless they are naturally occurring or derived from natural sources.

Organic gardening obviously requires more personal attention and human energy than the standard integrated techniques, and generally results in greater damage to the produce. For persons with the interest and time, the natural scheme of gardening is highly sensible. After all, our forefathers managed their gardens and field crops in this manner until a little over a generation ago.

A combination of several methods is usually needed for satisfactory control of the many garden pests. Since insect and disease control methods vary in their effectiveness, you may wish to select alternative methods to correspond with differences in plant development and productivity, insect damage, weather conditions and cultural practices.

CULTURAL METHODS

Cultural control alters routine production practices in ways that are detrimental to the biological success of insects, diseases and weeds. For instance, spading or plowing and cultivating the soil kills weeds, buries disease organisms, and exposes soil insects to adverse weather conditions, birds and other predators. Additionally, deep spading or plowing buries some insects, preventing their emergence.

Crop rotation is effective against insects and diseases that are fairly specific in their range of affected plants and especially against insects with short migration ranges. The movement of crops to different locations will isolate these pests from their food source. If additional land is not available for an alternate site, change the sequence of plants grown in the garden. Don't plant members of the same family in the same location in consecutive seasons, such as melons, squash and cucumbers following each other.

Intelligent use of fertilizers and water induces normal, healthy plant growth and enhances the plant's capability of tolerating insect and disease damage. Along this same line, zealous use of compost or manure in gardens makes them highly attractive to millipedes, white grubs, and certain mulch-inhabiting beetles, and promotes mildew and some stem blights.

Planting or harvesting times can be changed to reduce disease vulnerability or keep insect pests separated from susceptible stages of the host plant. Corn and bean seed damage by seed maggots can be greatly reduced by delaying planting until the soil is warm enough to result in rapid germination. This same technique reduces root and seedling diseases that attack slow-growing plants. Placing protective coverings over seedlings during early season not only preserves heat, but also protects plants from wind, hail, some insects, and certain cool weather diseases. In some situations a healthy transplant overcomes insect and disease damage more readily than plants developing from seed in the field.

The removal of crop remains and disposal of weeds and other volunteer plants eliminates disease organism sources, and food and shelter for many arthropod pests including cutworms, webworms, white grubs, aphids, millipedes and spider mites. When garden plants stop producing, spade them into the soil to convert them to compost. The exception is when rootknot nematode is present, for this merely increases its distribution in the garden.

INTERPLANTING

Interplanting is an orderly mixing of crop plants aimed at diversifying populations. There are many claims made about the ability of certain plants to protect other species from insect damage. Unfortunately, I have no hard research data to support the claims, but rather many literature sources and testimonials. Quite likely they work to some extent. and several are included as trial combinations:

1. Alternate plants of cucumbers with radishes or nasturtiums to reduce numbers of cucumber beetles.
2. Nasturtium is reported to repel aphids and squash bugs.
3. Interplant basil or borage with tomatoes to discourage tomato hornworms.
4. Mix plantings of beans with marigolds or summer savory to reduce attraction to Mexican bean beetles. Watch out for increased spider mites.
5. Interplant catnip with eggplant, tomato, or potato to repel flea beetles.
6. Interplanting datura or rue with plants to be protected is reported to repel Japanese beetles.
7. The weed, dead nettle, or interplanted beans are purported to prevent Colorado potato beetles from invading potatoes.
8. Chives or garlic interplanted with lettuce or peas may reduce aphids.
9. Flax or horseradish also have reputations for repelling potato beetles.
10. Geraniums planted among roses or grapes are reported to repel Japanese beetles.
11. Garlic planted among various vegetables is reported to repel Japanese beetles, aphids, the vegetable weevil, and spider mites.
12. Marigolds interplanted with curcurbits may help avoid cucumber beetles.

13. Mint, thyme or hyssop planted adjacent to most cole crops may reduce cabbage worms.
14. Potatoes interplanted with beans may reduce Mexican bean beetles.
15. Castor bean is repellent to moles and is said to repel aphids.
16. Pennyroyal is avoided by ants and aphids, when interplanted with susceptible hosts.
17. Rosemary and sage are reported to be repellent against cabbage butterfly, bean beetles and carrot flies.
18. Sassafras as an interplant is supposed to repel aphids, as does stinging nettle, a weed.
19. Tansy as an interplant is reported to ward off Japanese beetles, striped cucumber beetles, squash bugs and ants.

The principle involved here is to avoid planting all plants of one kind together. This is comparable to monoculture in agriculture, large fields planted to one crop. A patch or row of plants in a garden is much more attractive than one or two plants mingled in with other species. The interplanting principle literally causes vulnerable plants to lose their identity in the forest of other garden plants, thus losing their attractiveness to egg-laying by adult insects. It becomes a simple and logical way to avoid infestation, which surpasses controlling the pests after they nibble on the beautiful results of your hard work.

RESISTANT VARIETIES

Conscientious gardeners should make every effort to plant species or varieties which are resistant or tolerant to disease and insect damage. Resistance in plants is likely to be interpreted by the layman as meaning immune to damage. In reality, it distinguishes plant varieties that exhibit less insect or disease damage when compared to other varieties under similar circumstances. Some varieties may not taste as good to the pest, may not support disease organisms or may possess certain physical or chemical properties which repel or discourage insect feeding or egg-laying, or may be able to support insect populations with no appreciable damage or alteration in quality or yield.

When buying seeds or plants, check seed catalogs for information on resistant varieties that grow well in your area. Inquire as well at your County Agent's or seed dealers and nurserymen in your area. (The County Agent can be found by looking in the white pages of the telephone directory for Cooperative Extension Service listed under the local County Government.) Vegetable varieties that grow well and are resistant to certain pests in Ohio may be a flop in California or Montana. Experienced gardening neighbors may be the best information source of all! Resistant varieties are included in the tables of insect and disease control when known.

MECHANICAL AND PHYSICAL METHODS

Mechanical control actually aims activities directly toward diseased tissues, and inseits in various stages of development and is usually more practical for small gardens than for large. The methods can be used in combinations or singly as seems fitting.

Removing by hand or handpicking of diseased stems, leaves and fruit, or insects and insect egg masses insures immediate and positive control. This is especially effective with anthracnose and leaf spots and foliage-consuming insects as potato beetles, hornworms, or bean beetles.

Plant guards or preventive devices are easy to use against insects, although sometimes more ornamental than effective. These include: (1) cheesecloth screens for cold frames and hot beds to prevent insect egg-laying; (2) paper collars, paper cups or tin cans with the end removed to prevent cutworm damage; (3) sticky barriers on the trunks of trees and woody shrubs to prevent damage by crawling insects; (4) mesh covers for small fruit trees and berry bushes to screen out larger insects and birds; and (5) aluminum foil on the soil beneath plants to repel aphids and leafhoppers, referred to as aluminum foil mulch.

A stiff stream of water removes insects without injuring the plants or drenching the soil. Simple water sprays from a hose or pressure sprayer, sometimes with a small amount of soap or detergent added, will dislodge certain pests as aphids, mealybugs and spider mites.

TRAPS

Traps of various types have always been popular because of the visible results. Examples: (1) slugs and pillbugs can be trapped under boards on the ground; (2) earwigs are trapped in rolled up newspaper placed among the plants or other locations where they gather; (3) a 2-quart container half-filled with a 9:1 dilution of water to molasses will collect grasshoppers, moths and certain beetles, and (4) the famous beer pan trap for slugs and snails is made by placing a small can or pan flush with the soil and half-filling with beer to attract and drown them. Generally speaking, simple types are the easiest to set up and the most effective. Commercial traps are seldom better than those made at home.

Light traps, particularly blacklight or bluelight traps (special bulbs that emit a higher proportion of ultraviolet light that is highly attractive to nocturnal insects) are a good insect monitoring tool but provide little or no protection for the garden. True, they usually capture a tremendous number of insects, however, a close examination of light trap collections shows that they attract both beneficial and harmful insects that would ordinarily not be found in that area. Those insects attracted but not captured remain in the area, and the destructive ones may cause damage later. Also some species that are wingless and those active only during the day (diurnal, as opposed to nocturnal) are not caught in these traps. Consequently, the value of blacklight or simple light traps in protecting the home garden is generally of no benefit and in some instances detrimental.

BIOLOGICAL METHODS

Biological control is the use of parasites, predators or disease pathogens (bacteria, fungi, viruses, and others) to suppress pest populations to low enough levels to avoid economic losses. There are 3 categories of biological control: (1) introduction of natural enemies which are not native to the area, and which will have to establish and perpetuate themselves; (2) expanding existing populations of natural enemies by collecting, rearing and releasing additional parasites or predators, (inundative releases); and (3) conserving resident beneficial insects by the judicious use of insecticides and the maintenance of alternate host insects so that the beneficials can continue to reproduce and be available when needed. Figure 1 shows some of the beneficial insects found in the home garden and orchard.

The introduction of a parasite or predator does not guarantee its success as a biological control. However, certain conditions can indicate the potential value of such a natural enemy. The effectiveness of a predator or parasite is usually dependent on (1) its ability to find the host when host populations are small, (2) its ability to survive under all host-inhabited conditions, (3) its ability to utilize alternate hosts when the primary host supply runs short, (4) a high reproductive capacity and short life cycle (high biological potential), and (5) close synchrony of its life cycle with the host so the desirable host stage is available when needed by its natural enemy.

The garden and immediate surroundings are alive with many beneficial organisms that are there naturally; however, they may not be numerous enough to control a pest before its damage is done. Actually, parasites and predators are most effective when pest populations have stabilized or are relatively low. Their influence on increasing pest populations is usually minimal since any increase in parasite and predator numbers depends on an even greater increase in pest numbers. Disease pathogens, however, seem to be most effective when pest populations are large. Thus the nature of the host-enemy relationship makes it impossible to have an insect-free garden and simultaneously retain sizable populations of beneficials.

Several biological control agents are available to natural gardeners. There is a list of sources in the Appendix or your County Extension Agent or Extension Entomologist can supply a list of sources.

Predators

Praying mantids are usually sold in the form of egg cases and are readily available. Mantids are hungry hunters, are cannibalistic and will eat their brothers and sisters immediately after hatching; thus, few survive. They really don't live up to their reputations because they usually wait for prey rather than search them out. Naturally this determines the kinds of insects it captures. They prefer grasshoppers, crickets, wasps, bees and flies, none of which are particularly important garden pests.

Lady beetles or ladybird beetles are also readily available commercially since they can be collected by the millions in the California mountains during their summer hibernation. Adult lady beetles and larvae prefer aphids, though they will feed readily on mealy bugs, spider mites and certain other soft-bodied pests and eggs. They do not feed on caterpillars, grubs and other beetles. If an adequate food supply of aphids or other hosts is not available at the release point, lady beetles will move out until they find sufficient food. Because of their physiological conditions under which they were originally collected, they may leave the area regardless of the food supply, to complete their hibernation or aestivation cycle. So, don't be surprised if today's release is absent from roll call tomorrow.

Green lacewings (Chrysopa) are usually sold as eggs, since the adults are difficult to manage. The larvae, commonly referred to as aphid lions, feed on several garden pests including aphids, spider mites, thrips, leafhopper nymphs, moth eggs and small larvae. Adults feed on pollen, nectar and honeydew secreted by aphids. Introduced lacewings must also have an available food supply or they too will leave.

There are several other beneficial insects that are resident helpers and are not available commercially. With the ecological mistakes of man now becoming more apparent, it is reassuring to know that nature can establish certain controls which prevent some destructive insects from overpopulating the environment. The attentive gardener can encourage and prepare the conditions for an increase in insect predators by not using insecticides except where and when absolutely necessary.

Several true bugs occur naturally, are seen around the yard and garden, and are among the predaceous beneficial insects for they feed on insect pests. Bugs stab their hosts with piercing-sucking mouthparts and inject digestive enzymes then remove the liquid contents. Included in the true bug predators are assassin bugs, damsel bugs, spined soldier bugs, and the smaller bigeyed bugs and minute pirate bugs. Together they prey on caterpillars, lygus bugs, fleahoppers, spider mites, aphids, thrips, and leafhoppers. When food becomes scarce they may occasionally become cannabalistic and attack their own species. As adults the bugs are winged and fly readily. The nymphs, or immature stages, are wingless and cannot fly, though they resemble the adults in all other characteristics.

Common wasps haunt your trees and gardens looking for food. They may be the common paper wasp, hornet, yellow jacket, mud dauber, or wasps so tiny that they can be seen only after very close observation. Wasps vary widely in size, color, and general body structure. Some are parasites while others are predators. Tiny parasitic wasps lay eggs in bodies of insects and their developing offspring devour and kill their hosts from within. Large predaceous wasps — well known to every gardener — sting caterpillars to paralyze them, and feed them to their young. Mud daubers commonly seek out spiders to stock their nests for feeding of larvae. The author has found as many as 10 black widow spiders, male and female, in one larval cell of a mud dauber.

Ground beetles include hundreds of species and exhibit differences in size, shape and color. Most ground beetles are somewhat flattened, dark, shiny, and have visible mandibles or jaws with which they grasp their prey. They may be found under stones, logs, bark, debris or running about on the ground. Most of them hide by day and feed at night. Nearly all are predaceous on other insects and many are beneficial. There are several that feed primarily on snails.

Dragonflies are those large, biplane-like insects that soar and dart about near and over ponds and streams. They depend entirely on other insects for their food, both catching and eating their prey in flight. Mosquitoes and flies make up a large portion of their diet.

Syrphid flies are commonly called flower flies. They may be brightly colored and several resemble wasps and bees hovering over flowers. They do not sting because they are true flies. The larvae are the active predators, resemble maggots, and feed on aphids or the young of termites, ants, or bees. Most commonly they can be found in dense colonies of aphids.

Antlions, or doodlebugs, are larvae of large, clear-winged predatory insects, and are strange-looking creatures with long sickle-shaped mouthparts. They are really first-cousins to the lacewings or aphid lions mentioned earlier. Antlions are more commonly found in the South and Southwest, but there are a few species found throughout most of the U.S. The larvae hide in cone-shaped burrows in the ground waiting for an ant or other unlucky insect to stumble into its lair. Once inside, the insect is quickly dispatched.

Lightning bugs occur at evening during the early summer and late spring, and are conspicuous by their blinking yellow light. Most of the larvae are luminescent and are given the name, glow-worms. The larvae feed on various smaller insects and on snails.

Spiders make their contribution to insect control not only in the garden but on porches, roof overhangs, in trees and shrubs, and inside the home. Basically there are two types of spiders, the web-spinning varieties that capture flying insects, and the crab spiders that do not make webs but rather lie in wait for their unsuspecting prey. Spiders, unlike previously mentioned beneficial insects, are not very selective and will capture and feed on any insect within their size range, including the beneficial species.

Parasites

Trichogramma wasps are also available commercially "in season". These tiny wasps deposit their eggs in the eggs of other insects, namely moths, including armyworms, cutworms, fruitworms and many others found in orchards and gardens. These egg-parasitic wasps should be released when the moths are laying eggs, but a sequence of releases throughout the season is preferable to a single large release. The results of these releases will depend on timing, selection of *Trichogramma* species, and placement of wasps near host egg masses. A list of caterpillars whose eggs are parasitized by *Trichogramma* wasps is presented in Table 1.

Braconid wasps and other tiny wasps are seldom seen except as they form their little white cocoons on the outside of their hosts after killing it. There are many species of parasitic wasps most of which are quite small. Like the related Ichneumen wasps, they feed on the inner body fluids of their hosts. The most common ones are parasitic on sphinx moth caterpillars like the tobacco and tomato hornworms.

Parasitic flies are another important group of insects that help hold down populations of caterpillars. These are known as tachinid flies, are about the size of house flies, but are darker and covered with heavy bristles. They lay their eggs on, in or near caterpillars, and in a day or two the eggs hatch. The parasitic fly larva lives inside and feeds on the caterpillar until it pupates. Usually at this stage the caterpillar will die, though it continues to feed and appears normal while the parasite is developing inside.

Disease Organisms — Microbial Insecticides

Microbial insecticides obtain their name from microbes, or microorganisms, which are used by man to control certain insects. Insects, like mammals, also have diseases caused by fungi, bacteria, and viruses. In several instances, these have been isolated, cultured, and mass produced for use as pesticides.

The insect-disease-causing microorganisms do not harm other animals or plants. The reverse of this is also true. Vertebrate disease microorganisms do not harm insects. This method of insect control is ideal in that the diseases are usually rather specific. Undoubtedly the future holds many such materials in the arsenal of insecticides, since several new insect pathogens are identified each year. However, at the present only a few are produced commercially and approved by the Environmental Protection Agency for use on food and feed crops. There is still some

TABLE 1. Some of the important caterpillars whose eggs are parasitized by *Trichogramma* sps. wasps.

There are several species of the tiny, parasitic trichogramma wasps, all of which lay their eggs in the eggs of other insects. There are more than 200 known caterpillar pests whose moth eggs are attacked by trichogramma.

alfalfa caterpillar	tobacco budworm	luna moth
armyworm	inchworm	polyhemus moth
bagworm	(measuring worm)	promethea moth
European corn borer	(span worm)	regal moth
pink bollworm	bean leaf roller	rosy maple moth
peachtree borer	cotton leafworm	royal moth
squash vine borer	cabbage looper	tussock moth
imported cabbageworm	Angoumis grain moth	wax moth
fall cankerworm	browntail moth	nymphalid butterfly
eastern tent caterpillar	pecan nutcase bearer	prominents
forest tent caterpillar	oriental fruit moth	datanas
cutworm	carpenterworm	plum moth
corn earworm	codling moth	giant silkworm
(tomato fruitworm)	dagger moth	skippers
(cotton bollworm)	gypsy moth	swallowtail butterfly
grassworm	hawk moth	fall webworm
tobacco hornworm	hummingbird moth	California oakworm
tomato hornworm	io moth	overflow worm

FIGURE 1. Some of the beneficial insect predators found in the home garden and orchard.

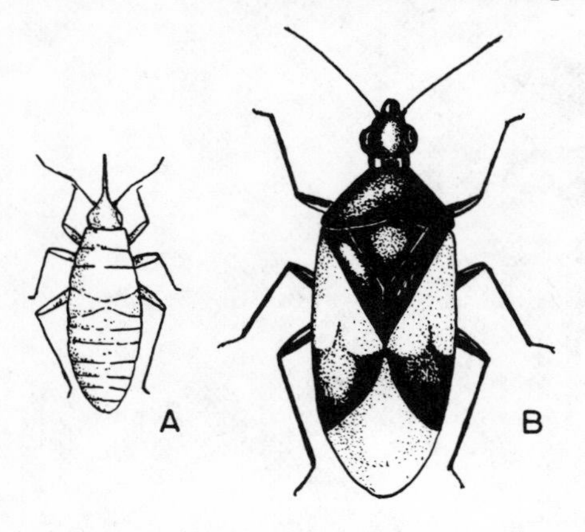

Minute Pirate Bug: *A*. Nymph,
B. Adult
(Univ. of Arizona)

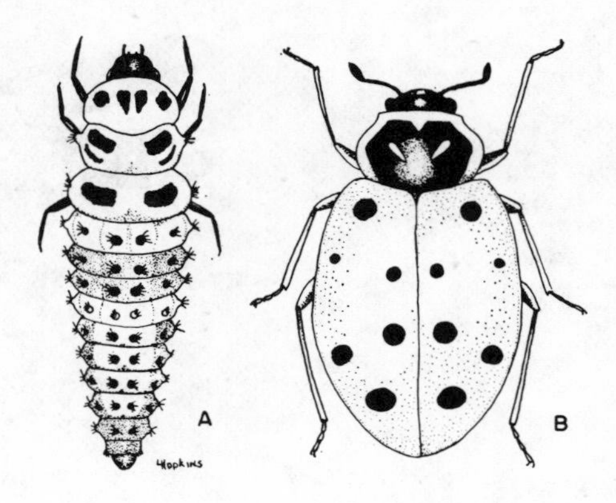

Convergent Lady Beetle.
A. Larva, *B*. Adult
(Univ. of Arizona)

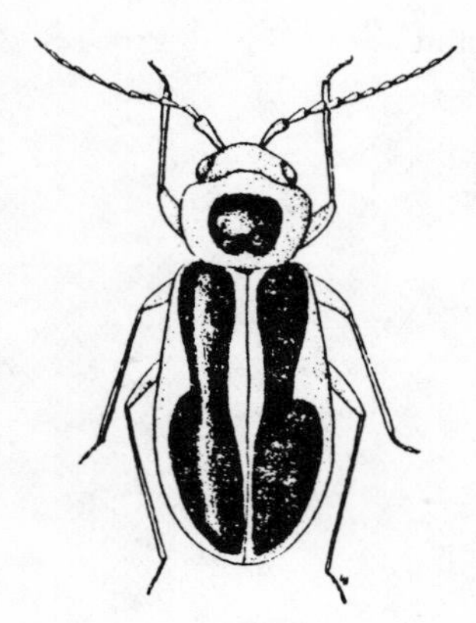

Striped Collops Beetle
(Univ. of Arizona)

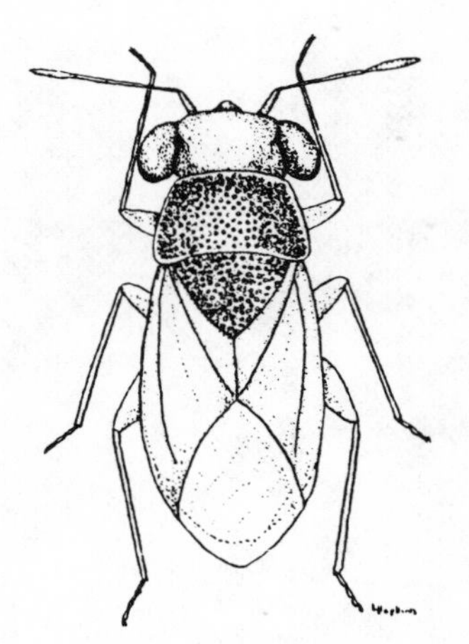

Big-eyed Bug
(Univ. of Arizona)

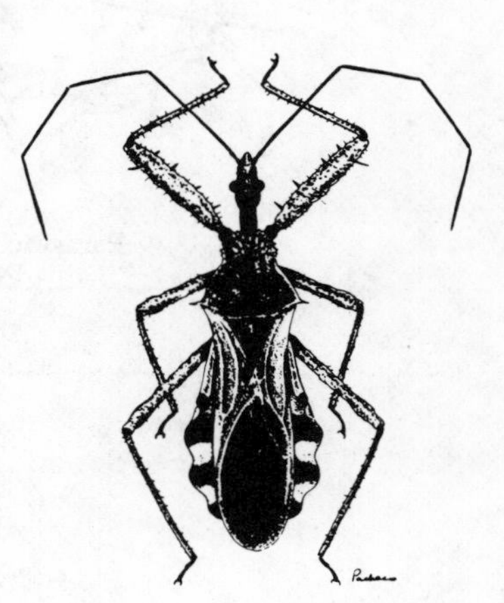

Spined Soldier Bug
(Univ. of Arizona)

Lacewing Adult, Eggs and Larva
(Univ. of Arizona)

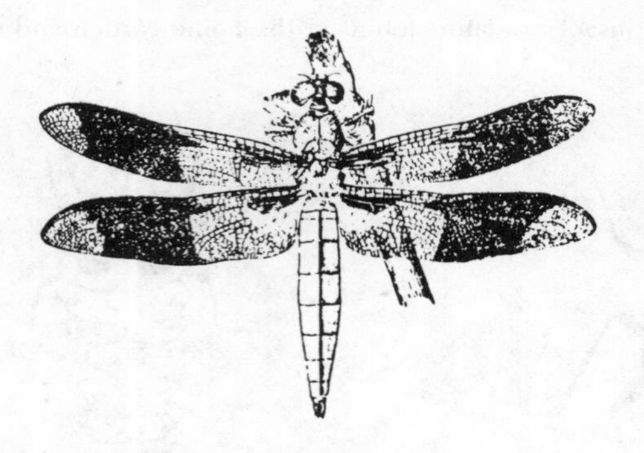

Dragonfly
(U.S.D.A.)

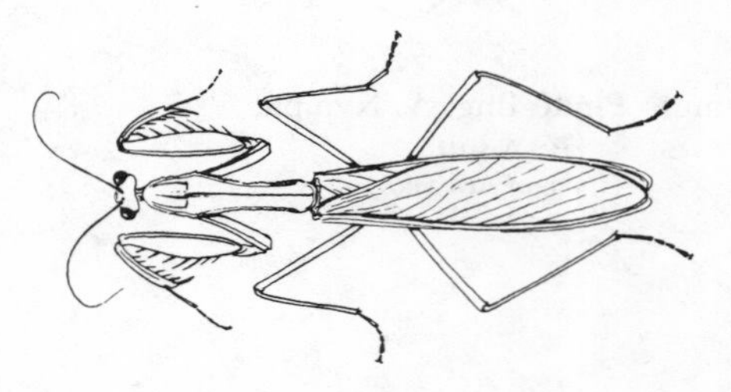

Praying Mantid
(U.S.D.A.)

Parasitic Wasp
(U.S.D.A.)

Trichogramma female stinging a moth egg and placing its own egg inside.
(U.S.D.A.)

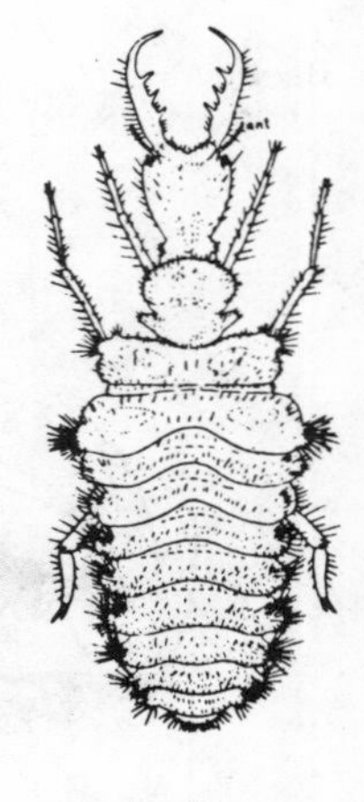

Antlion
(U.S.D.A.)

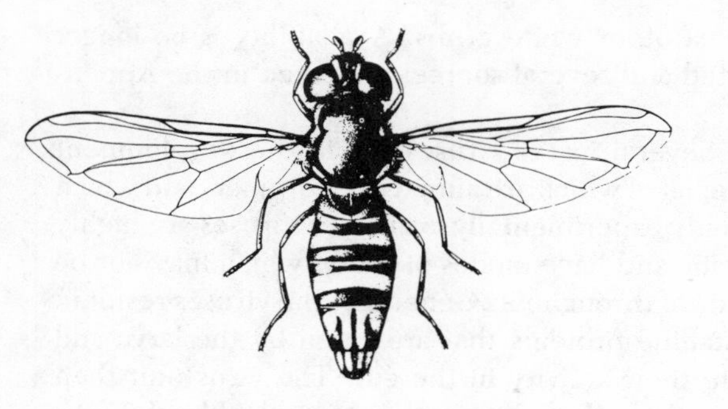

Syrphid or Flower Fly
(U.S.D.A.)

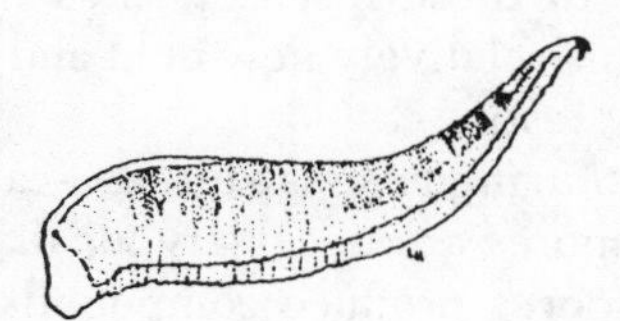

Syrphid Fly Maggot
(Univ. of Arizona)

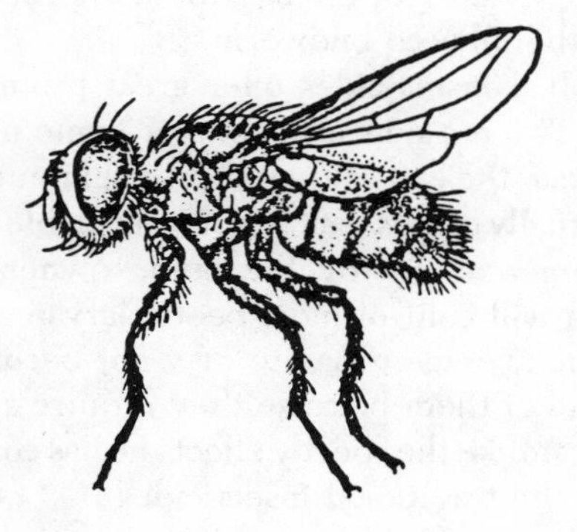

Tachinid Fly
(Univ. of Arizona)

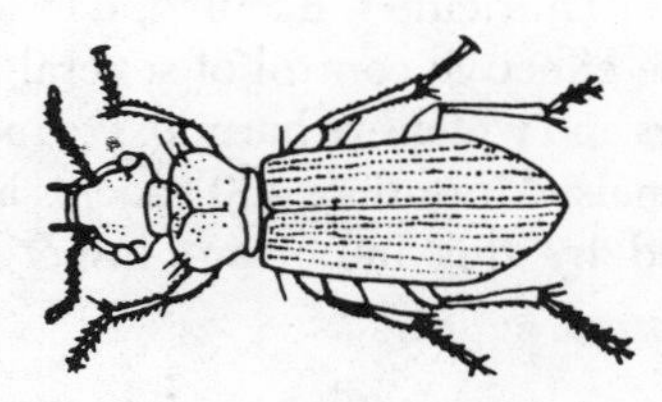

Ground Beetle
(U.S.D.A.)

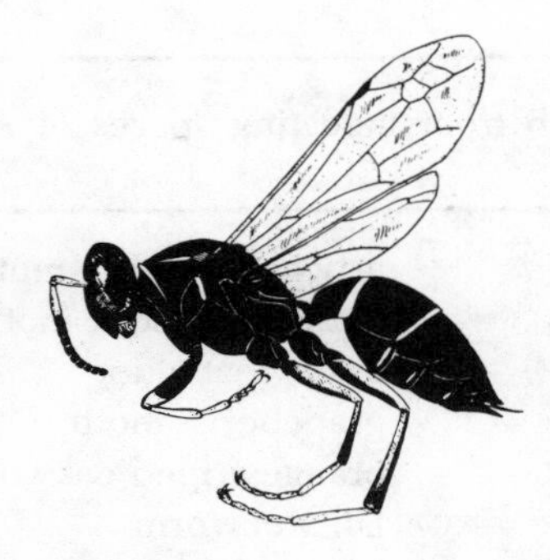

Dark Paper Wasp
(U.S. Public Health Service)

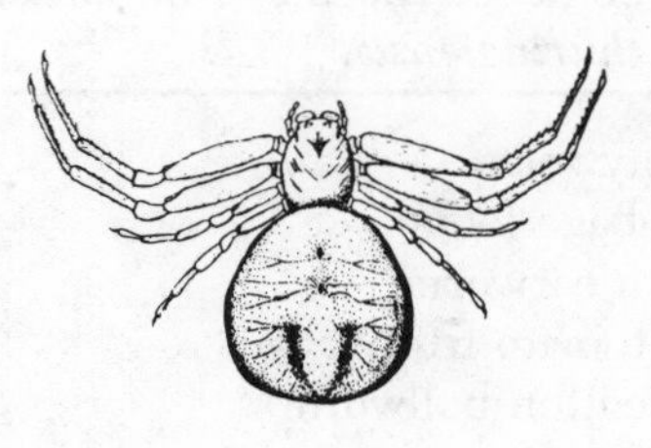

Crab Spider
(Univ. of Arizona)

concern regarding the very remote chance of man's susceptibility to these diseases, thus the slow advances into this relatively new field and exceptional precautionary testing.

Bacillus thuringiensis is a disease-causing bacterium whose spores are necessary for disease induction. These spores produce compounds that injure the gut of insect larvae in a way that invasion of the body cavity follows. The organism produces four substances toxic to insects: (1) a crystalline protein that paralyzes the gut of caterpillars, (2) a toxin that passes through the gut and kills certain fly maggots and pupae, (3) phospholipase C, an enzyme that breaks down the insect cell wall; and (4) another phospholipase that acts on phospholipids. The active agents of *Bacillus thuringiensis* are large protein molecules whose structures are not known.

Several formulations are available to the homeowner, Dipel®, Thuricide®, Bactur®, BT® and others which provide effective control of several leaf-feeding caterpillars and which are harmless to man and his domestic animals. More than 400 species are known to be affected by this important insect pathogen (Table 2).

Bacillus thuringiensis is quite slow in its action. For example, caterpillars that consume some of the spores will stop eating within 2 hours, but may continue to live and move around until they die, which may be as long as 72 hours. When this occurs the untrained gardener may assume the material was ineffective because of the continued pest activity and impatiently apply a chemical pesticide to correct his mistake.

Bacillus popillae, the milky disease bacteria, is closely related to the above and was originally developed to control the larval or grub stage of Japanese beetles; however, it appears to be effective against other white grubs. Availability is no longer limited and several sources are listed in the Appendix.

Several insect viruses are in the development stage, all of which are aimed at caterpillars and which are only experimentally available. Viruses are highly specific and have modes of action which may not be identical throughout. Generally, the viruses result in crystalline proteins that are eaten by the larva and begin their activity in the gut. The virus unit then passes through the gut wall and into the blood. From there, several possibilities occur, but the known aspects are that the units multiply rapidly and take over complete genetic control of the cells, causing their death. None of the viruses are yet available for home use, though one has become available for agricultural use (Elcar®) for the control of the cotton bollworm and the tobacco budworm.

The microbial insecticides offer great promise for the future. Two are now available for home use, while others are in the experimental or developmental stage. Eventually others may become available for lawn and ornamental use by the home owner as mixtures which will control most beetle larvae and caterpillars. The average urbanite may not become particularly fond of them because they require several days to kill, unlike the speedy effects he has come to expect from the traditional insecticides.

Birds

Far more important in preventing insect outbreaks than in controlling them are birds, birds of all kinds. All bird species feed on insects to some extent. The creepers, flycatchers, nuthatches, swallows,

TABLE 2. Some of the more important caterpillar pests that succumb from ingesting spores of *Bacillus thuringiensis*.

armyworm	tobacco hornworm	artichoke plume moth
cabbage looper	European corn borer	western tussock moth
corn earworm	celery leaf tier	grape leaffolder
(tomato fruitworm)	sod webworm	grapeberry moth
(cotton bollworm)	melonworm	orangestriped oakworm
eastern tent caterpillar	redbanded leafroller	fall webworm
saltmarsh caterpillar	imported cabbageworm	tomato hornworm
walnut caterpillar	gypsy moth	California oakworm
tobacco budworm	brown-tail moth	

vireos, warblers, and woodpeckers are almost entirely insectivorous, while blackbirds, crows, gulls, magpies, robins and even birds of prey, the hawks and owls, commonly feed on insects. To increase bird numbers near gardens it is necessary to encourage those species which feed largely on insects. If all species are encouraged, including those that damage gardens, such as blackbirds, robins, and starlings, trouble will develop. Insect-feeding birds can best be encouraged by providing cover, supplementary feed, and prevention of predation from cats.

Many insectivorous birds can be attracted to the home area by planting ornamentals which provide suitable bird cover and food. Especially valuable are the following trees and shrubs: apple, bittersweet, cherry, cotoneaster, crabapple, dogwood, elderberry, firethorn, gray birch, hawthorn, highbush cranberry, holly, male mulberry, mountain ash, plum, red cedar, Russian olive, sumac (aromatic and staghorn), tupelo, white oak, and wild plum or cherry. Additionally, sunflowers are especially attractive to insect-feeding birds.

CHEMICAL CONTROL — PESTICIDES

Frequently non-chemical methods of pest control will fail to do the job. At these times, after everything has been tried, the use of chemicals may be the only alternative. These chemicals are classed under the broad heading of *pesticides*. (A pesticide is anything which kills pests, thus *Bacillus thuringiensis*, found under Biological Control on p. 12, is a pesticide. Because it occurs naturally or is not synthesized in a chemical factory does not remove it from the pesticide classification.)

Because this book is about pests and their control, with and without chemicals, we must now discuss several aspects of the chemical tools. Pesticides, those "super chemicals" used to control pests around the home as well as in agriculture, have a prominent place in the day-to-day activities of our technologically-advanced society. Pesticides can't be ignored, and they won't go away. Like it or not, we find ourselves well into the last quarter of the 20th century, just beginning what historians will refer to as "The Chemical Age". Time, population density and technology have isolated us from the "good old days" (whenever they were), and we could no more return to the non-chemical, back-to-nature way of life, than we could park our automobiles and walk where we would normally drive.

Because pesticides are now a way of life, and are essential ingredients in our affluent lives both here and abroad, it becomes necessary for every educated person, every conscientious citizen, to know something about these valuable chemicals. Practically speaking, there are two classes of pesticides: Those naturally occurring minerals, chemical compounds and plant parts or plant extracts that we use as pesticides, and those man-made chemicals which are designed and used only as pesticides, synthetic pesticides.

Synthetic Pesticides

Synthetic pesticides, by their simplest definition, are those made or synthesized by man in chemical laboratories or factories. Examples of these include the insecticides such as malathion and methoxychlor, herbicides such as 2,4-D and dalapon, or the fungicides maneb and zineb. The first synthetic pesticide was probably kerosene emulsion sprayed on citrus to control scale insects in California in 1868. Several more complex synthetic pesticides were discovered in the 1920's and 1930's, some of which are still in use. The real surge of development, however, came with World War II, beginning with the discovery of DDT.

CHAPTER 3

The proposed decoctions and washes we are well satisfied, in the majority of instances, are as useless in application as they are ridiculous in composition . . .

Editorial, *Practical Entomologist*, Oct. 30, 1865.

THE PESTICIDES

Most pesticides are synthetic, though a few are produced naturally by plants. The U.S. Environmental Protection Agency (EPA) had more than 1200 pesticides registered in 1980. Of these 275 were herbicides, 400 were insecticides, 200 were fungicides and nematicides, 100 were rodenticides, and 225 were disinfectants. These are sold in the form of 30,000 products or formulations, and 5 pounds of pesticides are used each year to feed, clothe, and protect every man, woman, and child in the United States alone. Part of these 5 pounds are used at home by the homeowner as do-it-yourself pest control, and that is what part of this book is about.

Pesticides have become extremely beneficial tools to the urbanite, the home gardener. He depends on pesticides, perhaps more than he realizes: for algae control in the swimming pool, weed control in his lawn, flea collars and powders for pets, sprays for controlling a myriad of garden and lawn insects and diseases, household sprays for ants and roaches, aerosols for flies and mosquitoes, soil and wood treatment for termite protection by professional exterminators, baits for the control of mice and rats, woolen treatment at the dry cleaners for clothes moth protection, and repellents to keep off biting flies, chiggers, and mosquitoes when camping or fishing.

These chemical tools are used by the homeowner as intentional additions to his home and garden environment in order to improve environmental quality for himself, his animals, and his plants, giving him the advantage over his pest competitors. Pesticides are used in agriculture to increase the ratio of cost/benefit in favor of the grower and ultimately the consumer of food and fiber products — the public. Pesticides have contributed significantly to the increased productive capacity of the U.S. farmer, each of whom produced food and fiber for 3 persons in 1776, 37 in 1965, and 59 persons in 1980. Where can one find a more successful story of technology?

THE LANGUAGE OF PESTICIDES

To one person the word pesticide may suggest the insecticide malathion. To another it may conjure up the herbicide dalapon, and still another the garden fungicide maneb. All are correct, however only in part, for the uses and effects of these three materials are totally unrelated.

"Pesticide" is an all-inclusive, but nondescript word meaning "killer of pests". The various generic words ending in "-icide" (from the Latin, —cida, to kill) are classes of pesticides, such as fungicides and insecticides. In the table below are listed the various pesticides and other classes of chemical compounds not commonly considered pesticides. These others, however, are included among the pesticides as defined by federal and state laws.

Pesticides are legally classed as "economic poisons" in most state and federal laws and are defined as "any substance used for controlling, preventing, destroying, repelling, or mitigating any pest". Should you ever pursue the subject of pesticides from a legal viewpoint they would be discussed as economic poisons.

The pesticide vocabulary not only includes all of the commonly-used pesticides, but also includes groups of chemicals which do not actually kill pests, as shown in Table 3. However, because they fit rather practically as well as legally into this umbrella word, pesticides, they are included.

TABLE 3. A list of pesticide classes, their use and derivation.

PESTICIDE CLASS	FUNCTION	ROOT-WORD DERIVATION
Insecticide	controls insects	L.[a] insectum (insecre, to cut or divide into segments)
Herbicide	kills weeds	L. herba (an annual plant)
Fungicide	kills fungi	L. mushroom (Gr.[b] spongos)
Nematicide	kills nematodes	L. nematoda (Gr. nema = thread)
Rodenticide	kills rodents	L. rodere (to gnaw)
Bactericide	kills bacteria	L. bacterium (Gr. baktron, a staff)
Acaricide	kills mites	Gr.[b]akari (mite or tick)
Algicide	kills algae	L. alga (seaweed)
Miticide	kills mites	synonymous with Acaricide
Molluscicide	kills snails and slugs (May include oysters, clams, mussels)	L. molluscus (soft- or thin-shelled)
Pediculicide	kills lice (head, body, and crab lice)	L. pedis (a louse)
Avicide	controls or repels birds	L. avis (bird)
Slimicide	controls slimes	Anglo-Saxon slim
Piscicide	controls fish	L. piscis (a fish)
Ovicide	destroys eggs	L. ovum (an egg)
Predicide	kills predators (coyotes, etc.)	L. praeda (prey)
Chemicals classed as pesticides not bearing the -icide suffix:		
Disinfectants	destroy or inactivate harmful microorganisms	
Growth regulators	stimulate or retard plant growth	
Defoliants	remove leaves	
Desiccants	speed drying of plants	
Repellents	repel insects, mites and ticks or pest vertebrates (dogs, birds)	
Attractants	attract insects or pest vertebrates	
Pheromones	attract insects	
Chemosterilants	sterilize insects or pest vertebrates (birds, rodents)	

[a] Latin origin [b] Greek origin

THE NAMING OF PESTICIDES

A passing knowledge of pesticides involves, among other things, learning something of their names or nomenclature. For example, let us look at malathion, a commonly known household insecticide.

Malathion is the common name for the compound. Common names are selected officially by the appropriate professional scientific society and approved by the American National Standards Institute (formerly United States of America Standards Institute) and the International Organization for Standardization. Common names of insecticides are selected by the Entomological Society of America; herbicides by the Weed Science Society of America; and fungicides by the American Phytopathological Society. The proprietary name, Cythion®, trade name or brand name, for the pesticide is given by the manufacturer or by the formulator. It is not uncommon to find several brand- or trademark names given to a particular pesticide in various formulations by their formulators. To illustrate, Cythion is (or was) also known as Malaspray, Malamar, and Zithiol.

Common names are assigned to avoid the confusion resulting from the use of several trade names, as just illustrated.

The long chemical name, diethyl mercaptosuccinate, S-ester with 0,0-dimethyl phosphorodithioate, is the scientific or chemical name. It is usually presented according to the principles of nomenclature used in Chemical Abstracts, a scientific abstracting journal which is generally accepted as the world's standard for chemical names.

PESTICIDE FORMULATIONS

Pesticides are formulated to improve their properties of handling, application, effectiveness, safety, and storage. After a pesticide is manufactured in its relatively pure form, the technical grade material, whether herbicide, insecticide, fungicide or other classification, the next step is formulation. It is processed into a usable form for direct application, or for dilution followed by application. The formulation is the final physical condition in which the pesticide is sold for use. The technical grade material may be formulated by its basic manufacturer or sold to a formulator. The formulated pesticide will be sold under the formulator's brand name or it may be custom-formulated for another firm.

Formulation is the processing of a pesticidal compound by any method which will improve its properties of storage, handling, application, effectiveness, or safety. The term is usually reserved for commercial preparation prior to actual use, and does not include the final dilution in application equipment.

The real test for a pesticide is acceptance by the user. And, to be accepted for use by the home gardener or pest control operator, a pesticide must be effective, safe and easy to apply, but not necessarily economical, especially from the home gardener's viewpoint. The urbanite commonly pays 5 to 25 times the price that a grower may pay for a given weight of a particular pesticide, depending to a great extent on the formulation. For instance, the most expensive formulation of an insecticide is the pressurized aerosol.

Pesticides, then, are formulated into many usable forms for satisfactory storage, for effective application, for safety to the applicator and the environment, for ease of application with readily available equipment, and for economy. This is not always simply accomplished, due to the chemical and physical characteristics of the technical grade pesticide. For example, some materials in their "raw" or technical condition are liquids, others solids; some are stable to air and sunlight, whereas others are not; some are volatile, others not; some are water soluble, some oil soluble, and others for example, the insecticide carbaryl (Sevin®) may be insoluble in either water or oil. These characteristics pose problems to the formulator, since the final formulated product must meet the standards of acceptability by the user.

About 98% of all pesticides used in the United States in 1980 are manufactured in the formulations appearing in the simplified classification presented in Table 4. Familiarity with the more important formulations is essential to the well-informed home gardener. We will now examine the major formulations used for the home, lawn and garden, as well as those employed in structural pest control and agriculture.

SPRAYS

Emulsifiable Concentrates. (EC). Formulation trends shift with time and need. Traditionally, pesticides have been applied as water sprays, water suspensions, oil sprays, dusts, and granules. Spray formulations are prepared for insecticides, herbicides, miticides, fungicides, algicides, growth regulators, disinfectants, repellents, and molluscicides. Consequently, more than 75% of all pesticides are applied as sprays. The bulk of these are currently applied as water emulsions made from emulsifiable concentrates.

Emulsifiable concentrates are concentrated oil solutions of the technical grade material with enough emulsifier added to make the concentrate mix (emulsify) readily with water for spraying. The emulsifier is a detergent-like material that makes possible the suspension of microscopically small oil droplets in water to form an emulsion.

When an emulsifiable concentrate is added to water, the emulsifier causes the oil to disperse immediately and uniformly throughout the water, with agitation, giving it an opaque or milky appearance. This oil-in-water suspension is a normal emulsion. There are a few rare formulations of invert emulsions, which are water-in-oil suspensions, and are opaque in the concentrated forms, resembling salad dressing or face cream. These are employed almost exclusively as herbicide formulations for agri-

TABLE 4. Common Formulations of Pesticides Available for Do-It-Yourself.

1. Sprays (insecticides, herbicides, fungicides)
   a. Emulsifiable concentrates (also emulsible concentrates)
   b. Water miscible liquids
   c. Wettable powders
   d. Water soluble powders, e.g. pre-packaged tank drop-ins
   e. Oil solutions, e.g. house and garden or barn and corral ready-to-use sprays
   f. Soluble pellets for water-hose attachments
   g. Flowable or sprayable suspensions
2. Dusts (insecticides, fungicides)
   a. Undiluted toxic agent
   b. Toxic agent with active diluent, e.g. sulfur
   c. Toxic agents with inert diluent, e.g. home garden insecticide-fungicide combination in pyrophyllite carrier, flea powder
   d. Aerosol "dust", e.g. silica in aerosol form
3. Aerosols (insecticides, disinfectants or "germicides")
   a. Push-button
   b. Total release
4. Granulars (insecticides, herbicides, algicides)
   a. Inert carrier impregnated with pesticide
   b. Soluble granule (pool chlorine)
5. Fumigants (insecticides, nematicides)
   a. Stored products and space treatment, e.g. liquids, gases, and moth crystals
   b. Soil treatment liquids which vaporize
6. Impregnates (insecticides, fungicides, herbicides)
   a. Plastic strips containing a volatile insecticide, e.g. pet collars
   b. Shelf papers, strips, cords containing a volatile or contact insecticide
   c. Mothproofing agents for woolens (insecticides)
   d. Wood preservatives (fungicides, insecticides)
   e. Wax bars (herbicides)
7. Fertilizer combinations with herbicides, insecticides, or fungicides
8. Baits
   a. Insecticides, e.g. ants, roaches, wasps, crickets, grasshoppers, fruit flies
   b. Molluscicides, e.g. slugs, snails
   c. Rodenticides, e.g. mice, rats, gophers
   d. Avicides, e.g. pigeons, starlings, blackbirds
9. Slow-release insecticides
   a. Encapsulated materials for agriculture and mosquito abatement
   b. Paint-on lacquers
   c. Adhesive tapes
10. Insect repellents
    a. Aerosols
    b. Rub-ons (liquids, cloths, and "sticks")
    c. Vapor-producing candles, torch fuels, smoldering wicks
11. Insect attractants
    a. Food (Japanese beetle traps)
    b. Sex lures (pheromones for agricultural and forest pests)
12. Animal systemics (insecticides, parasiticides)
    a. Oral (pre-measured capsules or liquids)
    b. Dermal (pour-on or sprays)
    c. Feed-additive

This list is incomplete, and contains only the more common formulations that are available in most areas.

cultural and industrial use. The thickened sprays result in reduced drift and can be applied in sensitive situations.

If properly formulated, emulsifiable concentrates should remain suspended without further agitation for several days after dilution with water. A pesticide concentrate that has been held over by the home gardener from last year can be easily tested for its emulsifiable quality by adding one tablespoon to a quart of water and allowing the emulsion to stand after shaking. The material should remain uniformly suspended for at least 24 hours with no precipitate. If a precipitate does form the same condition may occur in your spray tank, resulting in a clogged nozzle and uneven application. This can be remedied by adding two tablespoons of a quality liquid dishwashing detergent to each pint of concentrate and mixing thoroughly. The bulk of pesticides available to the homeowner are formulated as emulsifiable concentrates and generally have a shelf life of about 3 years. Emulsifiable concentrates should be stored where they will not freeze, which usually causes separation of ingredients and failure to remain emulsified when diluted.

Water Miscible Liquids readily mix with water. The technical grade material may be water-miscible initially or it may be alcohol-miscible and formulated with an alcohol to become water-miscible. These formulations resemble the emulsible concentrates in viscosity and color, but do not become milky when diluted with water. Few of the home and garden pesticides are sold as water miscibles since few of the pesticides that are safe for home use have these physical characteristics.

Wettable Powders, (WP) are essentially concentrated dusts containing a wetting agent to facilitate the mixing of the powder with water before spraying. The technical material is added to the inert diluent, in this case a finely ground talc or clay, in addition to a wetting agent, similar to a dry soap or detergent, and thoroughly ground together in a ball mill. Without the wetting agent, the powder would float when added to water and the two would be almost impossible to mix. Because wettable powders usually contain from 50% to 75% clay or talc, they sink rather quickly to the bottom of spray tanks unless the tank is shaken repeatedly during use. Many of the insecticides sold for garden use are in the form of wettable powders because there is very little chance that this formulation will be phytotoxic, that is burn foliage, even at high concentrations. This is not true for emulsible concentrates, since the original carrier is usually an aromatic solvent, which alone in relatively moderate concentrations can cause foliage burning at high temperatures (90°F. and above).

Water Soluble Powders (SP) are properly titled and self explanatory. Here, the technical grade material is a finely ground water-soluble solid and contains nothing else to assist its solution in water. It is merely added to the proper amount of water in the spray tank where it dissolves immediately. Unlike the wettable powders and flowables, these formulations do not require repeated agitation; they are true solutions and do not settle to the bottom.

Oil Solutions in their commonest form are the ready-to-use, household and garden insecticide sprays, sold in an array of bottles, cans, and plastic containers, all usually equipped with a handy spray atomizer. Not to be confused with aerosols, these sprays are intended to be used directly on pests or where they frequent. Oil solutions may be used as weed sprays, for fly control around stalls and corrals, for standing pools to control mosquito larvae, in fogging machines for mosquito and fly abatement programs, or for household insect sprays purchased in supermarkets. For commercial use they may be sold as oil concentrates to be diluted with kerosene or diesel fuel before application, or as the dilute, ready-to-use form. In either case, the compound is dissolved in a light-weight oil and is applied as an oil spray; it contains no emulsifier or wetting agent.

Soluble Pellets, despite their seeming convenience and ease of handling with a water hose, are not very effective. They are sold in kits, including the waterhose attachment, fertilizer, fungicide, insecticide, and even a car wash detergent and wax pellets. The actual amount of active ingredient is very small, and uniform distribution with a watering hose is difficult.

Flowable or Sprayable Suspensions (F or S) exemplify an ingenious solution to a formulation problem. Earlier it was stated that some pesticides are soluble in neither oil nor water. They are soluble in one of the exotic solvents, however, making the formulation quite expensive and perhaps pricing it out of the marketing competition. To handle the problem, the technical material is blended with one of the dust diluents and a small quantity of water in a mixing mill, leaving the pesticide-diluent mixture finely ground with a custard-like appearance. This formulation mixes well with water as a suspension and can be sprayed, but with the same tank-settling characteristics as the wettable powders.

Fogging Concentrates are the formulations sold strictly for public health use in the control of nuisance or disease vectors, such as flies and mosquitoes, and to pest control operators. Fogging machines generate droplets whose diameters are usually less than 10 microns but greater than 1 micron. They are of two types. The thermal fogging device utilizes a flash heating of the oil solvent to produce a visible vapor or smoke. Unheated foggers atomize a tiny jet of liquid in a venturi tube through which passes an ultra-high velocity air stream. The materials used in fogging machines depend on the type of fogger. Thermal foggers use oil only, whereas unheated generators use water, emulsions, or oils.

DUSTS

Historically, dusts (D) have been the simplest formulations of pesticides and the easiest to apply. Examples of the undiluted toxic agent are sulfur dust used on ornamentals for some disease and mite control, and one of the older household roach dusts, sodium fluoride. An example of the toxic agent with active diluent would be one of the garden insecticides or fungicides having sulfur dust as its carrier or diluent. A toxic agent with an inert diluent is the most common type of dust formulation in use today, both in the home garden and in agriculture. Insecticide-fungicide combinations are applied in this manner, with the carrier being an inert clay such as talc or pyrophyllite. The last type, the aerosol dust, is a finely ground silica in a liquefied gas propellant that can be directed into crevices of homes and commercial structures for insect control. The aerosol dust never became popular for home use.

Despite their ease in handling, formulation, and application, dusts are the least effective, and ultimately the least economical, of the pesticide formulations. The reason is that dusts have a very poor rate of deposit on foliage, unless it is wet from dew or rain. For instance, in agriculture an aerial application of a standard dust formulation of pesticide will result in 10% to 40% of the material reaching the crop. The remainder drifts upward and downwind. Psychologically, dusts are annoying to the nongrower who sees great clouds of dust resulting from an aerial application, in contrast to the grower who believes he is receiving a thorough application for the very same reason. The same statement may be relevant to the avid user of dusts in his garden and the abstaining neighbor! Under similar circumstances, a garden hand sprayer application of water emulsion spray will deposit 70 to 80% of the pesticide on target plants or turf.

AEROSOLS

Most of us have been raised in the aerosol culture: bug bombs, hair sprays, underarm deodorants, home deodorizers, oven sprays, window sprays, repellents, paints, garbage can and shower-tub disinfectants, and supremely, the anti-itch remedies for foot and crotch. Among pesticides the insecticides are dominant. Developed during World War II for use by military personnel living in tents, the push-button variety was used as space sprays to knock down mosquitoes. More recently the total-release aerosol has been designed to discharge its entire contents in a single application. They are now available for home owners as well as commercial pest control operators. In either case, the nozzle is depressed and locked into place, permitting the aerosol total emission while the occupants leave and remain away for a few hours. Aerosols are effective only against resident flying and crawling insects, and provide little or no residual effect as do conscientiously-applied sprays. How do aerosols work? Essentially the active ingredients must be soluble in the highly volatile, liquified hydrofluorocarbon[1] gas in its pressurized condition. When the liquified gas is released it evaporates almost instantly, leaving the micro-sized droplets of toxicant suspended in air. I should point out that aerosols commonly produce droplets well below 10 microns in diameter, which are respirable, meaning that they will be absorbed by alveolar tissue in the lungs rather than impinging in the bronchioles, as do larger droplets. Consequently, aerosols of every type should be handled with discretion and breathed as little as possible.

FERTILIZER COMBINATIONS

Fertilizer-pesticide combinations are fairly common formulations to the home gardener who has purchased a lawn or turf fertilizer containing a herbicide for crabgrass control, insecticide for grubs and sod webworms, and/or a fungicide for numerous lawn diseases. Fertilizer-insecticide mixtures have been made available over the years to growers, particularly in the corn belt, by special order with the fertilizer distributor. The fertilizer and insecticide can then be applied to the soil during planting in a single, economical and energy-saving operation.

[1]The EPA has ruled that all aerosols not contain hydrofluorocarbons as the propellent by 1980.

GRANULAR PESTICIDES

Granular pesticides (G) overcome the disadvantages of dusts in their handling characteristics. The granules are small pellets formed from various inert clays and sprayed with a hot oil solution of the toxicant to give the desired content. After the solvent has evaporated, the granules are packaged for use. Granular materials range in size from 20 to 80 mesh, which refers to the number of grids per linear inch of screen through which they will pass. Only insecticides, and a few herbicides, are formulated as granules. They range from 1% to 15% active ingredient, including some systemic insecticides as granules available for garden, lawn, and ornamentals. Granular materials may be applied at virtually any time of day in winds up to 20 mph without problems of drift, an impossible task with sprays or dusts. They also lend themselves to soil application in the drill at planting time to protect the roots from insects or to introduce a systemic to the roots for transport to above-ground parts in certain garden vegetables, lawns, and ornamentals.

FUMIGANTS

Fumigants are a rather loosely defined group of formulations. Plastic insecticide-impregnated pest strips and pet collars of the same materials are really a slow-release formulation, permitting the insecticide to work its way slowly to the surface and volatilize. Moth crystals and moth balls (paradichlorobenzene and napthalene) are crystaline solids that evaporate slowly at room temperatures, exerting both a repellent as well as an insecticidal effect. (They can also be used in small quantity to keep cats and dogs off of or away from their favorite parking places.) Soil fumigants are used in horticultural nurseries, greenhouses, and on high-value cropland to control nematodes, insect larvae and adults, and sometimes diseases. Depending on the fumigant, the treated soils may require covering with plastic sheets for several days to retain the volatile chemical, allowing it to exert its maximum effect. The latter are not likely to be used in a home garden situation and are not recommended in this book.

IMPREGNATING MATERIALS

Impregnating materials mentioned here will include only treatment of woolens for mothproofing and timbers against wood-destroying organisms. For several years woolens and occasionally leather garments have been mothproofed in the final stage of dry-cleaning (using chlorinated solvents). The last solvent rinse contained in ultra-low concentration of the insecticide dieldrin, which had fantastic residual qualities against moths and leather-eating beetle larvae. Railroad ties, telephone and light poles, fence posts, and other wooden objects that have close contact with or are actually buried in the ground, soon begin to deteriorate as a result of attacks from fungal decay microorganisms and insects, particularly termites, unless treated with fungicides and insecticides. Such treatments permit poles to stand for 40 to 60 years, that would otherwise have been replaced in 5 to 10 years.

IMPREGNATED SHELF PAPERS

Shelf papers, strips and cords containing insecticides are in a rapid state of decline. Thoroughly effective against stored products insect pests, they usually contained one of the chlorinated insecticides to give long residual activity. These insecticides, along with most others, cannot be used where food and food utensils are stored, according to regulations established by the Environmental Protection Agency, resulting in the decline of impregnated materials.

IMPREGNATED WAX BARS

Impregnated wax bars contain herbicides which are selective against broad-leaf plants. When dragged over grass lawns in a uniform pattern enough is rubbed off on weeds to eliminate them, leaving the grass unaffected. This type of application is very selective, represents a spot application that is not disruptive to the environment, and is the type that should be strongly encouraged for home use.

BAITS

Several baits can be purchased for home and garden, such as for snails and slugs, and wasps and hornets. They contain low levels of the toxicant incorporated into materials that are relished by the target pests. Here is another example where spot application, placing the bait in selected places accessible only to the target species, permits the use of very small quantities of often times highly toxic materials in a totally safe manner, with no environmental disruption.

SLOW-RELEASE INSECTICIDES

Slow-release insecticides are relatively new and not yet available to the homeowner. The principle, as mentioned under the paragraph on fumigants, involves the incorporation of the insecticide in a permeable covering, which permits its escape at a reduced, but effective rate. Several agricultural insecticides have been encapsulated into extremely small plastic spheres which are then sprayed on crops. The insecticide escapes through the sphere wall over an extended period, preserving its effectiveness much longer than if formuated as an emulsible concentrate or wettable powder.

An invisible paint-on lacquer is probably the most recent innovation in household pest control (Pest Aid "180", Positive Control, and Farm Aid "180"). Here the insecticide is dissolved in a special solvent containing small quantities of dissolved plastics and lacquers. Following its paint-on application as a spot treatment in homes, the solvent evaporates leaving the insecticide incorporated in the thin film of transparent lacquer. With time the insecticide "blooms" at a constant rate presenting a fresh, thin insecticide surface to crawling insects at all times. Because of the safe-handling characteristics it should become a part of the arsenal in household pest control.

The new insecticidal adhesive tapes work essentially by the same principle but are perhaps related more to the plastic strips mentioned under fumigants. The adhesive back is exposed by removing a protective strip, and the tape is attached beneath counters, under shelves, and in other protected places. These two new formulations have just recently become available to homeowners.

In closing this section of pesticides, it might appear that there are no limits to the different forms in which a pesticide can be prepared. This is almost the case. While reading the last entry on slow-release formulations, your imagination may have been stimulated to think about formulations of the future. If not through economy, then by efforts of the Environmental Protection Agency, we will learn to formulate and apply pesticides in extremely conservative ways, to preserve our health, our resources and the environment that at times appears to be in some jeopardy. In summary, pesticides are formulated to improve their properties of handling, application, effectiveness, safety, and storage.

THE INSECTICIDES

As the facts fall into place, man has been on earth somewhere between 1 and 2 million years, a figure that most of us have difficulty in conceiving. In contrast, however, try to visualize 250,000,000 years, the period insects are known to have existed. Despite their head start, man has been able to carve his niche and to forge a path through their devastations. He has learned to live and to compete with them. There is no way to determine when insecticides became a tool, but we can guess that the first materials used by our primitive ancestors were mud and dust spread over his skin to repel biting and tickling insects, a habit resembling those of water buffalo, pigs, and elephants. In fact, we could speculate that their first use was around the home tree or cave, making them domestic or urban repellents. Fictitious as it may seem, this may have been the origin of urban pest control.

It was not until 1690 that the first truly insecticidal material was used — water extracts of tobacco sprinkled on garden plants to kill sucking insects. Then about 1800 a louse powder was being used in the Napoleonic Wars which was made by grinding the flower heads of a chrysanthemum. You know the active ingredient as pyrethrins. These naturally occurring or botanical insecticides were the first real step forward in man's perpetual war against insects.

Natural or Botanical Insecticides

The botanical insecticides are of great interest to many gardeners, especially organic gardeners because they are "natural" insecticides, toxicants derived from plants. Historically, the plant materials have been in use longer than any other group of insecticides with the possible exception of sulfur. Tobacco, pyrethrum, derris, hellebore, quassia, camphor, and turpentine were some of the more important plant products in use before the organized search for synthetic insecticides had begun.

Some of the most widely used insecticides have come from plants. The flowers, leaves, or roots have been finely ground and used in this form, or the toxic ingredients have been extracted and used alone or in mixtures with other toxicants. There are five natural or botanically-derived insecticides that will be of interest to gardeners in general, but especially to the organic gardener: Pyrethrins, rotenone, sabadilla, ryania and nicotine. All except nicotine are exempt from the requirement of a tolerance when applied to growing fruit and vegetables. That is, they can be eaten anytime after application, but these botanical insecticides must be used according to label directions (See Appendix for pesticides that can be made at home).

Pyrethrins. Pyrethrins are extracted from the flowers of a chrysanthemum grown in Kenya, Africa, and Ecuador, South America. It has an oral LD_{50} of approximately 1,500 mg/kg and is one of the oldest household insecticides available (LD_{50} means a dose lethal to 50 percent of the organisms treated, usually the white laboratory rat, and the higher the figure the safer the pesticide). (See p. 232 for more details regarding the meaning of LD_{50}) The ground, dried flower heads were used back in the 19th century as the original louse powder to control body lice in the Napoleonic Wars. Pyrethrin acts on insects with phenomenal speed causing immediate paralysis, thus its popularity in fast knock-down household aerosol sprays. However, unless it is formulated with one of the synergists most of the paralyzed insects recover to once again become pests. Pyrethrins are formulated as household sprays and aerosols and are available as spray concentrates and dusts for use on vegetables, fruit trees, ornamental shrubs and flowering plants at any stage of growth. Vegetables and fruit sprayed or dusted with pyrethrins may be harvested or eaten immediately. In other words, there is no waiting interval required between application and harvest of the food crop.

Because of its general safety to man and his domestic animals and its effectiveness against practically every known crawling and flying insect pest, pyrethrins have more uses approved by the Environmental Protection Agency than any other insecticide, numbering literally in the thousands!

Rotenone. Rotenoids, the rotenone-related materials, have been used as crop insecticides since 1848, when they were applied to plants to control leaf-eating caterpillars. However, they have been used for centuries before that (at least since 1649) in South America to paralyze fish, causing them to surface.

Rotenoids are produced in the roots of two genera of the legume (bean family), *Derris*, grown in Malaya and the East Indies, and *Lonchocarpus* (also called cubé), grown in South America. Rotenone has an oral LD_{50} of approximately 350 mg/kg and has been used for generations as the ideal general garden insecticide. It is harmless to plants, highly toxic to fish and many insects, especially caterpillars, moderately toxic to warm-blooded animals, and leaves no harmful residues on vegetables. There is no waiting interval between application and harvest of a food crop.

It is both a contact as well as a stomach poison to insects and is sold as spray concentrates and ready-to-use dust. It kills insects slowly, but causes them to stop their feeding almost immediately. Like all the other botanical insecticides its life in the sun is short, 1 to 3 days. It is useful against caterpillars, aphids, beetles, true bugs, leafhoppers, thrips, spider mites, ants, rose slugs, whiteflies, sawflies, bagworms, armyworms, cutworms, leafrollers, midges, and a host of other pests.

Next to pyrethrins, rotenone is probably second in the number of approved uses, exceeding well over 1,000.

Rotenone is the most useful piscicide (fish control chemical) available for reclaiming lakes for game fishing. It eliminates all fish, closing the lake to reintroduction of rough species. After treatment the lake can be restocked with the desired species. Rotenone is a selective piscicide in that it kills all fish at dosages that are relatively non-toxic to fish food organisms. It also breaks down quickly leaving no residues harmful to fish used for restocking. The recommended rate is 0.5 part of rotenone to one million parts of water (ppm), or 1.36 pounds per acre-foot of water.

Sabadilla. Sabadilla is extracted from the seeds of a member of the lily family. Its oral LD_{50} is approximately 5,000 mg/kg, making it the least toxic to warm blooded animals of the five botanical insecticides discussed. It acts as both a contact and stomach poison for insects. It is irritating to the eyes of humans and causes violent sneezing in some sensitive individuals. It deteriorates rapidly in sunlight and can be used safely on food crops with no waiting interval required by the Environmental Protection Agency. Sabadilla will probably be the most difficult of the five botanical insecticides to purchase, simply because there was a period of about 15 years during which there was hardly any demand for it.

Sabadilla is registered for most commonly grown vegetables and will control caterpillars, grasshoppers, beetles, leafhoppers, thrips, chinch bugs, stink bugs, harlequin and squash bugs, other true bugs and potato psyllids. It is not very useful against aphids and will not control spider mites.

Ryania. Ryania is another botanical or plant-derived insecticide that is quite safe for man and his animals — so safe, that no waiting is required between the time of application to food crops and harvest, as there is for most other insecticides. Ryania is made from the ground roots of the ryania shrub grown in Trinidad, and like nicotine, belongs to the chemical class of alkaloids. It has an oral LD_{50} of approximately 750 mg/kg. It is a slow-acting insecticide, requiring as much as 24 hours to kill. Insects exposed to ryania usually stop their feeding almost immediately, making it particularly useful for caterpillars.

The preferred uses for ryania are against fruit and foliage-eating caterpillars on fruit trees, especially the codling moth on apple trees. However, it is useful against almost all plant-feeding insects, making it an ideal material for small orchards of deciduous fruits. It is not effective against spider mites.

Ryania is exempt from a waiting period between application and harvest, and is registered by the Environmental Protection Agency for the control of a host of insect pests on a wide variety of plants, shrubs and trees. Vegetable garden pests include: aphids, cabbage loopers, Colorado potato beetle, corn borers, cucumber beetles, diamond back moth, flea beetles, leafhoppers, Mexican bean beetle, spittle bugs, and tomato hornworms.

It is registered for deciduous fruit trees to control aphids (except the woolly aphid), codling moth, Japanese beetle, and cherry fruit fly. It has long been used to control citrus thrips on all citrus. Though not very effective, ryania can be used in the home to control ants, silverfish, cockroaches, spiders and crickets. On ornamentals it is registered for aphids and lace bugs. It can be used on roses against aphids, Japanese beetle, thrips and whiteflies. On brambles it is used for aphids and raspberry fruitworm and sawfly. And it can be used on grapes for aphids (except the woolly aphid) and the Japanese beetle.

Ryania is difficult to obtain since its importation into the United States has been discontinued by its major distributor, S. B. Penick Company. See Appendix for sources.

Nicotine. Smoking tobacco was introduced to England in 1585 by Sir Walter Raleigh. As early as 1690 water extracts of tobacco were reported as being used to kill sucking insects on garden plants. As early as about 1890, the active principle in tobacco extracts was known to be nicotine, and from that time on, extracts were sold as commercial insecticides for home, farm, and orchard.

Today, nicotine (nicotine sulfate) is commercially extracted from tobacco by steam distillation or solvent extraction. Black Leaf "40", which has long been a favorite house plant and garden spray, is a concentrate containing 40% nicotine sulfate.

Nicotine is an alkaloid and has prominent physiological properties. Other well-known alkaloids, which are not insecticides, are caffeine (found in tea and coffee), quinine (from chinchona bark), morphine (from the opium poppy), cocaine (from coca leaves), ricinine (a poison in castor oil beans), strychnine (from *Strychnos nux vomica*), conine (from spotted hemlock, the poison that killed Socrates), and finally, LSD (from the ergot fungus attacking grain), one of the banes of our 20th cenury culture.

Nicotine sulfate, as it is commonly sold, is highly toxic to all warm-blooded animals, as well as insects, having an LD_{50} of 50-60 mg/kg. This makes it the most hazardous of the botanical insecticides to home gardeners. The chemical is usually sold as a 40% nicotine sulfate concentrate, to be diluted with water and used as a spray for plants. Again, it has been used with great success since before the turn of the century.

Dusts are not available for garden use because they are too toxic to man. Nicotine sulfate is used primarily for piercing-sucking insects such as leafhoppers, aphids, scales, thrips and whiteflies, but it can kill all insects and spider mites on which it is sprayed directly. It is more effective during warm weather, but degrades quickly. It is registered for use on flowering plants, and ornamental shrubs and trees to control aphids, mealybugs, scales and thrips, lace bugs, leafminers, leafhoppers, rose slugs, and spider mites. Its use for most greenhouse pests is also acceptable.

A homemade nicotine spray can be made by soaking one or two shredded, cheap cigars in 1 gallon of water overnight at room temperature. Remove the tobacco parts and add 1 teaspoon of household detergent. When used as a foliar spray against aphids and other small insects it is usually as effective as sprays made from the commercial product. Caution! This product is more concentrated, thus more toxic, than the commercial preparations. This homemade concentrate may also infect your tomatoes and certain other plants with tobacco virus.

Nicotine sulfate is registered for use on a variety of vegetables and fruit trees, but it cannot be used in most cases within 7 days of harvest, as required by the Environmental Protection Agency. There are several ornamental plants that are sensitive to nicotine, such as roses. The label should identify those sensitive plants.

Nicotine is also registered for use as a dog and cat repellent out-of-doors as well as for furniture in the home. Additionally, tobacco dust is acceptable as a dog and rabbit repellent out of doors.

When using nicotine or nicotine sulfate, it is important to follow the instructions on the label and wear the necessary protective clothing as indicated, since nicotine can be absorbed through the skin.

In closing, let me point out that these botanical insecticides are chemicals and no safer than most of the currently available synthetic insecticides. Their only distinction is that they are produced by plants and extracted for use by man. (See Appendix for list of pesticides that can be made at home).

ORGANOCHLORINES

Even as recently as 1940, our insecticide supply was still limited to lead or calcium arsenate, petroleum oils, nicotine, pyrethrum, rotenone, sulfur, hydrogen cyanide gas, and cryolite. And then, World War II opened the Chemical Era with the introduction of a totally new concept of insecticide control chemicals — synthetic organic insecticides, the first of which was DDT.

The advantages of the synthetic insecticides over the botanical or natural insecticides were efficacy, cost, and persistence. Less of the synthetic materials was needed to kill, the cost was usually much less, and their effectiveness lasted longer, sometimes as much as months. These synthetics were so easy to use and so economical that they soon replaced the botanicals around the home as well as in agriculture.

The organochlorine insecticides should be familiar because of the notoriety given them in recent years by the press. The organochlorines are insecticides that contain carbon (thus the name organo-), chlorine, and hydrogen. They are also referred to by other names: "chlorinated hydrocarbons," "chlorinated organics," "chlorinated insecticides," "chlorinated synthetics," and perhaps others. They include chlordane, lindane, dicofol (Kelthane) methoxychlor, endosulfan (Thiodan), DDT and dieldrin.

DDT and Related Insecticides. DDT was first known chemically as *d*ichloro *d*iphenyl *t*richloroethane, thus ddt or DDT. It is available only as an insecticide for louse control on humans through a physician's prescription.

There are two other relatives of DDT which should be mentioned because they are all recommended for home pest control in this book: Methoxychlor and dicofol (Kelthane). The latter is not really an insecticide, but rather an acaricide (miticide) and will be found recommended for several species of mites on fruit, vegetables and ornamentals. Methoxychlor is an excellent substitute for DDT, because of its long residual action and low toxicity to man and his domestic animals. It controls a wide variety of insects on fruit and shade trees, vegetables, and domestic animals. It is seldom phytotoxic (damaging to plants).

DDT is a persistent insecticide, which is one reason that its registrations were cancelled by EPA. Persistence, as used here, implies a chemical stability giving the products long lives in soil and aquatic environments, and in animal and plant tissues. They are not readily broken down by microorganisms, enzymes, heat or ultraviolet light. From the insecticidal viewpoint these are good characteristics. From the environmental viewpoint they are not. Using these qualities, the other relatives of DDT are considered nonpersistent.

How does DDT kill? The mode of action, or type of biological activity, has never been clearly worked out for DDT or any of its relatives. It does affect the neurons or nerve fibers in a way that prevents normal transmission of nerve impulses, both in insects and mammals. Eventually the neurons fire impulses spontaneously, causing the muscles to twitch; this may lead to convulsions and death. There are several valid theories for DDT's mode of action, but none has been clearly proved.

DDT is relatively stable to the ultraviolet of sunlight, not readily broken down by microorganisms in the soil or elsewhere, stable to heat, acids, and unyielding in the presence of almost all enzymes. In other words, it is poorly biodegradable. Next, it has practically zero water solubility. DDT has been reported in the chemical literature to be probably the most water insoluble compound ever synthesized. Its water solubility is actually somewhere in the neighborhood of 6 parts per billion parts of water (ppb). On the other hand, it is quite soluble in fatty tissue, and as a consequence of its resistance to metabolism, it is readily stored in fatty tissue of any animal ingesting DDT alone or DDT dissolved in the food it eats, even when it is part of another animal.

If it is not readily metabolized, and thus not excreted, and if it is freely stored in body fat, it should come as no surprise that it accumulates in every animal that preys on other animals. It also accumulates in animals that eat plant tissue bearing even traces of DDT. Here we aim at the dairy and beef cow. The dairy cow excretes (or secretes) a large share of the ingested DDT in its milk fat. Man drinks milk and eats the fatted calf. Guess where the DDT is now.

The same story is repeated time and again in food chains ending in the osprey, falcon, golden eagle, seagull, pelican, and so on.

The explanation of these food chain oddities is this: Any chemical which possesses the characteristics of stability and fat solubility will follow the same biological magnification (condensed to biomagnification) as DDT. Other insecticides incriminated to some extent in biomagnification, belonging to the organochlorine group are TDE, DDE (a major metabolite of DDT), dieldrin, aldrin, several isomers of BHC, endrin, heptachlor and mirex. And, of course, they all possess these two crucial prerequisites.

Lindane. HCH (BHC), hexachlorocyclohexane, was first discovered in 1825. But like DDT, was not known to have insecticidal properties until 1940, when French and British entomologists found the material to be active against all insects tested.

Lindane has a higher vapor action than most insecticides, and is recommended in instances where the vaporized insecticide can reach the insect, such as with borers in the trunks of fruit trees and woody ornamentals.

Since the Gamma isomer was the only active ingredient, methods were developed to manufacture a product containing 99% Gamma isomer, Lindane, which was effective against most insects, but also quite expensive, making it impractical for crop use.

Lindane is odorless and has a high degree of volatility and is recommended only for outdoor, non-food plants, such as for borers in the trunks of trees and woody ornamentals.

Cyclodienes. The cyclodienes are a prominent and extremely useful group of insecticides, also known as the diene-organochlorine insecticides.

They were developed after World War II, and are, therefore, of more recent origin than DDT (1939) and HCH (1940). The 3 compounds listed below were first described in the scientific literature or patented in the year indicated: chlordane, 1945; dieldrin, 1948; and endosulfan (Thiodan®), 1956.

The cyclodienes are persistent insecticides, except endosulfan, and are stable in soil and relatively stable to the ultraviolet action of sunlight. Consequently, chlordane and dieldrin are now used only as soil insecticides for the control of termites. Because of their persistence, the use of cyclodienes on crops was restricted; undesirable residues remained beyond the time for harvest. To suggest the effectiveness of cyclodienes as termite control agents, structures treated with chlordane and dieldrin in the year of their development are still protected from damage. This is 35 and 32 years, respectively. It would be elementary to say that these insecticides are the most effective, long-lasting, economical, and safest termite control agents known.

ORGANOPHOSPHATES

This next group, the chemically unstable organophosphate (OP) insecticides, has virtually replaced the persistent organochlorine compounds. This is especially true with regard to their use around the home and garden.

The OPs have several commonly used names, any of which are correct: Organic phosphates, phosphorus insecticides, nerve gas relatives, phosphates, phosphate insecticides, and phosphorus esters or phosphoric acid esters. They are all derived from phosphoric acid, and are generally the most toxic of all pesticides to vertebrate animals except those recommended for the home which are quite safe to use.

The OPs have two distinctive features. First, they are generally much more toxic to vertebrates than the organochlorine insecticides, and second, they are chemically unstable or nonpersistent and break down rather quickly and easily. It is this latter quality which brings them into general agricultural use as substitutes for the more persistent organochlorines.

Malathion, one of the oldest and safest organophosphate insecticides, appeared in 1949. It is highly recommended and commonly used in and around the home with little or no hazard either to man or his pets. It controls a wide variety of pests including aphids, spider mites, scale insects, mosquitoes, house flies, and a broad spectrum of other sucking and chewing insects attacking fruit, vegetables, ornamentals, and stored products. It is safe to use on most pets as a dust. Malathion is sold to the homeowner as emulsifiable concentrates, wettable powders, and dusts.

Trichlorfon (Dylox®) is a chlorinated OP, which has been useful for crop pest control and fly control around barns and other farm buildings. It is registered for several vegetables and ornamentals but will not be found often as a recommended insecticide in this book.

Naled (Dibrom®) is an organophosphate that contains bromine in addition to chlorine, making it somewhat unique. Even though it is registered with the Environmental Protection Agency for several vegetable crops, it is recommended only for use on ornamentals in this book. It has a broad spectrum of activity, much as malathion does and can be particularly useful in fly control around barns, stables, poultry houses, and kennels.

One of the old standbys is diazinon, which appeared first in 1952. With the exception of DDT, more diazinon has been used in and around homes than any other insecticide. It is a relatively safe OP that has an amazingly good track record around the home, used effectively for practically every conceivable use: Insects in the home, lawn, garden, ornamentals, around pets (not on) and for fly control in stables and pet quarters.

Chlorpyrifos (Dursban®) has become the most frequently used insecticide by pest control operators in homes and restaurants for controlling cockroaches and other household insects. Chlorpyrifos is also sold as ready-to-use oil sprays for insect pest control in the home. This is an insecticide that is uniquely sold under another name, Lorsban®, when formulated for agricultural use. You will find chlorpyrifos recommended in this book only for household and ornamentals use.

Ethion is one of the OPs useful both as an insecticide and acaricide, but not among the very common materials used in the home garden. It controls aphids, scales, thrips, leafhoppers, maggots, foliage-eating caterpillars, and mites, including Eriophyid mites, on a large number of fruit, vegetables, and ornamentals.

Dichlorvos (Vapona®, DDVP) is not only a contact and stomach poison but also acts as a fumigant. It has been incorporated into vinyl plastic pet collars and pest strips from which it escapes slowly as a fumigant. It may last up to several months and is very effective for insect control in the home, other closed areas, and on pets.

SYSTEMIC ORGANOPHOSPHATES

Systemic insecticides are those that are taken into the roots of plants and translocated to the aboveground parts, where they are toxic to any sucking insects feeding on the plant juices. Normally caterpillars and other plant tissue-feeding insects are not controlled, because they do not ingest enough of the systemic-containing juices to be affected.

Contained among the several plant systemics are dimethoate (Cygon®), disulfoton (Di-Syston®) and acephate (Orthene®), all of which can be used safely by the homeowner and are the only ones recommended in this book.

Dimethoate is used as a residual wall spray in farm buildings for fly control and for control of a wide range of mites and insects on ornamentals, vegetables, apples and pears, citrus and melons.

Disulfoton can give up to 5-6 weeks of control from seed treatment or application in the seed furrow or as a side dressing, controlling many species of insects and mites, and especially sucking insects, such as aphids.

Acephate (Orthene®) is one of the more recent additions to the systemic insecticides available for home use, though it also acts as a contact insecticide. It has moderate persistence with 10 to 15 days of systemic activity. Around the home it controls aphids, cabbage loopers, bagworms, tent caterpillars, gypsy moths, lace bugs, leaf miners and rollers, leafhoppers, thrips and webworms. It has a few garden vegetable uses and more on ornamentals. Check the label.

Another group of systemic insecticides is available for animal use. Ronnel (Korlan®) is the only one recommended here for pet use. When administered orally to livestock it controls cattle grubs, lice, horn flies, face flies, screwworms, ticks, sheep keds and wool maggots. The animal systemics are truly remarkable and useful insecticides.

ORGANOSULFURS

The organosulfurs, as the name suggests, are synthetic organic insecticides built around sulfur. Dusting sulfur alone is a good acaricide (miticide), particularly in hot weather. The organosulfurs, however, are far superior, requiring much less material to achieve control. Of greater interest, however, is that the organosulfurs, even though toxic to mites, have very low toxicity to insects. As a result they are normally used only for mite control.

This group has one other valuable property: They are usually ovicidal (kill the eggs) as well as being toxic to the young and adult mites.

Tetradifon (Tedion®) is one of the older acaricides and safe for home use. It is registered for use on most fruit trees, including citrus and nuts, vegetables and ornamentals.

Oxythioquinox (Morestan®) is an amazing chemical. It has the distinction of being all things to all pests — an insecticide, acaricide, fungicide, and ovicide! It is found frequently as a recommended material for control of pests on ornamentals, but not for the vegetable garden. It gives residual control of mites, mite eggs, powdery mildew and pear psylla. It can be used as a pre-bloom spray on most deciduous fruits and in both pre- and post-bloom sprays on apples and pears. It is also registered for non-bearing citrus and walnuts, and ornamentals.

CARBAMATES

Considering that the organophosphate insecticides are derivatives of phosphoric acid, then the carbamates must be derivatives of carbamic acid. The carbamates kill insects in ways similar to the organophosphates.

Carbaryl (Sevin®), the first successful carbamate, was introduced in 1956. More of it has been used world over than all the remaining carbamates combined. Two distinctive qualities have resulted in its popularity: Very low mammalian oral and dermal toxicity and a rather broad spectrum of insect control. This has led to its wide use as a lawn and garden insecticide.

Carbaryl is registered for insect control on more than 100 different crops. These include vegetables, citrus, deciduous fruit, forage crops, field crops, forests, lawn and turf, nuts, ornamentals, rangeland, shade trees, poultry and most pets.

Another carbamate, propoxur (Baygon®), is highly effective against cockroaches that have developed resistance to the organochlorines and organophosphates. Propoxur is used by most structural pest control operators for roaches and other household insects in restaurants, kitchens, and homes. It is also formulated as bottled sprays for home use.

SYNERGISTS OR ACTIVATORS

Synergists are not in themselves considered toxic or insecticidal but are materials used with insecticides to synergize or enhance the activity of the insecticides. They are added to certain insecticides in the ratio of 8:1 or 10:1, synergist:insecticide. The first synergist was introduced in 1940 to increase the effectiveness of the plant-derived insecticide, pyrethrin. Since then many materials have been introduced, but only a few have survived because of cost and ineffectiveness. Synergists are found in practically all of the "bug-bomb" aerosols to enhance the action of the fast knock-down insecticide, pyrethrin, against flying insects.

The synergists are usually used in sprays prepared for the home and garden, stored grain, and on livestock, particularly in dairy barns. Synergists and the insecticides they synergize, such as natural pyrethrins, are quite expensive, thus seldom if ever used on crops. The new synthetic pyrethrins (or pyrethroids) however, are stable to sunlight and weather and are being used on field crops and some vegetables.

The first synergist was discovered in sesame oil and later named sesamin. Four other compounds you may find listed on the labels of certain insecticide sprays and aerosols are piperonyl butoxide, Sesamex, Sulfoxide, and MGK-264. MGK-264 was discovered in 1944 and has been used in great quantity for livestock and pet sprays.

MICROBIALS

The insect disease pathogens, known as microbial insecticides, are discussed earlier in the section on Biological Methods (page 8).

INSECT REPELLENTS

Historically repellents included smokes, plants hung in dwellings or rubbed on the skin as the fresh plant or brews, oils, pitches, tars, and various earths applied to the body. Camel urine sprinkled on the clothing has been of value, though questionable, in certain locales. Camphor crystals sprinkled among woolens has been used for decades to repel clothes moths. Prior to man's more edified approach to insect olefaction and behavior it was assumed that if a substance was repugmant to man it would likewise be repellent to annoying insects.

Prior to World War II there were only 4 principal repellents: (1) oil of citronella, discovered in 1901, used also as a hair-dressing fragrance by certain Eastern cultures; (2) dimethyl phthalate, discovered in 1929; (3) Indalone, introduced in 1937, and (4) Rutgers 6-12, which became available in 1939.

With the onslaught of World War II, and the introduction of American military personnel into new environments, particularly the tropics, it became necessary to find new repellents which would survive both time and dilution by perspiration. The ideal repellent would be nontoxic and nonirritating to man, nonplasticizing, and long lasting (12 hours) against mosquitoes, biting flies, ticks, fleas and chiggers.

Unfortunately, the ideal repellent is still being sought. Some have unpleasant odors, require massive dosages, are oily or effective only for a short time, irritate the skin, or soften paint and plastics.

Insect repellents come in every conceivable formulation — undiluted, diluted in a cosmetic solvent with added fragrance, aerosols, creams, lotions, treated cloths to be rubbed on the skin, grease sticks, powders, suntan oils, and clothes-impregnating laundry emulsions. Despite the formulation, their periods of protection vary with the chemical, the individual, the general environment, insect species, and avidity of the insect.

What happens to repellents once applied? Why are they not effective longer than 1 or 2 hours? No single answer is satisfactory but generally they evaporate, are absorbed by the skin, are lost by abrasion of clothing or other surfaces, and diluted by perspiration. Usually it is a combination of them all, and the only solution is a fresh application.

Deet (Delphene®) and Rutgers 6-12 are those most commonly found in today's repellents. Of these, deet is by far superior to all others against biting flies and mosquitoes.

FUMIGANTS

The fumigants are small, volatile, organic molecules that become gases at temperatures above 40°F. They are usually heavier than air and commonly contain one or more of the halogens (chlorine, bromine or fluorine). Fumigants, as a group, are narcotics, that is, they induce narcosis, sleep or unconsciousness, which in effect is their action on insects. The fumigants are fat-soluble and their effects are reversible.

Fat solubility appears to be an important factor in the action of fumigants, since these narcotics lodge in fat-containing tissues, which are found in the nervous system. Only two materials are recommended for home use, naphthalene and paradichlorobenzene (PDB). You know them as moth balls and moth crystals.

INORGANICS

Inorganic insecticides are those which do not contain carbon. Usually they are white and crystalline, resembling the salts. They are stable chemicals, do not evaporate, and are frequently soluble in water. Neither of the two discussed here are white salts, and one is not water soluble — sulfur.

Sulfur. Sulfur dusts are especially toxic to mites, such as chiggers and spider mites of every variety, thrips, newly-hatched scale insects, and as a stomach poison for some caterpillars. Sulfur dusts and sprays, made from wettable sulfur, are fungicidal, especially against powdery mildews.

Lime-sulfur is an old dormant-spray material, used on fruit trees since about 1886. It is effective against many scale insects and fungus diseases and as a foliage spray on fruit trees for fungus diseases, thrips, and various mites. Lime-sulfur solutions can be made at home by cooking together sulfur and unslaked lime (quicklime) in water. It has a disagreeable odor and is caustic to the skin. As with any other insecticide, it can be used safely and effectively if handled with the usual element of caution and common sense. (See Recipes for Home-made Insecticides in the Appendix).

Cryolite is a naturally occurring mineral that is mined in its pure form and is used as a stomach poison insecticide for chewing insects, especially caterpillars. Its low toxicity hazard and lack of effect on most predators make this a very useful material for the home garden, orchard and ornamentals. The insecticidal properties of cryolite were discovered in the 1920's, and it was used heavily until the mid-1950's when it was displaced by the more effective synthetic organic insecticides. It is registered for use on most garden crops, fruit crops and ornamentals, and provides control for several days since it does not evaporate or deteriorate under sunlight. I am acquainted with only one product that is currently available to the homeowner, Kryocide®, manufactured by the Pennwalt Corporation.

Boric Acid powder and tablets used indoors are especially effective against all species of cockroaches.

THE FUNGICIDES

Fungicides, strictly speaking, are chemicals used to kill or halt the development of fungi. Most plant diseases can be controlled to some extent with today's fungicides. Among those that cannot be controlled with chemicals are *Phytophthora* and *Rhizoc-*

tonia root rots, *Fusarium*, *Verticillium*, and bacterial wilts, and the viruses. The difficulties with these diseases are that they occur either below ground, and beyond the reach of fungicides, or they are systemic within the plant.

The fungal diseases are basically more difficult to control with chemicals than insects because the fungus is also a plant living in close quarters with its host. This explains the difficulty in finding chemicals which kill the fungus without harming the plant. Also, fungi that can be controlled by fungicides may undergo secondary cycles rapidly and produce from 12-25 "generations" during a 3-month growing season. Consequently, repeated applications of protective fungicides may be necessary due to plant growth dilution and removal by rain and other weathering.

It is necessary that fungicides be applied to protect plants during stages when they are vulnerable to inoculation by pathogens, before there is any evidence of disease. Fungicides can help to control certain diseases after the symptoms appear, referred to as chemotherapeutants. Also, protective fungicides are commonly used, even after symptoms of disease have appeared. Eradicant fungicides are usually applied directly to the pathogen during its "over-wintering" stage, long before disease has begun and symptoms have appeared. In the case of crops, however, if their sale depends on appearance, such as lettuce and celery, then the fungicide must be applied as a protectant spray in advance of the pathogens to prevent the disease.

There are about 150 fungicidal materials in our present arsenal, most of which are recently discovered organic compounds. And most of these act as protectants, preventing spore germination and subsequent fungal penetration of plant tissues. Protectants are applied repeatedly to cover new plant growth and to replenish the fungicide that has deteriorated or has been washed off by rain.

The application principle for fungicides differs from that of herbicides and insecticides. Only that portion of the plant which has a coating of dust or spray film of fungicide is protected from disease. Thus a good, uniform coverage is essential. Fungicides, with several new exceptions, are not systemic in their action. They are applied as sprays or dusts, but sprays are preferable since the films stick more readily, remain longer, can be applied during any time of the day, and result in less off-target drift.

Thanks to modern chemistry, many of the serious diseases of grain crops are controlled by seed treatment with selective materials. Others are controlled with resistant varieties. Diseases of fruit and vegetables are often controlled by sprays or dusts of fungicides.

Historically, fungicides have centered around sulfur, copper, and mercury compounds, and, even today, most of our plant diseases could be controlled by these groups. However, the sulfur and copper compounds can retard growth in sensitive plants, and, as a result, the organic fungicides were developed. These sometimes have greater fungicidal activity and usually have less phytotoxicity.

The general-purpose fungicides for home use include inorganic forms of copper, sulfur, and a variety of organic compounds. The general purpose lawn and garden fungicides are few in number and are usually organic compounds.

THE INORGANIC FUNGICIDES

SULFUR

Sulfur in several forms is probably the oldest effective fungicide known to man and still a very useful home-garden fungicide. There are three physical forms or formulations of sulfur used as fungicides. The first is finely ground sulfur dust which contains 1 to 5% clay or talc to assist in dusting qualities. The sulfur in this form may be used as a carrier for another fungicide or an insecticide. The second is flotation or colloidal sulfur, which is so very fine that it must be formulated as a wet paste in order to be mixed with water. In its original microparticle size, it would be impossible to mix with water, and would merely float. Wettable sulfur is the third form; it is finely ground with a wetting agent so that it mixes readily with water for spraying. The easiest to use, of course, is dusting sulfur when plants are slightly moist with the morning dew.

Some pathogenic organisms and most mites are killed by direct contact with sulfur, and also by its fumigant action at temperatures above 70°F. The fumigant effect is, however, somewhat secondary at marginal temperatures and under windy conditions. It is quite effective in controlling powdery mildews of plants that are not unduly sensitive to sulfur. Unlike those of any other fungus, spores of powdery mildews will germinate in the absence of a film of water and penetrate the leaf tissue. Its fumigant effect — acting at a distance — is undoubtedly important in killing spores of powdery mildews. In its fungicidal action sulfur is eventually converted to hydrogen sulfide (H_2S), which is a toxic entity to the cellular proteins of the fungus.

COPPER

Most inorganic copper fungicides are practically insoluble in water, and are the pretty blue, green, red, or yellow powders sold for plant disease control. Included in these is Bordeaux mixture, named after the Bordeaux region in France where it originated. Bordeaux is a chemically undefined mixture of copper sulfate and hydrated lime, which was accidentally discovered when sprayed on grapes in Bordeaux to scare off "freeloaders". It was soon observed that downy mildew, a disease of grapes, disappeared from the treated plants. From this unique origin began the commercialization of fungicides.

Bordeaux mixture is the oldest recognized fungicide, and is readily available to home gardeners in the apple and grape producing districts of the country. Bordeaux is one of two do-it-yourself fungicides that can be made at home. The other is lime-sulfur, and both recipes can be found in the Appendix.

While primarily a fungicide, Bordeaux is also repellent to many insects, such as flea beetles, leafhoppers, and potato psyllid, when sprayed over the leaves of plants.

The copper ion, which becomes available from both the highly soluble and relatively insoluble copper salts, provides the fungicidal as well as phytotoxic and poisonous properties. A few of the many inorganic copper compounds used over the years are presented in Table 5.

Protective fungicides have very low solubilities but, in water, some toxicant does go into solution. That small quantity absorbed by the fungal spore is then replaced in solution from the residue. The spore accumulates the toxic ion and "commits suicide" so to speak. Except for powdery mildews, water permits spore germination and solubilizes the toxic portion of the fungicidal residue.

The copper ion is toxic to all plant cells, and must be used in discrete dosages or in relatively insoluble forms to prevent killing all or portions of the host plant. This is the basis for the use of relatively insoluble or "fixed" copper fungicides, which release only very low levels of copper, adequate for fungicidal activity, but not enough to become toxic to the host plant.

Copper compounds are not easily washed from leaves by rain since they are relatively insoluble in water, thus give longer protection against disease than do most of the organic materials. They are relatively safe to use and require no special precautions during spraying. Despite the fact that copper is an essential element for plants, there is some danger in an accumulation of copper in agricultural soils resulting from frequent and prolonged use. In fact, certain citrus growers in Florida have experienced a serious problem of copper toxicity after using fixed copper for disease control for many years.

TABLE 5. Some of the Inorganic Copper Compounds Used as Fungicides.

NAME	USES
Cupric sulfate	Seed treatment and preparation of Bordeaux mixture
Copper dihydrazine sulfate	Powdery mildews Black spot of roses
Copper oxychloride	Powdery mildews
Copper oxychloride sulfate	Many fungal diseases
Copper zinc chromates	Diseases of potato, tomato, cucurbits, peanuts and citrus.
Cuprous oxide	Powdery mildews
Basic copper sulfate	Seed treatment and preparation of Bordeaux mixture
Cupric carbonate	Many fungal diseases

THE ORGANIC FUNGICIDES

A host of synthetic sulfur and other organic fungicides have been developed over the past 30 years to replace the more harsh, less selective inorganic materials. Most of them have had no measurable build-up effect on the environment after many years of use. The first of the organic sulfur fungicides was thiram, discovered in 1931. This was followed by many others. Then came other new classes, the dithiocarbamates and dicarboximides (zineb and captan) introduced in 1943 and 1949, respectively. Since then, the organic synthesists have literally opened the doors with now more than 125 fungicides of all classes in use and in various stages of development.

The newer organic fungicides possess several outstanding qualities. They are extremely efficient, that is, smaller quantities are required than of those used in the past; they usually last longer; and they are safer for crops, animals and the environment. Too, most of the newer fungicides have very low phytotoxicity, many showing at least a 10-fold safety factor over the copper materials. And, most of them are readily degraded by soil microorganisms, thus preventing their accumulation in soils.

DITHIOCARBAMATES

These are the "old reliables", thiram, maneb, ferbam, ziram, Vapam® (SMDC), and zineb, all developed in the early 1930's and 1940's. They probably have greater popularity, including home garden use, than all other fungicides combined. Except for systemic action, they are employed collectively in every use known for fungicides.

Thiram is used as a seed protectant for certain fungus diseases of apples, peaches, strawberries and tomatoes. As a turf fungicide it is applied for control of large brown patch and dollar spot of turf. And it acts as a repellent for rabbit and deer, when used on fruit trees, shrubs, ornamentals and nursery stock.

Maneb is used for the control of early and late blights on potatoes and tomatoes and many other diseases of fruits and vegetables.

The principal uses of ferbam are in the control of apple scab and cedar apple rust and tobacco blue mold. It is also applied as a protective fungicide to other crops against many fungus diseases.

Zineb is used on a variety of fruits and vegetables, especially on potato seed pieces only, and on tomatoes for blight.

SUBSTITUTED AROMATICS

This is a somewhat arbitrary classification assigned to the benzene derivatives that possess long-recognized fungicidal properties. Pentachloronitrobenzene (PCNB) was introduced in the 1930's as a fungicide for seed treatment and selected foliage applications. Chlorothalonil first became available in 1964 and has proved to be a very useful, broad-spectrum foliage protectant fungicide. It has wide use in the garden, including beans, cole crops, carrot, sweet corn, cucumber, cantaloupe, muskmelon, honeydew, watermelon, squash, pumpkin, onion, potato, tomato and turf.

THIAZOLES

The thiazoles contain among others, the ever popular Terrazole®, which is used as a soil fungicide. Terrazole is recommended as a turf fungicide for the control of pythium diseases (pythium blight, cottony blight, grease spot, spot blight, and damping off). It is also effective against the seedling disease complex (*Pythium*, *Fusarium*, and *Rhizoctonia*) of several garden and field crops. Application can be either by seed treatment or in-furrow at time of planting.

TRIAZINES

The triazines are usually thought of as herbicides, however, there is one that is a well known fungicide. Anilazine (Dyrene®) was introduced in 1955, and has received wide use on vegetables, for potato and tomato leaf spots along with fungus diseases of lawn and turf grass.

DICARBOXIMIDES

Two extremely useful foliage protectant fungicides belong to this group: captan appeared in 1949 and is undoubtedly the most heavily used fungicide around the home of all classes; folpet appeared in 1962. They are used primarily as foliage dusts and sprays on fruits, vegetables, and ornamentals.

Captan controls scabs, blotches, rots, mildew, and other diseases on fruit, vegetables, and flowers. It is used also as a dust or slurry seed treatment. Folpet controls apple scab, cherry leaf spot, rose black spot, and rose mildew. It is also useful as seed and plant bed treatments.

The dicarboximides are some of the safest of all pesticides available and are recommended for lawn and garden use as seed treatments and as protectants for mildews, late blight, and other diseases. Remember the old garden center adage, "When in doubt use captan".

SYSTEMIC FUNGICIDES

Systemic fungicides, those which penetrate the plant cuticle and which are translocated to growing points, have become available only in recent years, and very few are yet available. Most systemic fungicides have eradicant properties which stop the progress of existing infections. They are therapeutic in that they can be used to cure plant diseases. A few of the systemics can be applied as soil treatments and are slowly absorbed through the roots to give prolonged disease control.

Systemics offer much better control of diseases than is possible with a protectant fungicide that requires uniform application and remains essentially where it is sprayed onto the plant surfaces. There is, however, some redistribution of protective fungicidal residues on the surfaces of sprayed or dusted plants, giving them longer residual activity than would be expected.

Benzimidazoles. The benzimidazoles, represented by benomyl which was introduced in 1968, have received wide acceptance as systemic fungicides against a broad spectrum of diseases. Benomyl has the widest spectrum of fungitoxic activity of all the newer systemics, including control of *Sclerotinia, Botrytis, Rhizoctonia,* powdery mildews, and apple scab. Thiophanate was introduced in 1969, and although not a benzimidazole initially, it is converted to one by the host plant and the fungus through their metabolism. Thiophanate has a fungitoxicity similar to that of benomyl. Both compounds have been used in foliar applications, seed treatment, dipping of fruit or roots, and soil application.

Methyl thiophanate controls brown patch, fusarium blight, dollar spot, red thread, stripe smut and powdery mildew on turfs.

Systemic fungicides cover susceptible foliage and flower parts more efficiently than protectant fungicides because of their ability to translocate through the cuticle and across leaves. They bring into play the perfect method of disease control by attacking the pathogen at its site of entry or activity, and reduce the risk of contaminating the environment by frequent broadcast fungicidal treatments. Undoubtedly, as newer and more selective systemic molecules are synthesized, they will gradually replace the protectants which comprise the bulk of our fungicidal arsenal.

DISEASE RESISTANCE TO FUNGICIDES

Disease resistance to chemicals other than the heavy metals occurs commonly in fungal and on rare occasions in bacterial plant disease pathogens. Several growing seasons after a new fungicide appears, it becomes noticeably less effective against a particular disease. As our fungicides become more specific for selected diseases, we can expect the pathogens to become resistant. This can be attributed to the singular mode of action of a particular fungicide which disrupts only one genetically controlled process in the metabolism of the pathogen. The result is that resistant populations appear suddenly, either by selection of resistant individuals in a population or by a single gene mutation. Generally, the more specific the site and mode of fungicidal action, the greater the likelihood for a pathogen to develop a tolerance to that chemical.

DINITROPHENOLS

The basic dinitrophenol molecule has a broad range of toxicities. Compounds derived from it are used as herbicides, insecticides, ovicides, and fungicides. They act by preventing the utilization of nutritional energy.

One dinitrophenol has been used since the late 1930s, dinocap (Karathane®), both as an acaricide and for powdery mildew on a number of fruit and vegetable crops and ornamentals. Dinocap undoubtedly acts in the vapor phase since it is quite effective against powdery mildews whose spores germinate in the absence of water. This also is a popular home fungicide.

ALIPHATIC NITROGEN COMPOUNDS

Dodine (Cyprex®), a fungicide introduced in the middle 1950s, has proved effective in controlling certain diseases such as apple, pear and pecan scab, and cherry leaf spot, foliar diseases of strawberries, blossom brown rot on peaches and cherries (western states only), peach leaf curl (western states only), bacterial leafspot on peaches, leaf blight of sycamores and black walnuts. It has disease-specificity, and slight systemic qualities and is taken up rapidly by fungal cells.

ANTIBIOTICS

The antibiotic fungicides are substances produced by microorganisms, which in very dilute concentrations inhibit growth and even destroy other microorganisms. The largest source of antifungal antibiotics is the actinomycetes, a group of the lower plants. Within this group is one amazing species, ***Streptomyces griseus***, from which are derived streptomycin and cycloheximide.

Streptomycin is the only antibiotic recommended to the home gardener and is used as dusts, sprays and seed treatment. It will control bacterial diseases such as blight on apples and pears, soft rot on leafy vegetables, and some seedling diseases. It is also effective against a few fungal diseases.

THE HERBICIDES

Herbicides, or chemical weedkillers, have largely replaced mechanical methods of agricultural and industrial weed control in the past 30 years, especially in intensive and highly mechanized agriculture. Herbicides provide a more effective and economical means of weed control than cultivation, hoeing, and hand pulling (except in the home garden). Without the use of herbicides in agriculture, it would have been impossible to mechanize fully the production of cotton, sugar beets, grains, potatoes, and corn.

Other locations where herbicides are used extensively include areas such as industrial sites, roadsides, ditch banks, irrigation canals, fence lines, recreational areas, railroad embankments, and power lines. Herbicides remove undesirable plants that might cause damage, present fire hazards, or impede work crews. They also reduce costs of labor for mowing.

Herbicides are classed as selective when they are used to kill weeds without harming the crop and nonselective when the purpose is to kill all vegetation.

Selective and nonselective materials can be applied to weed foliage or to soil containing weed seeds and seedlings depending on the mode of action. True selectivity refers to the capacity of an herbicide, when applied at the proper dosage and time, to be active only against certain species of plants but not against others. But selectivity can also be achieved by placement, as when a nonselective herbicide is applied in such a way that it contacts the weeds but not the crop.

The classification of herbicides would be a simple matter if only the selective and nonselective categories existed. However there are multiple-classification schemes which may be based on selectivity, contact vs. translocated, timing, area covered, and chemical classification. One method of classifying the herbicides is presented in Table 6.

Each herbicide affects plants either by contact or translocation. Contact herbicides kill the plant parts to which the chemical is applied and are most effective against *annuals*, those weeds that germinate

TABLE 6. Herbicide classification chart.

Herbicide (Common Name)	Selective			Nonselective		
	Contact	Foliage Translocated	Soil Residual	Contact	Foliage Translocated	Soil Residual
Amitrol				X	X	
Benefin			X			X
Bensulide			X			X
Bromoxynil				X	X	
Cacodylic acid				X		
CDEC						X
Dalapon		X			X	
DCPA						X
Dichlobenil						X
Diphenamid			X			X
Diquat				X		
DSMA		X		X	X	
EPTC						X
Glyphosate					X	
MSMA		X		X	X	
Petroleum oils				X		
Trifluralin						X
2,4-D amine		X				

from seeds and grow to maturity each year. Complete coverage is essential in weed control with contact materials.

The translocated herbicides are absorbed either by roots or aboveground parts of plants and are then moved within the plant system to distant tissues. Translocated herbicides may be effective against all weed types, however their greatest advantage is seen when used to control established *perennials*, those weeds that continue their growth from year to year. Uniform application is needed for the translocated materials, whereas complete coverage is not required.

Another method of classification is the timing of herbicide application with regard to the stage of crop or weed development. The three categories of timing are preplanting, preemergence, and postemergence.

Preplanting applications for control of annual weeds are made to an area before the crop is planted, within a few days or weeks of planting. Preemergence applications are completed prior to emergence of the crop or weeds, depending on definition, after planting. Postemergence applications are made after the crop or weed emerges from the soil.

Broadcast applications cover the entire area, including the crop. Spot treatments are confined to small areas of weeds. Directed sprays are applied to selected weeds or to the soil to avoid contact with the crop.

THE INORGANIC HERBICIDES

The first chemicals utilized in weed control were inorganic compounds. These were brine, and a mixture of salt and ashes, both of which were used as early as Biblical times by the Romans to sterilize the land. In 1896, copper sulfate was used selectively to kill weeds in grain fields.

The borate herbicides are a class of inorganics available to homeowners. For example, sodium tetraborate or sodium metaborate can be purchased for removal of all plant growth in an area. Use with caution, because they are leached to roots below the weed level. The amount of boron or boric acid determines their effectiveness. Borates are absorbed by plant roots, translocated to above-ground parts, and are nonselective, persistent herbicides. Boron accumulates in the reproductive structures of plants, causing them to become stunted and die. These are still used to give a semi-permanent form of sterility to areas where no vegetation of any sort is wanted, such as driveways, gravel lawns, alleys and other barren areas.

Though organic herbicides are not superior to inorganic ones, intensive EPA restrictions have been placed on some of the inorganics because of their persistence in soils. The inorganic herbicides are not wise choices for use around the home except with great care when removing all vegetation from an area.

THE ORGANIC HERBICIDES

PETROLEUM OILS

Petroleum oils were the earliest organic herbicides. They are a complex mixture of long-chain hydrocarbons containing traces of nitrogen- and sulfur-linked compounds.

The petroleum oils are effective contact herbicides for all vegetation. Homeowners today can make use of old crank case oil, gasoline, paint thinner, kerosene, Stoddard's solvent and diesel oil for spot treatments. This is not, however, an environmentally sound selection, in view of the efforts to save energy and avoid adding hydrocarbons to our atmosphere. Though only temporary in effect, they are fast-acting and probably the safest of the materials available to use around the home. The petroleum oils exert their lethal effect by penetrating and disrupting plasma membranes, almost immediately giving treated plants a frost-bitten appearance.

ORGANIC ARSENICALS

Widely used as agricultural herbicides, and in some instances on turf, are the organic arsenicals, namely the arsinic and arsonic acid derivatives. Cacodylic acid (dimethylarsinic acid), a derivative of arsinic acid, is used for the seasonal killing of lawns for their conversion to winter or summer lawns. Salts of arsonic acid, disodium methanearsonate (DSMA) and monosodium acid methanearsonate (MSMA), can be used for perennial weed control in Bermudagrass turf. The organic arsenicals are much less toxic to mammals than the inorganic forms, are crystalline solids, and are relatively soluble in water.

Arsonates are absorbed and translocated to underground tubers and rhizomes, making them extremely useful against Johnsongrass and nut sedges, when applied as spot treatments.

PHENOXY HERBICIDES

An organic herbicide introduced in 1944, later to be known as 2,4-D was the first of the "phenoxy herbicides," "phenoxyacetic acid derivatives," or "hormone" weed killers. These were highly selective for broadleaved weeds and were translocated throughout the plant. 2,4-D provided most of the impetus in the commercial search for other organic herbicides in the 1940s. There are several compounds belonging to this group, of which 2,4-D, 2,4,5-T and Silvex are the most familiar.

The phenoxy herbicides cause responses in broadleaf plants resembling those of auxins (growth hormones). They affect cellular division, and result in unusual and rapid growth in treated plants.

2,4-D and 2,4,5-T have been used for years in gargantuan volume world wide with no adverse effects on human or animal health. Recently, however, 2,4,5-T used mainly for control of woody perennials, became the subject of heavy investigation, particularly on its use in Vietnam in the form of "Agent Orange". Certain samples were found to contain excessive amounts of a highly toxic impurity, tetrachlorodioxin, commonly referred to as dioxin. Slight alterations in manufacturing procedures brought this dioxin content down to acceptable levels after the source was determined. The uses of 2,4,5-T and Silvex were banned in 1979, thus, they are no longer available for home use.

SUBSTITUTED AMIDES

The amide herbicides are relatively simple chemical compounds that are easily degraded by plants and soil. CDEC (Vegadex®) inhibits the germination or early seedling growth of most annual grasses.

It is well suited for use in sandy soils and controls many annual grasses and broadleaf weeds in vegetable gardens (Table 28), ornamentals and shrubbery.

Diphenamid is used as a preemergence soil treatment and has little contact effect. Most established plants are tolerant to diphenamid, because it affects only seedlings. It persists in the soil from 3 to 12 months and can be used for the control of annual grass and broadleaf weeds in dichondra, iceplant, and Bermudagrass lawns and for other ornamental plants.

NITROANILINES

Examples of this group, which have high preemergence herbicidal action, are trifluralin and benefin. Trifluralin has very low water solubility, which minimizes leaching and movement away from the target. The nitroanilines inhibit both root and shoot growth when absorbed by roots. Benefin is effective in the control of annual grasses and broadleaf weeds in established turf.

SUBSTITUTED UREAS

The substituted ureas are a group of compounds used primarily as selective preemergence herbicides. The only one of this group recommended for use around the home is siduron. The ureas are strongly absorbed by the soil, then absorbed by roots. They inhibit photosynthesis, the production of plant sugars with the energy from light.

Siduron is used only for the establishment of lawns. It is applied as a preemergence herbicide for the control of annual weed grasses, such as crabgrass, foxtail, and barnyardgrass, in newly seeded or established plantings of bluegrass, fescue, redtop, smooth brome, perennial ryegrass, orchard grass, and certain strains of bentgrass.

CARBAMATES

The esters of carbamic acid are physiologically quite active. As we have seen, some carbamates are insecticidal, while others are fungicidal. Some carbamates also are herbicidal. Discovered in 1945, the carbamates are used primarily as selective preemergence herbicides, but some are also effective as postemergence ones.

EPTC belongs to the carbamates and is the only one of this category recommended for use around the home. In this instance, EPTC is recommended only in the granular formulation for nutsedge control in and around woody ornamentals.

TRIAZINES

The triazines are strong inhibitors of photosynthesis. Their selectivity depends on the ability of tolerant plants to degrade or metabolize the herbicide whereas the susceptible plants do not. Triazines are applied to the soil primarily for their post-emergence activity.

There are many triazines on the market today, two of which are used around the home. Amitrole may be used in the home garden, with great care, for complete weed control, and simazine is used where all vegetation needs to be controlled, for example, around patios, walks and driveways.

ALIPHATIC ACIDS

One heavily used herbicide in this group is dalapon, used against grasses, particularly our old enemies (or friends, depending on your view), quackgrass and Bermudagrass. Dalapon is widely used around homes to control Johnsongrass, Bermudagrass and other perennial grasses. It is also effective against cattails and rushes. It is translocated to the roots of most grasses and acts as a growth regulator.

ARYLALIPHATIC ACIDS

A number of these materials are employed as herbicides, and are applied to the soil against germinating seeds and seedlings.

DCPA (Dacthal®) produces auxin-like growth effects in plants similar to those of 2,4-D. It is approved for use on turf and ornamentals, and is effective against smooth and hairy crabgrass, green and yellow foxtails, fall panicum, and other annual grasses in addition to the perennial witchgrass. It is also useful against certain broadleaf weeds such as carpet weed, dodder, purslane, and common chickweed.

SUBSTITUTED NITRILES

Nitriles are organic compounds containing a cyanide group. There are several substituted nitrile herbicides, which have a wide spectrum of uses against grasses and broadleaf weeds. Their actions are broad, involving seedling growth and potato sprout inhibition. They are fast acting which is attributed primarily to rapid permeation.

Dichlobenil is used for weed control around woody ornamentals, nurseries and in fruit orchards. Bromoxynil is used on industrial sites, vacant lots, driveways, and other areas to be freed from weeds for post-emergent control of weeds such as blue mustard, fiddleneck, corn gromwell, cowcockle, field pennycress, green smartweed, groundsel, lambsquarters, London rocket, shepherdspurse, silverleaf nightshade, tartary buckwheat, tarweed, tumble mustard, wild buckwheat, and wild mustard.

BIPYRIDYLIUMS

There are two important herbicides in this group, diquat and paraquat. Both are contact herbicides that damage plant tissues quickly, causing the plants to appear frostbitten because of cell membrane destruction. Both materials reduce photosynthesis and are more effective in the light than in the dark. Neither material is active in soils. Diquat is the only one recommended and available to the amateur.

MISCELLANEOUS HERBICIDES

One of the better turf herbicides, especially for the control of crabgrass, is bensulide. It is an organophosphate but is considered one of the less toxic herbicidal materials. Bensulide acts by inhibiting cell division in root tips. It is used as a preemergence herbicide in lawns to control certain grasses and broadleaf weeds, But it fails as a foliar spray because it is not translocated.

Glyphosate (Roundup®) belongs to the organophosphate classification, though it is indeed not an insecticide, but a very popular herbicide. It is useful for the control of many annual and perennial grasses and broadleaf weeds. A foliarly applied, translocated herbicide, it may be applied in spring, summer, or fall. It is ideal for general nonselective weed control in noncrop areas such as in industrial areas and in turfgrass establishment or renovation.

Many chemical and use classes of herbicides are available. Those discussed here are but a cross-section of the existing herbicides since they are available for use around the home (except paraquat). We can expect to see new and different classes develop in the future just as in the past.

The home gardener is urged to review the compound label carefully to gain an understanding of why a certain herbicide is used in a particular way and of how to use it in ways that will not injure or kill wanted plants.

THE MOLLUSCICIDES

Metaldehyde has been used in baits for the control of slugs and snails commercially and around the home since its discovery in 1936. Its continued use and success can be attributed both to its attractant and toxicant qualities. Many other materials have been used as baits, sprays, fumigants, and contact toxicants, with but little success.

The other three toxicants are all carbamate insecticides that have proved very successful as bait formulations. Carbaryl (Sevin®) was the first successful carbamate insecticide and has very low mammalian oral and dermal toxicity.

Mexacarbate (Zectran®) has been prepared as a snail and slug bait for use in ornamental plantings to be applied only to the soil.

Mercaptodimethur (Mesurol®) is another successful insecticide registered for use against snails and slugs on ornamentals. It is highly effective against these pests and has also demonstrated repellency to several bird species.

THE NEMATICIDES

Nematodes are covered with an impermeable cuticle which provides them with considerable protection. Chemicals with outstanding penetration characteristics are required, therefore, for their control.

Nematicides are seldom used by the homeowner unless he has a greenhouse or cold frames. For the most part we can say that nematicides are not and should not be used by the layman due mainly to their hazard.

Those that are available commercially fall into four groups: (1) halogenated hydrocarbons (some of which were described earlier), (2) isothiocyanates, (3) organophosphate insecticides, and (4) carbamate or oxime insecticides. None of the latter two groups are available to the homeowner.

Most of today's nematicides are soil fumigants, volatile halogenated hydrocarbons. To be successful they must have a high vapor pressure to spread through the soil and contact nematodes in the water films surrounding soil particles.

HALOGENATED HYDROCARBONS

The nematicidal properties of a mixture of dichloropropane and dichloropropene (DD) were discovered in 1943 and effectively launched the use of volatile nematicides on a field-scale basis. Previously only seed beds, greenhouse beds, and potting soil were treated with the materials like chloropicrin, carbon disulfide, and formaldehyde. These were very expensive, in some instances explosive, and usually required a surface seal because of their relatively high vapor pressures.

DD is injected into the soil several days before planting to kill nematodes, eggs and soil insects. Dibromochloropropane (DBCP) was once available to homeowners, but is no longer because of a recent decision by the Environmental Protection Agency.

ISOTHIOCYANATES

In this classification belongs sodium methyldithiocarbamate (SMDC). It is a dithiocarbamate mentioned under the fungicides, but it is activated in the soil and is effective against all living matter in the soil, including nematodes and their eggs.

THE RODENTICIDES

Most of the methods used to control rodents are aimed at destroying them. Poisoning, shooting, trapping, and fumigation are among the methods selected. Of these, poisoning is most widely used and probably the most effective and economical. Because rodent control is in itself a diverse and complicated subject, we will mention only those more commonly used rodenticides.

Rodenticides differ widely in their chemical nature. Strange to say, they also differ widely in the hazard they present under practical conditions, even though all of them are used to kill animals that are physiologically similar to man.

COUMARINS (ANTICOAGULANTS)

The most successful group of rodenticides are the coumarins, represented classically by Warfarin®. There are five compounds belonging to this classification, all of which have been very successful rodenticides. Their mode of action is twofold: (1) inhibition of prothrombin, the material in blood responsible for clotting, and (2) capillary damage, resulting in internal bleeding. The coumarins require repeated ingestion over a period of several days, leaving the unsuspecting rodents growing weaker daily. The coumarins are thus considered relatively safe, since repeated accidental ingestion would be required to produce serious illness. In the case of most other rodenticides a single accidental ingestion could be fatal to man.

Of the several types of rodenticides available, only those with anticoagulant properties are safe to use around the home.

Warfarin was released in 1950 by Wisconsin Alumni Research Foundation (thus its name WARF coumarin, or WARFarin). It was immediately successful as a rat poison because rats did not develop "bait shyness" as they did with other baits during the required ingestion period of several days.

Fumarin®, is another commonly used material, and has had more recent success, following Warfarin.

Two new coumarin rodenticides are available, brodifacoum (Talon®) and bromadiolone (Maki®). These differ from the earlier coumarins in that, although they are anticoagulant in their mode of action, they require but a single feeding for rodent death to occur within 4 to 7 days. They are both effective against rodents that are resistant to conventional anticoagulants.

To protect children and pets from accidental ingestion of these highly toxic anticoagulants, the treated baits must be placed in tamper-proof boxes or in locations not accessible to children.

These rodenticides are relatively specific for rodents in that they are offered as rodent-attractive, premixed baits inaccessible to pets and domestic animals, and their relative toxicity to other warm-blooded animals is low. For instance, the rat is 2 times more sensitive to brodifacoum than the pig and dog, 36 times more than the chicken, and 90 times more sensitive than the cat.

INDANDIONES

Two compounds, diphacinone (Promar®, Ramik®) and pindone (Pival®), belong to this class, have the same remarkable anticoagulant property as Warfarin, and have replaced Warfarin where rodent avoidance behavior (bait shyness) made it ineffective. Sold as baits containing up to 0.0075% of the toxicant, they must be ingested for several consecutive days before they become effective. Because of this characteristic, the anticoagulants provide a definite safety factor for children and animals.

RODENT REPELLENTS

"Rodent" is rather all-inclusive and perhaps a bit deceptive, because all rodent repellents are not registered for all rodents. This requires the accurate identification of the particular pest and selection of one of the following materials or combinations which clearly indicates that pest on its label: Biomet-12 (tri-n-butyltin chloride) naphthalene, paradichlorobenzene, polybutanes, polyethylene, R-55 (tert-butyl dimethyltrithioperoxycarbamate), and thiram.

THE AVICIDES

The old, general purpose poison, strychnine is registered by EPA as an avicide for the control of English sparrows and pigeons when used as a 0.6% grain bait. Strychnine acts very quickly leaving the treated area strewn with dead birds which should be removed at regular intervals. Prebaiting for several days is necessary before distributing the treated bait. Treated grain should be colored before distribution for bird protection and uneaten bait removed.

Ornitrol® is a new and apparently humane approach to bird control. It is a chemosterilant, a birth control agent for pigeons, designed to control population growth rather than eradicate. It causes temporary sterility to pigeons after a 10-day feeding exposure by inhibiting egg production. It has little, if any, activity on mammals, and is selective toward pigeons when impregnated on whole corn grains too large for other species to feed on. It is formulated as a 0.1% bait.

BIRD REPELLENTS

Bird repellents can be divided into three categories: (1) olfactory (odor), (2) tactile (touch), and (3) gustatory (taste). In the first category, only naphthalene granules or flakes are registered by the EPA. It should come as no surprise that naphthalene is repellent to all domestic animals as well.

Tactile repellents are made of various gooey combinations of castor oil, petrolatum, polybutane, resins, diphenylamine, pentachlorophenol, quinone, zinc oxide and aromatic solvents applied as thin strips or beads to roosts, window ledges and other favorite resting places.

The taste repellents are varied and somewhat surprising in certain instances, since they have other uses. The fungicides captan and copper oxalate are examples and are used as seed treatments to repel seed-pulling birds. Two other popular seed treatments are anthraquinone and Glucochloralose®. Turpentine, an old stand-by with multiple uses, can also be used as a seed treatment.

THE ALGICIDES

COPPER COMPOUNDS

Rather for industrial, public water systems and agriculture but not for the swimming pools are the organic copper complexes. Most commonly used among these are the copper-triethanolamine materials. Generally they can be used as surface sprays for filamentous and planktonic forms of algae in potable water reservoirs, irrigation water storage and supply systems, farm, fish and fire ponds, lake and fish hatcheries, and aquaria. The treated water can be used immediately for its intended purpose. The same generalities apply also to another organic copper compound, copper ethylenediamine complex.

QUATERNARY AMMONIUM HALIDES

The quaternary ammonium (QA) compounds include a host of algicides. The QA's are characterized as containing a chlorine or bromine and nitrogen.

They are general purpose antiseptics, germicides and disinfectants, ideal for algae control in the greenhouse as pot dips, and wall, bench and floor sprays, swimming pool and recirculation water systems. Algae control lasts up to several months.

The alkyldimethylbenzylammonium chlorides can be used to give long-lasting control of algae and bacteria in swimming pools, cooling systems, air conditioning systems and glass houses, and compose the bulk of the quaternary ammonium halides. They are toxic to fish and consequently cannot be used in fish ponds or aquaria. Dimanin C® (sodium dichloroisocyanurate) also gives long-lasting control of algae in swimming pools and can be used as a disinfectant.

Aquazine®, a herbicide introduced earlier as simazine, is also registered by EPA as an algicide. It performs well at 1.0 ppm against a broad spectrum of algae and may be used in fish ponds but not in swimming pools. It is only an algicide and does not purify pools by killing bacteria.

Dichlone, an agricultural fungicide, is also registered as an algicide for use against blue-green algae in lakes and ponds, including your fish pond, It too, is not a disinfectant.

THE DISINFECTANTS

Disinfectants are an almost limitless number of chemical agents for controlling microorganisms, and new ones appear on the market regularly. A common problem confronting persons who must utilize disinfectants or antiseptics is which one to select and how to use it. There is no single ideal or all-purpose disinfectant, thus the compound to choose is the one that will kill the organisms present in the shortest time, with no damage to the contaminated substrate.

The EPA has registered currently approximately 440 active ingredients as disinfectants. There are 23 major use categories further divided into 200 use sites. Thus, the reader quickly sees the difficulty in making specific, precise recommendations.

Before exploring the various chemicals referred to as disinfectants, it is necessary to make a quick distinction between two confusing words, antisepsis and sanitation. Antisepsis is the disinfection of skin and mucous membranes, while sanitation is the disinfection of inanimate surfaces. Consequently, much more severe treatment can be used for sanitation than for antisepsis. We will direct our presentation only to the use of disinfectants used in sanitation.

THE PHENOLS

Phenol can probably be called the oldest recognized disinfectant, because Dr. Lister used phenol in the 1860's as a germicide in the operating room. When greatly diluted, its deadly effect is due to protein precipication. It is used as the standard for comparison of the activities of other disinfectants in terms of phenol coefficients. Phenol and the cresols have very distinct odors, which change little with modification of their chemical structures. Lysol® and its mimics belong to this group. The addition of chlorine or a short chain organic compound increases the activity of the phenols. For instance, hexachlorophene is one of the most useful of the phenol derivatives but is in a state of some disrepute at this writing. After much public debate, it was removed from all across-the-counter items, including soaps and shampoos, and can be dispensed only with a physician's prescription.

THE HALOGENS

These are compounds containing chlorine, iodine, bromine or fluorine, both organic and inorganic. Generally, the inorganic halogens are deadly to all living cells.

Chlorine. Chlorine was first used as a deodorant and later as a disinfecting agent. It is a standard treatment for drinking water in all communities of the United States. Hypochlorites are those most commonly used in disinfecting and deodorizing procedures because they are relatively safe to handle, colorless, good bleaches and do not stain. Several organic chlorine derivatives are used for the disinfection of water, particularly for campers, hikers and the military. The most common of these are halazone and succinchlorimide. Chlorine is the dominant element in disinfectants, in that roughly 25% of those registered with EPA contain one or more atoms of this important halogen.

Hypochlorites. Calcium hypochlorite and sodium hypochlorite mentioned earlier as algicides, are popular compounds, widely used both domestically and industrially. They are available as powders or liquid solutions and in varying concentrations depending on the use. Products containing 5 to 70% calcium hypochlorite are used for sanitizing dairy equipment and eating utensils in restaurants. Solutions of sodium hypochlorite are used as a household disinfectant; higher concentrations of 5 to 12% are also used as household bleaches and disinfectants and for use as sanitizing agents in dairy and food-processing establishments.

Chloramines. Another category of chlorine compounds used as disinfectants, or sanitizing agents are the chloramines. Three chemicals in this group are monochloramine, Chloramine-T®, and azochloramide. One of the advantages of the chloramines over the hypochlorites is their stability and prolonged chlorine release.

The germicidal action of chlorine and its compounds comes through the formation of hypochlorous acid when free chlorine reacts with water. Similarly, hypochlorites and chloramines also form hypochlorous acid. The hypochlorous acid formed in each instance is further decomposed releasing oxygen. The oxygen released in this reaction (nascent oxygen) is a strong oxidizing agent, and through its action on cellular constituents, microorganisms are destroyed. The killing of cells by chlorine and its compounds is also due in part to the direct combination of chlorine with cellular materials.

Iodine. Water or alcohol solutions of iodine are highly antiseptic and have been used for decades before surgical procedures. Several metallic salts, such as sodium and potassium iodide, are registered as disinfectants, but the number of compounds containing iodine nowhere approaches those containing chlorine.

THE DETERGENTS

These are organic compounds which have two chemical ends or poles. One end is hydrophilic and mixes well with water while the other is hydrophobic and does not. As a result, the compounds orient themselves on the surfaces of objects with their hydrophilic poles toward the water. Basically, these may be classed as ionic or non-ionic detergents. The ionic are either anionic (negatively charged) or cationic (positively charged). The anionic detergents are only mildly bactericidal. The cationic materials, which are the quaternary ammonium compounds, are extremely bactericidal, especially for *Staphylococcus*, but do not affect spores. Hard water, containing calcium or magnesium ions, will interfere with their action, and they also rust metallic objects. Even with these disadvantages, the cationic detergents are among the most widely used disinfecting chemicals, since they are easily handled and are not irritating to the skin in concentrations ordinarily used.

THE ALDEHYDES

Combinations of formaldehyde and alcohol are outstanding sterilizing agents, with the exception of the residue remaining after its use. A related compound, glutaraldehyde, in solution is as effective as formaldehyde. Most organisms are killed in 5 minutes exposure to glutaraldehyde, while bacterial spores succumb in 3 to 12 hours.

There are other classes of disinfectants including a number of dyes, acids and alkalies, alcohols, peroxides, and fumigants (ethylene oxide and methyl bromide) which are very effective under certain conditions. However, due to the scope of this book, they are only mentioned in passing.

Remember that no single chemical antimicrobial agent is best for any and all purposes. This is not surprising in view of the variety of conditions under which agents may be used, differences in modes of action, and the many types of microbial cells to be destroyed.

DEODORANTS

There are frequent occasions following the control of small mammal pests when there arises the need for a deodorant, for example, when a mouse or rat dies in some unreachable void within the walls or in a crawl space, or bat guano becomes pungent. The mere presence of rodents in large numbers will result in a musky, offensive odor. The odors from small home fires, cigarette and cigar smoke, and cooking odors can be equally offensive. Deodorants can provide partial or complete relief.

To handle unpleasant odors, isobornyl acetate, Neutroleum Alpha®, (see Skunks) Super Hydrasol®, quaternary ammonium compounds, Styamine 1622, Zephiran chloride, and Bactine® can be used as an aerosol, mist spray, or in a bowl or bottle with a tissue or cotton wick. These are more than masking agents. They are deodorants, in that they react chemically with the odors. These are the materials the professionals use.

Several of the old familiar masking agents also can be used. These are oil of pine, oil of peppermint, oil of wintergreen, formalin, and anise. For example, 10 drops of pine oil in one gallon of water can be applied with an atomizer or with a mist sprayer to provide relief. Unless you like the fragrance of the men's gym, I recommend not using oil of wintergreen.

CHAPTER 4

A good gardener never blames his tools.

American Proverb.

THE EQUIPMENT

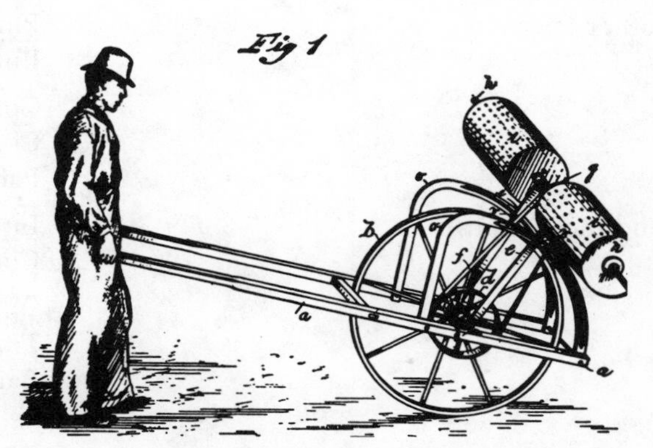

"Original 1881 Patent Drawing of a Manually Operated Row-Crop Duster"

EQUIPMENT FOR HOME PEST CONTROL

Pesticide application equipment must be used for the purposes and formulations for which it was intended to produce the maximum effect at minimum cost and to avoid hazards to humans, domestic animals, pets, plants and other nontarget organisms. Most application equipment distributes pesticides, whether sprays, dusts, granules, or aerosols, to produce uniform coverage. That describes the scientific aspects of using application equipment. The art depends on the user for good judgment and smooth even movements of the "business end" of the equipment.

SELECTING THE PROPER EQUIPMENT

Much could be written about the selection and proper use of equipment for a particular pest control job. But in the good old American tradition, we "make do" with what we have. Almost any kind of pesticide application could be carried out with any one of the sprayers or dusters described, with of course, a couple of miniature-sized exceptions. Table 7 and Figure 1 contain a list of various home pests and the choices of equipment available to do the job.

In summary, the basic arsenal for the homeowner would be (1) a 2 or 3 gallon compressed air sprayer with 2 interchangeable nozzles, a cone-type and flat fan, and (2) a plunger duster. The following figure provides a general guide for selecting the proper equipment for the job. In the left hand column select the outdoor housekeeping task pertinent to your interest. Filled squares at the right indicate types of applicators suggested for that use. Consider all your various jobs to find the type of equipment that will do a wide range of work for you. Applications to lawn grasses can be done with sprayers only. In all other cases your choice can be sprayers or dusters, as you prefer.

TABLE 7. Pest Problems and Equipment to Use.

PESTS	USEFUL EQUIPMENT
Inside the Home	
1. Crawling insects and mites (ants, roaches silverfish, clover mites)	Continuous hand sprayer Paint brush Plunger duster
2. Fleas	Same as above
3. Ticks	Same as above
4. Pests on houseplants	Same as above
5. Clothes moths and carpet beetles infesting carpet and other fabrics	Continuous hand sprayer Plunger duster Bulb duster
6. Spiders	Continuous hand sprayer Compressed-air sprayer Paint brush
7. Flying insects (flies and mosquitoes)	Intermittent hand sprayer Continuous hand sprayer Aerosol bomb
8. Mice	Traps Bait Stations
Outside the Home	
1. Insects and plant diseases	Compressed-air sprayer Knapsack sprayer Hose-end sprayer Power sprayer Plunger duster Rotary duster Knapsack duster Granule spreader
2. Annoying insects a. On porch, patio, or terrace	Continuous hand sprayer Compressed-air sprayer Hose-end sprayer Aerosol bombs
b. Refuse disposal areas (garbage cans, compost pile, lawn clippings)	Compressed-air sprayer Hose-end sprayer Plunger duster Rotary duster
3. Weeds (dandelions, plantain, crabgrass)	Compressed-air sprayer with fan nozzle Power sprayer with fan nozzle Killer cane (spot treatment) Granules spreader
4. Rodents and other small mammals	Traps Bait stations

FIGURE 1. Suggested Pesticide Application Equipment Guide for Home Pest Control Jobs.

			Shaker Can	Pressure Can	Trigger Sprayer	Contin-uous Sprayer	Hose-End Sprayer	Bucket Pump Sprayer	Slide Pump Sprayer	Compress-ed Air Sprayer	Knapsack Sprayer	Wheel-barrow Sprayer	Power Sprayer	Mist Blower	Plunger Duster	Crank Duster	Drop Spreader	Centrif-ugal Spreader	Root Irrigator
Lawn Weeds		Small lots			X		X	X	X	X							X	X	
		Large lots								X	X	X	X				X	X	
Houseplants				X	X	X													
House Interior		Fliers		X															
		Crawlers	X	X	X	X				X					X				
Lawn Insects & Diseases		Small	X				X			X							X	X	
		Large					X			X	X		X				X	X	
Shrubs & Evergreens		24 or less		X	X		X		X	X					X	X			X
		25-50						X	X	X	X	X	X	X	X	X			
		50 or more									X	X	X	X		X			
Gardens Flowers & Vegetables		500 sq ft			X	X	X	X	X	X	X				X	X			
		500-1000							X	X	X	X	X		X	X			
		Over 1000								X	X	X	X	X		X			
Trees & Fruit Trees		1-5						X	X	X				X					X
		Over 5										X	X	X					X
Outdoor Comfort	SMALL LOTS	Mosquitoes					X		X	X				X		X			
		Flies Wasps		X					X	X					X				
		House Invaders	X		X					X					X	X			
	LARGE LOTS	Mosquitoes							X	X	X		X	X		X			
		Flies							X	X				X					
		House Invaders	X	X	X					X	X	X	X			X			

(Modified from Cornell University's Misc. Bull. 74.)

HAND SPRAYERS

There are basic as well as novelty tools for applying pesticides, but sprayers are the most efficient means, and hand sprayers are the most important types of application equipment the homeowner should have. They vary in size and weight from small, hand-operated "flit guns" containing one-half pint to larger, heavier pieces of equipment holding several gallons.

Compressed Air or Tank-Type Sprayer. If you could own but one piece of equipment, it should be this, particularly if you own fruit trees. It is the mainstay of home pest control and can be used by anyone. It will last a lifetime, by replacing the various parts that pop, split and crack. It is truly the universal pesticide application unit, as seen in Figure 2. This unit offers a continuous spray pattern expelled by a head of compressed air and has interchangeable or adjustable nozzles. It does require rinsing, draining by inverting, and thorough drying after use to preserve its serviceable life.

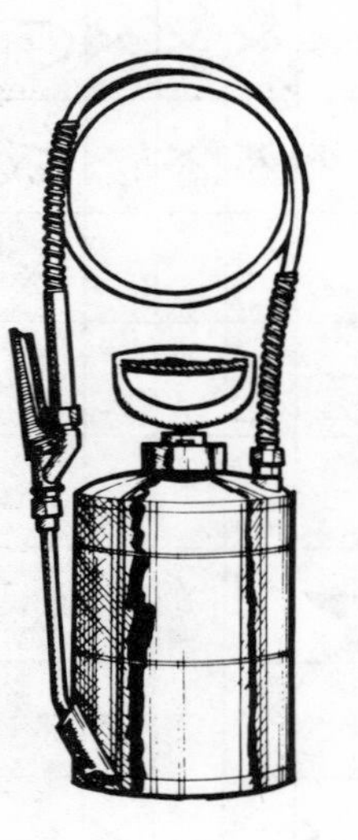

FIGURE 2. Compressed Air Sprayer
(U.S. Public Health Service)

Hand Pump Atomizer. These come in two styles. The first is the old-fashioned flit gun that emits a puff of spray with each plunger action, the intermittent sprayer. The second is a continuous sprayer that forces air into the tank to develop and maintain a constant pressure and deliver a continuous atomized spray. Most continuous hand sprayers have interchangeable or adjustable nozzles, and can deliver a fine-, medium-fine-, or coarse-droplet spray. These are designed basically to control flying insects in the home, particularly flies and mosquitoes. (Fig. 3).

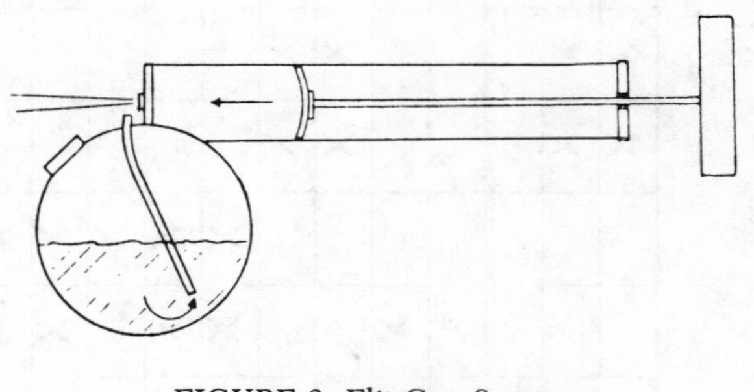

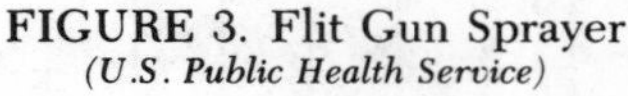

FIGURE 3. Flit Gun Sprayer
(U.S. Public Health Service)

Knapsack Sprayer. These are carried on the back of the operator, and have shoulder straps so that weight can be distributed evenly on both shoulders comfortably (Fig. 4). A diaphragm or piston pump and a mechanical agitator are mounted inside the tank and operated by a lever worked with the right hand. The pesticide is under liquid pressure during each stroke of the pump; thus, it produces a surging or pulsed spray. Knapsack sprayers are used chiefly for treatment of gardens, shrubs, and small trees. Adjustable nozzles are standard equipment. To extend the life of the pump and tank, rinse, drain, and dry after each use. This is not the most useful sprayer for the homeowner.

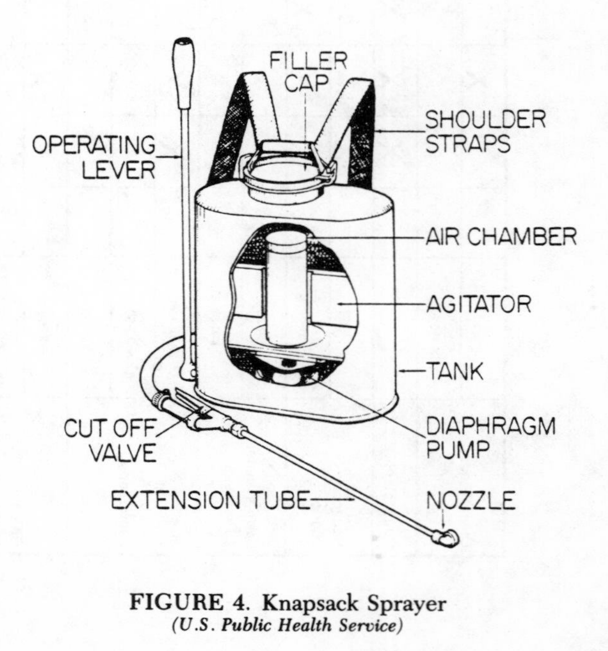

FIGURE 4. Knapsack Sprayer
(U.S. Public Health Service)

Side Pump or Trombone Sprayer. This bucket-mounted sprayer has a plunger and cylinder moving on each other with a trombone-like action, requiring the use of both hands. It, too, produces a surging or pulsed spray. Some types can throw a stream of spray 20 to 30 feet vertically and are useful for treating medium sized fruit and shade trees (Fig. 5). Again, this would not be the recommended sprayer for the homeowner.

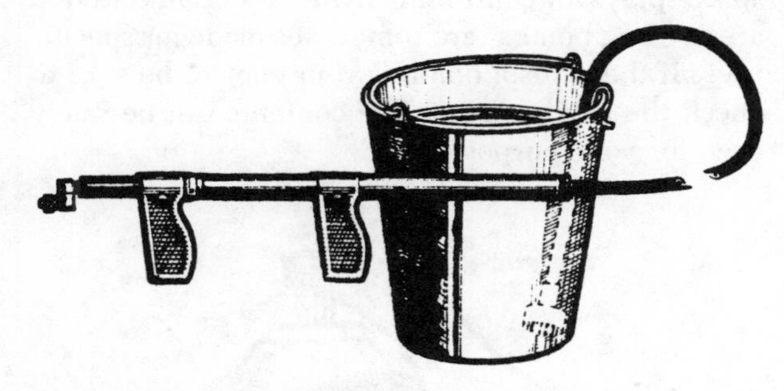

FIGURE 5. Trombone Sprayer
(U.S. Public Health Service)

Garden Hose Sprayer. These are designed to connect to a garden hose and utilize the household water supply and water pressure for dispersing the pesticide. It consists of a jar for holding concentrated spray material, a spray gun attached to the lid, and a suction tube from the gun to the bottom of the jar (Fig. 6). The gun siphons the spray concentrate from the jar and mixes it with the water flowing from the garden hose through the gun to produce a dilute spray. Although the primary design of this sprayer is for garden, lawn, and shrubbery insects, it is quite useful for controlling fleas, ticks, and chiggers in yards and for applying insecticides near the foundations of homes for cockroach, earwig and clover mite control. Its limitation is the length of the garden hose. Be certain the model you purchase contains a device to prevent back-siphoning. This is essential to prevent contamination of the water supply.

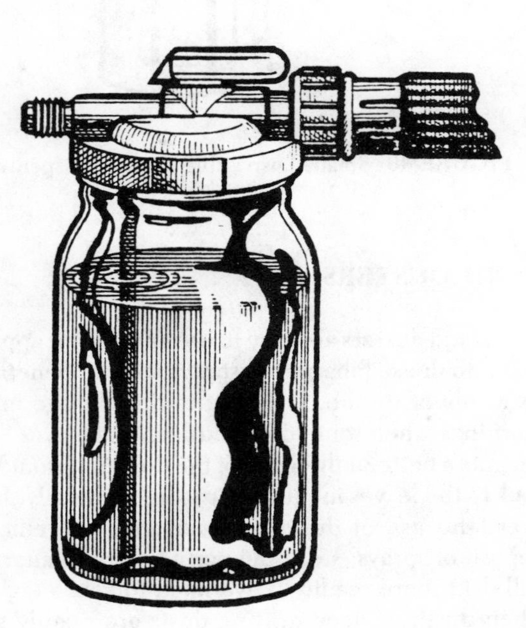

FIGURE 6. Garden Hose Sprayer
(U.S. Public Health Service)

Bucket Pump Sprayer and Wheelbarrow Sprayer. These are essentially the same, having hand pumps that utilize liquid pressure during each stroke of the pump, thus producing surging sprays. The only difference between the two is the mobility of the wheelbarrow and its increased tank capacity. They are useful mainly for spraying shrubs and small fruit and shade trees. The wheelbarrow sprayer can actually be used to treat practically all outdoor situations if a second person can be coerced into pushing the wheelbarrow as you pump and spray. These normally come equipped with interchangeable nozzles (Fig. 7).

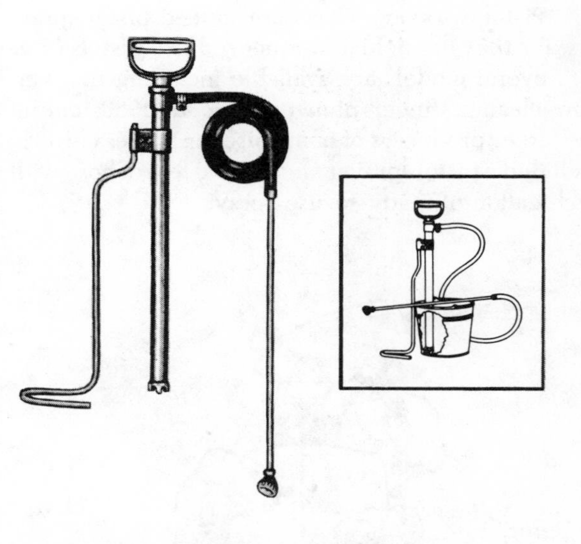

FIGURE 7. Bucket Pump Sprayer
(H. D. Hudson Co.)

Power Sprayers. Power sprayers take just a bit of the sport out of spraying around the home, because of their speed and ease of use. They can be purchased with gasoline or electrically powered pumps (Fig. 8).

Personally, I'll take the electric model. (Imagine getting prepared to spray your fruit trees with a dormant spray, first of the year, in mid-March, with no wind, and the gasoline engine refuses to start!) Power sprayers are usually piston pumps with a small head of air pressure, which reduces the surging to a minimum. They do require special care, cleaning and preparation for winter storage.

FIGURE 8. A Piston Pump Power Sprayer
(H. D. Hudson Co.)

Pistol Sprayer. These are indeed pistol sprayers, for they are held and squeezed like pistols (Fig. 9). Several models are available including the window-cleaner, finger-plunger type. In 1980, one of the larger producers of home-use pesticides cleverly included a pistol sprayer and several feet of hose with each gallon of ready-to-use spray.

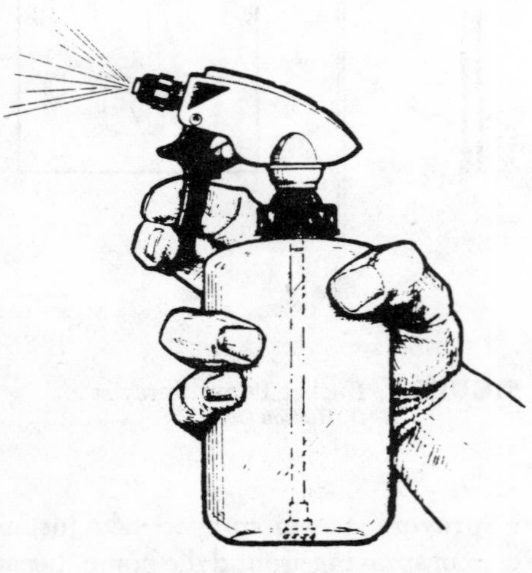

FIGURE 9. Pistol Sprayer
(U.S. Public Health Service)

Aerosol Bombs. Pressure cans and aerosol bombs (Fig. 10) containing pesticides are the most convenient of the spray applicators and also the most expensive. Insecticide aerosol cans totaled 132 million sold in 1978 (This accounted for only 5% of the total aerosol market for all uses in 1978.) (Fowler and Mahan, 1980). They are useful for making spot treatments to plants, to walls, cracks, and crevices both inside and outside the home and as atomizing space sprays for controlling flying insects in confined areas. "Bug bombs" are sometimes made for specific uses. If the aerosol bomb is your choice, be sure to check the label to see if the contents can be safely used for your purposes.

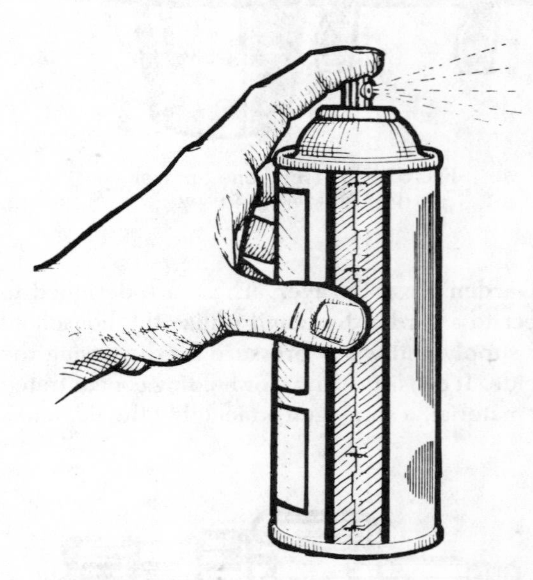

FIGURE 10. Small, Disposable Aerosol Dispenser
(U.S. Public Health Service)

HAND DUSTERS

Hand dusters are easy to use and permit application into dense foliage that sprays will not penetrate. As a rule of thumb, it is best to dust foliage in the mornings when some dew is still on the leaves. This permits a better adherence of the dust that would not stick to the leaves in mid-afternoon. Generally, however, the use of dusts does not give the efficient deposit of sprays, since the particles are smaller and will drift more readily. Give heed to your neighbor when dusting, since drifting dusts are readily seen and may be objectionable. Incidentally, many of the cans of dusts available may be used as self-contained dusters.

Plunger Duster. This duster is as important to dusts as the compressed air sprayer is to sprays. It consists of an air pump with a reservoir into which the air blast is directed to disperse the pesticide as a fine cloud or as a solid blast (Fig. 11). If the duster is turned so the delivery tube is beneath the dust, very heavy dust patterns will be produced, as needed for ant control. This duster can also be used to apply insecticides to control chiggers, ticks, fleas, and clover mites in and around lawns.

FIGURE 11. Hand Plunger Duster
(U.S. Public Health Service)

Knapsack Duster. This is a large bellows duster which is designed to be carried on the back, similar to the knapsack sprayer (Fig. 12). It consists of a large circular tank, the top of which forms the bellows operated by the up-down motion of a side lever. Dust is inserted through a cover on the back of the machine. It is used for treating flower and vegetable gardens, and shrubs. Because it emits puffs of dust rather than a uniform stream, it is not well suited for lawn and turf applications.

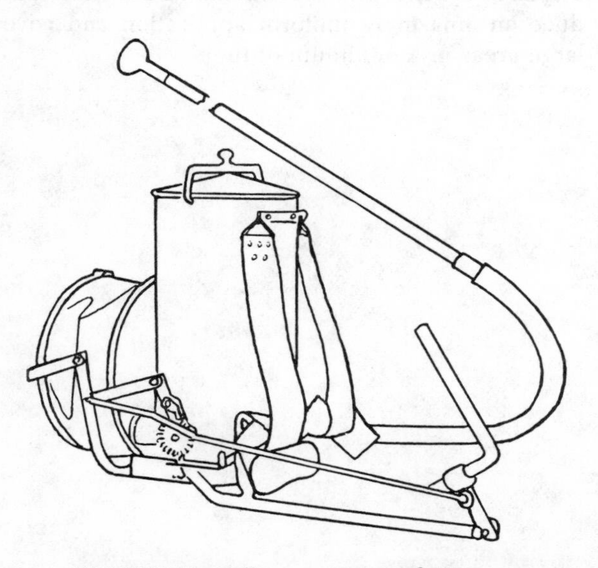

FIGURE 12. Bellows or Knapsack Duster
(USDA)

Rotary Hand Duster. This unit has a 5 to 10 pound capacity hopper from which dust is fed into a fan case, producing a steady stream of dust when the crank is turned (Fig. 13). Rotary dusters can be adjusted to deliver from 5 to 20 pounds of dust per acre with average walking speeds of 2 to 2½ mph. The advantage these have over knapsack models is the continuous dust pattern which permits their use on row crops and lawns. Most come equipped with a detachable fan-tip to produce a broad dust band for area treatment or broad plants such as melons and cucumbers.

FIGURE 13. Rotary Hand Duster
(U.S. Public Health Service)

Bulb Duster. This handy little gadget is designed for careful indoor placement of dusts. A 4-inch rubber bulb is equipped with a screw-cap that holds a small-orifice dust nozzle (Fig. 14). After the bulb is filled with dust, and the cap replaced, hand pressure on the bulb dispenses the dust. It is used where careful placement and neatness are required, such as crevices where cockroaches and silverfish hide, or behind baseboards where carpet beetles are found.

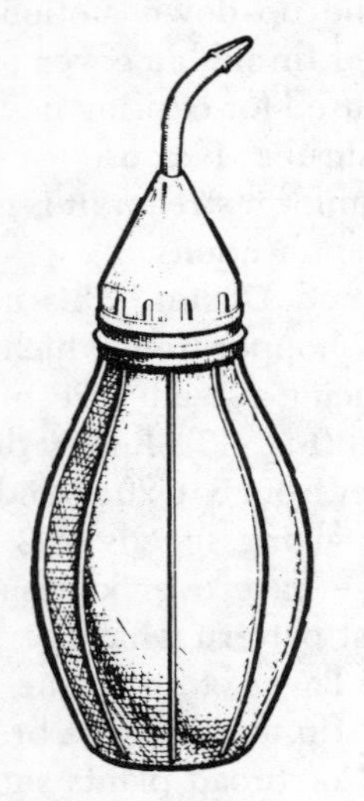

FIGURE 14. Bulb Duster
(U.S. Public Health Service)

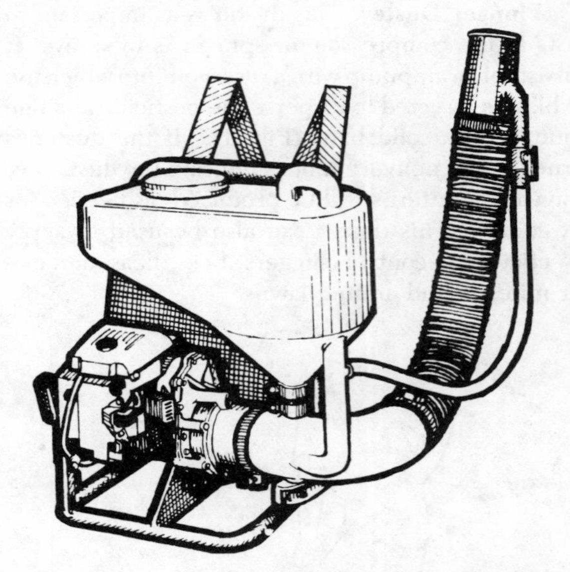

FIGURE 15. Small Portable Power Duster-Mister
(U.S. Public Health Service)

Power Duster. This light-weight unit is designed to be carried on the back of the operator and weighs from 50 to 60 pounds fully loaded, including 10 to 20 pounds of dust. It consists of a hopper, a small gasoline engine which operates a radial fan, and an air discharge nozzle into which is fed a metered amount of dust. Some models are mist/dust blowers (Fig. 15) and can be used to apply liquids, dusts or granules.

GRANULAR APPLICATORS

Granular applicators have the advantage of serving dual purposes, the application of fertilizers and pesticides. There are two main types used around the home, the two-wheeled fertilizer distributor and the cyclone grass seeder.

Fertilizer Distributor. Little need be said about this common lawn applicator. It has an agitator within the hopper activated by wheel rotation and an adjustable orifice plate to control the rate of flow. Its use is limited to continuous areas such as lawns and flat garden areas not yet planted or thrown up in rows.

Cyclone Seeder. These are available in the larger field-seeding models down to the smaller, plastic, 1-quart-capacity units for home lawn use. They produce an amazingly uniform application and cover large areas in a minimum of time.

CHAPTER 5

THE PROBLEMS – AND SOLUTIONS

> The biggest problem in the world could have been solved when it was small.
>
> Witter Bynner, *The Way of Life According to Laotzu.*

DIAGNOSIS AND IDENTIFICATION OF PLANT PEST PROBLEMS

To know the causes of poor plant growth and the associated symptoms, such as leafspot and discolored leaves, requires years of experience. However, generalizations can be made that will help you become proficient in diagnosing the ills of plants.

They may be injured by animals such as insects and rabbits; by other plants such as fungi and bacteria; by natural causes such as drought and nutritional disorders; and by chemical injury such as phytotoxic symptoms from sprays and air pollution.

Careful observation often reveals the cause. For example, a canker disease may have girdled a twig or limb. The symptoms might include wilted or dead leaves, but one must look carefully along the twig or limb to find the cause. Or an insect may be eating the roots of a plant — the symptoms appear first on the leaves, but until one examines the roots the cause cannot be established.

Two or more causes might produce the same kind of symptom, and there are hundreds of causes. With careful observation, knowledge of the plant's history, and a general knowledge of possible causes, plant ills can often be diagnosed without the aid of a plant pathologist or entomologist.

Ways in Which Insects Injure Plants

1. Chewing: devouring or notching leaves, eating wood, bark, roots, stems, fruit, seeds, mining in leaves. Symptoms: ragged leaves, holes in wood and bark or in fruit and seed, serpentine mines or blotches, wilted or dead plants, or presence of larvae.
2. Sucking: removing sap and cell contents and injection of toxins into plant. Symptom: usually off-color, misshapen foliage and fruit.
3. Vectoring Diseases: by carrying diseases from plant to plant, e.g. elm bark beetle and Dutch elm disease; various aphids and virus diseases. Symptoms: wilt, dwarf, off-color foliage.
4. Excreting Honeydew: deposits lead to the growth of sooty mold and the leaves cannot perform their manufacturing functions, which results in a weakened plant. Symptoms: sooty black leaves, twigs, branches and fruit.
5. Forming Galls: galls may form on leaves, twigs, buds, and roots. They disfigure plants, and twig galls often cause serious injury.
6. Scarring by Egglaying: scars formed on stems, twigs, bark, or fruit. Symptoms: scarring, splitting, breaking of stems and twigs, misshapen and sometimes infested fruit.

Ways in Which Diseases Injure Plants

1. Interfere with the supply lines by clogging water-conducting cells. Examples: late blight of tomato and potato. Dutch elm disease. Symptom: wilt.
2. Destroy chlorophyll. Examples and symptoms: blotch, scab, black spots on leaves, brown patch disease of turf.

3. Destroy water- and mineral-collecting tissues. Examples and symptoms: Fusarium wilt, root rot, general stunting of plant.
4. Gall-formation disrupts normal cellular organization. Symptom: unusual growths on flowers, twigs, and roots.
5. Produce flower and seed rots. Examples: fireblight, bacterial rot of potato.

SELECTING THE PROPER PESTICIDE

Ideally, each pest problem should be specifically identified and a particular pesticide chosen for the problem, but this is seldom practical because there are about 10,000 insect pests, 1,500 plant diseases, and 600 weeds. For the homeowner or amateur gardener, the purchase of a pesticide for each individual problem is impractical because of cost, lack of storage space, and increased danger of misuse and accidental poisoning.

Modern pesticides make pest control relatively simple, and only a few are needed around the home. Most of these control a rather wide range of pests, allowing generalizations to be made about control of pest groups such as aphids, mites, beetles, caterpillars, leaf spots, blights, turf diseases. The problem can be even further simplified to protect some groups of plants by use of "multipurpose" or "all purpose" mixtures of pesticides, which contain an insecticide, miticide, and fungicide in a single mixture and are effective against insects, mites, and diseases rather than against a few insects or a few diseases.

In either case, to select the proper pesticide or pesticide mixture, refer to the proper tables which list the most commonly observed pest problems for that particular problem area. These tables list alphabetically the host, the common name of the pest, the materials and methods of control, with and without chemicals, and the approximate time of treatment. The dilution rate for spray chemicals is found in Table 8. Read the label for more specific details and to make certain that the material is safe for the situation and effective for the pest to be controlled.

APPLYING THE PESTICIDE PROPERLY

A basic principle of pest control is the proper application of the pesticide. Generally, two intelligent efforts are involved. The first is to give good uniform coverage of the area to be treated, whether for insect, disease or weed control, or with sprays or dusts. The second effort probably should be first, the proper dilution of the pesticide before application.

Using the proper pesticide dilution is a fairly serious matter. Using too little material will obviously give less than hoped for control of the pest, regardless of its nature. Using too strong a spray mixture could cause plant damage, leave excess residues at time of harvest, or be hazardous to nontarget elements such as pets, sensitive plants, and beneficial insects. Using an excessive spray concentration will not kill more pest insects, control the disease longer, or kill more weeds. The recommended dilution has been demonstrated by research to be the most effective for that specific purpose. Heading the list of reasons for not using a heavier concentration than called for is cost. The extra amount of pesticide used is truly wasted. So, read and follow the label directions to obtain the best results.

Table 8 has been designed to aid the home gardener in making one gallon of the proper spray mixtures from wettable powders (WP) or from liquid concentrates (EC). To use Table 8 when mixing a spray of a prescribed percentage of the actual chemical (active ingredient), you need to know only the concentration of the formulation. First find the percentage of actual chemical wanted, then match it with the formulation.

Example: a 1% diazinon spray is recommended for cockroach control. To make a 1% diazinon spray using a 25% emulsifiable concentrate, add 10 tablespoonfuls of the diazinon concentrate to one gallon of water.

Example: a 0.25% Dyrene spray is recommended for powdery mildew control. To make 0.25% spray using a 15% Dyrene wettable powder, add 7 tablespoonfuls of the Dyrene powder to one gallon of water.

TABLE 8. Pesticide Dilution Table for Home and Garden

(Amount of pesticide formulation for each one gallon of water)

Pesticide Formulation	Percentage of Actual Chemical Wanted								
	.0313%	0.0625%	0.125%	0.25%	0.5%	1.0%	2.0%	3.0%	5.0%
					Wettable Powder (WP)				
15% WP	2½ tsp.	5 tsp.	10 tsp.	7 tbsp.	1 cup	2 cups	4 cups	6 cups	10 cups
25% WP	1½ tsp.	3 tsp.	6 tsp.	12 tsp.	8 tbsp.	1 cup	2 cups	3 cups	5 cups
40% WP	1 tsp.	2 tsp.	4 tsp.	8 tsp.	5 tbsp.	10 tbsp.	1¼ cups	2 cups	3¼ cups
50% WP	¾ tsp.	1½ tsp.	3 tsp.	6 tsp.	4 tbsp.	8 tbsp.	1 cup	1½ cups	2½ cups
75% WP	½ tsp.	1 tsp.	2 tsp.	4 tsp.	8 tsp.	5 tbsp.	10 tbsp.	1 cup	2 cups
					Emulsifiable Concentrate (EC)				
10%-12% EC 1 lb. actual/gal.	2 tsp.	4 tsp.	8 tsp.	16 tsp.	10 tbsp.	⅔ pt.	1⅓ pt.	1 qt.	3¼ pt.
15%-20% EC 1.5 lb. actual/gal.	1½ tsp.	3 tsp.	6 tsp.	12 tsp.	7½ tbsp.	½ pt.	1 pt.	1½ pt.	2½ pt.
25% EC 2 lb. actual/gal.	1 tsp.	2 tsp.	4 tsp.	8 tsp.	5 tbsp.	10 tbsp.	⅔ pt.	1 pt.	1¾ pt.
33%-35% EC 3 lb. actual/gal.	¾ tsp.	1½ tsp.	3 tsp.	6 tsp.	4 tbsp.	8 tbsp.	½ pt.	¾ pt.	1⅓ pt.
40%-50% EC 4 lb.-actual/gal.	½ tsp.	1 tsp.	2 tsp.	4 tsp.	8 tsp.	5 tbsp.	10 tbsp.	½ pt.	⅘ pt.
57% EC 5-lb.-actual/gal.	7/16 tsp.	⅞ tsp.	1¾ tsp.	3½ tsp.	7 tsp.	4½ tbsp.	9 tbsp.	14 tbsp.	1½ cups
60%-65% EC 6 lb. actual/gal.	⅜ tsp.	¾ tsp.	½ tbsp.	1 tbsp.	2 tbsp.	4 tbsp.	8 tbsp.	12 tbsp.	1½ cups
70%-75% EC 8 lb. actual/gal.	¼ tsp.	½ tsp.	1 tsp.	2 tsp.	4 tsp.	8 tsp.	5 tbsp.	7½ tbsp.	13 tbsp.

gal. = gallon lb. = pound pt. = pint tbsp. = tablespoon tsp. = teaspoon
3 level teaspoonfuls = 1 level tablespoonful 2 tablespoonfuls = 1 fluid ounce 8 fluid ounces or 16 tablespoonfuls = 1 cupful
2 cupfuls = 1 pint 1 quart = 2 pints or 32 fluid ounces 1 gallon = 4 quarts or 128 fluid ounces

INSECT PESTS

Upon his painted wings,
the butterfly Roam'd,
a gay blossom of the sunny sky.
Willis G. Clark

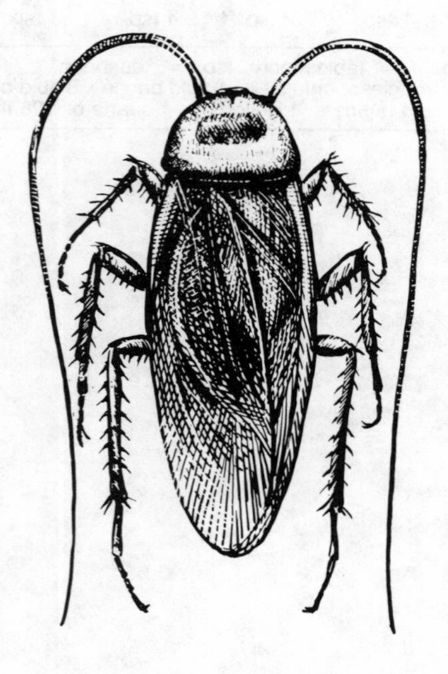

INSECT PESTS

Of the 1 million or more species of insects, only about 1000 are of economic pest status over the entire earth, and as you might guess, most of these fall in the agricultural pest class. Less studied and counted are those that are insect pests around and in man's structures, the domestic or urban insect pests. There may be as many as 600 of these but it will probably narrow down to around 200 species that are the normal, everyday type of damaging or simply annoying insects broadly classed as "pests".

But what is an insect pest? We could say that a pest is a pest "in the eyes of the beholder". For what may "drive one person up the wall" may not even have caught the attention of another. A good example of a pest insect that most people recognize immediately is the earwig, scurrying across the floor. To some, this is a hideous, despicable nuisance. To others they are only amusing, for the latter know that these are only small, harmless guests sharing their home, usually accidentally, but certainly with no malicious intent.

Honey bees are pests to some individuals; especially those who are allergic to bee stings, as well as those who spend long hours digging, transplanting and weeding in the flower garden. The fear of being stung usually becomes exaggerated once stung, and all efforts are made to avoid the nectar-foraging friends. Wasps and hornets are feared even more than the lowly bee. Add to this, bumble bees, and syrphid flies. Because syrphid flies resemble bees and wasps to the untrained eye, they are frequently mistaken for them and suffer the same classification, and as a consequence, fate, as the look-alike pests.

The best way to handle the subject of insects around the home and garden is to break them down into classes: household, vegetable garden, fruit trees, pets, lawn, and so on. Diagrams of these insect pests will be used when they are believed useful for gross identification.

VEGETABLE GARDEN

Insects and Mites in the Vegetable Garden

A Gallup Poll taken in 1978 showed that 41% of U.S. households have a vegetable garden, and 22% of school children have a high interest in gardening. The home vegetable garden not only helps reduce the food bill but also provides the gardener with good outdoor exercise and sources of the essential vitamins, nutrients and minerals. To produce vegetables without having them attacked by a wide variety of insects is next to impossible. Thus a control program is essential. This chapter presents important facts to the home gardener about insects and insecticides that will help him control, safely and effectively the more troublesome pests. Table 9 and Figure 16 contain guides to the types of insect injury commonly found in the garden and their probable cause.

Throughout the nation there are approximately 150 different species of injurious vegetable pests, and we never know which ones will be a problem from one season to the next. Nor can we guess how severe they will be. Some insects, for instance bean beetles, are usually a problem every year. Others become problems only occasionally. When an insect appears in the garden varies with different insects and the weather. Some insects spend the winter in or near the garden and become active as soon as warm days occur. Other insects fly in and out of the garden at random, while still others appear later in the summer. In other words, insects appear in the home vegetable garden from the time it is planted until it is leveled by frost in the fall.

TABLE 9. Guide to Insect Injury in the Garden — Plant Part and Kind of Injury

SEEDS (sprouting)
- Eaten or tunneled (chewing mouthparts)
  - seed corn maggot (beans, corn, melons)

SEEDLINGS
- Stems cut off or girdled (chewing mouthparts)
  - crickets
  - cutworms
  - darkling beetles
  - seed corn maggot (beans, corn, melons)
- Eaten or skeletonized (chewing mouthparts)
  - ants
  - caterpillars
    - armyworms
    - cabbage loopers
  - pillbugs
  - slugs and snails
- Wrinkled or withered (sucking mouthparts)
  - aphids
  - false chinch bugs
  - thrips

STEMS AND VINES
- Chewed or eaten
  - beetles

STEMS AND VINES (continued)

Chewed or eaten (continued)
- caterpillars
- crickets

Tunneling (chewing)
- cornstalk borer (corn)
- potato tuberworm (potato)
- corn borer (corn)
- squash vine borer (pumpkin, squash)
- stalk borers (eggplant, potato, tomato)

Discolored or withered (sucking)
- aphids
- plant bugs
- potato psyllid (potato, tomato)
- spider mites

LEAVES

Eaten, partly or totally (chewing)
- ants
- caterpillars
- blister beetles (eggplant, pepper, tomato)
- carrot beetle (larvae and adults)
- celery leaf tier (celery)
- crickets
- cucumber beetles (cucumber, melons)
- flea beetles
- grasshoppers
- hornworms (tomato)
- June beetles
- May beetles
- leaf beetles (beans)
- Mexican bean beetles (beans)
- leafrollers
- potato beetle (eggplant, potato, tomato)
- slugs and snails
- squash beetle (squash)
- tortoise beetles (peppers)
- vegetable weevil
- webworms

Tunneled or mined (chewing)
- potato tuberworm (potato)
- leaf miners (cucurbits, eggplant, pepper, tomato)
- tomato pinworm (tomato)

Discolored, wrinkled, withered or peppered
- aphids
- false chinch bug
- fleahoppers (melons)
- harlequin bug
- leafhoppers
- potato psyllid (potato, tomato)
- spider mites (melons, beans, tomato)
- squash bugs
- thrips (onion)

FRUIT

Chewed (chewing mouthparts)
- cucumber beetles (cucumber, melons)
- grasshoppers
- tomato fruitworm (tomato)

Tunneled
- corn earworm (corn)
- tomato pinworm
- pepper weevil (pepper, eggplant)

Discolored, wrinkled, withered (sucking mouthparts)
- aphids
- fleahoppers
- plant bugs
- squash bug (cucurbits)
- stink bugs (tomato)
- spider mites (tomato)

Secondary feeders in injuries
- Sap beetles (corn)

ROOTS

Chewed or gouged (chewing mouthparts)
- white grubs
- several beetle and weevil larvae

ROOTS OR TUBERS

Tunneled (chewing mouthparts)
- flea beetle larvae (potato)
- potato tuberworm (potato)
- wireworms (potato)

INSECT INJURY TO PLANTS

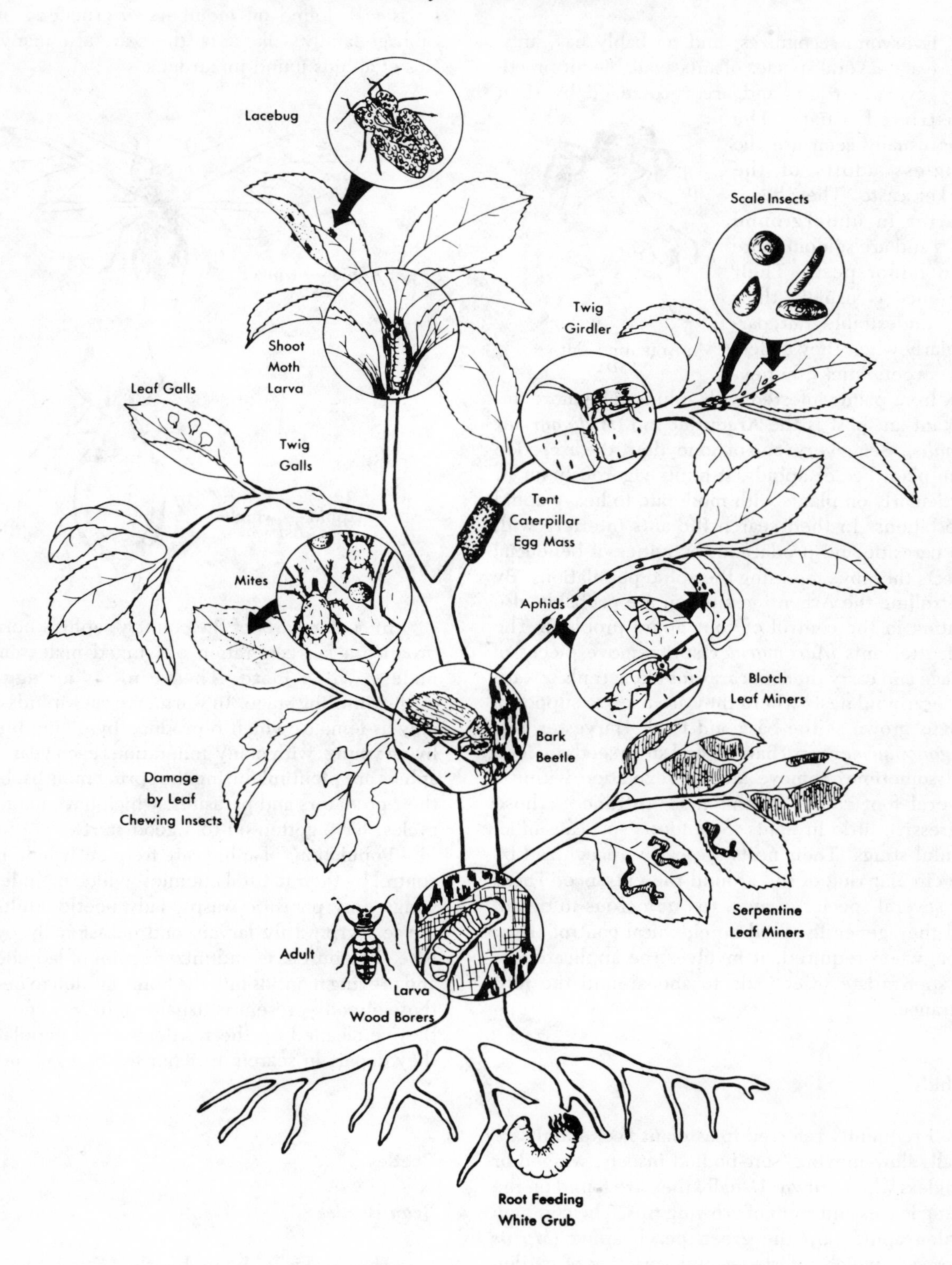

FIGURE 16. Insect Injury to Plants.

Ants

Everyone recognizes, and probably has, ants. There are several species of ants available for practically every garden, and are recognized by their constricted waists. The ones usually seen are the wingless adults of the worker caste. They headquarter in underground nests and are seldom more than minor pests. Their presence is usually their most undesirable trait, particularly when they collect the sweet, sticky honeydew from plants infested with aphids. The more important ant pest is the Argentine ant (*Iridomyrmex humilis*). Being very fond of honeydew, the excretion from plant lice or aphids, it hunts vigorously for it, particularly on plants with moderate to heavy aphid infestations. In their search the ants interfere with the parasitic and predaceous activities of beneficial insects thereby increasing the aphid populations. By controlling the Argentine ant you are probably also assiting in the control of your aphid problem. The leafcutter ants (*Acromyrmex* sp.) remove pieces of foliage and carry them in caravans along trails toward underground nests where they are used to support a fungus grown as the basic ant food. Harvester ants (*Pogonomyrmex* spp.) harvest and store seeds for food and sometimes remove all the vegetation within a several-foot radius of the nest entrance. Those aggressive little fire ants (*Solenopsis* sp.) can inflict painful stings. Their nests are easily recognized by the circular ring of soil around the entrance. There are several species of ants, too numerous to detail, and they generally need no chemical control. However, when required, it involves the application of an appropriate insecticide to and around the nest entrance.

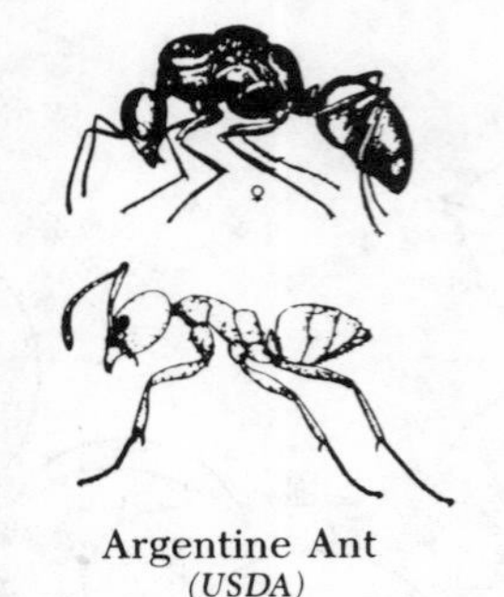

Argentine Ant
(USDA)

Aphids

Frequently referred to as plant lice, aphids are small, slow-moving, soft-bodied insects, winged or wingless when grown. Usually they are found on the young leaves and stem of growing tips. The common garden aphids are the green peach aphid (*Myzus persicae*), which infests several varieties of garden plants, the cotton or melon aphid (*Aphis gossypii*), which is commonly found on melons, squash, and cucumbers, and the cabbage aphid (*Brevicoryne brassicae*), found on members of crucifers, or the cabbage family. Like ants, there are also many species of aphids found in gardens.

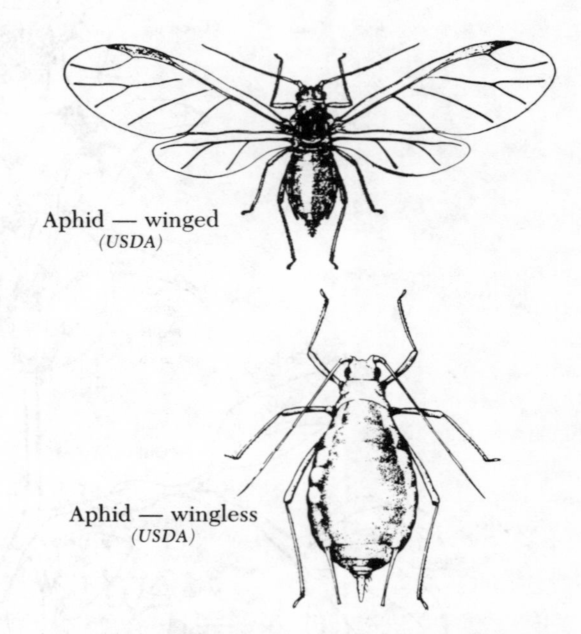

Aphid — winged
(USDA)

Aphid — wingless
(USDA)

In cooler parts of the country, aphids normally produce a fall generation of winged males and females, which mate. The result is an egg, the over-wintering stage. In warmer areas, aphids occur only as females which reproduce by giving birth to living young, with many generations each year. Most aphids are plentiful during the spring months, before their predators and parasites, which have longer life cycles, have gotten off to a good start.

Populations of aphids are frequently held under control by their natural enemies, which include lacewings, tiny parasitic wasps, lady beetle adults and larvae, syrphid fly larvae, and occasionally by disease. The purchase and introduction of lady beetles and preying mantids into the home garden to beef-up those already present is usually of little value. After they've cleaned up the resident aphid populations, they depart in search of richer feeding grounds.

Beetles

Bean Beetles

The Mexican bean beetle (*Epilachna varivestris*) is practically a universal pest of beans, including lima, bush or string, and pinto. The adults

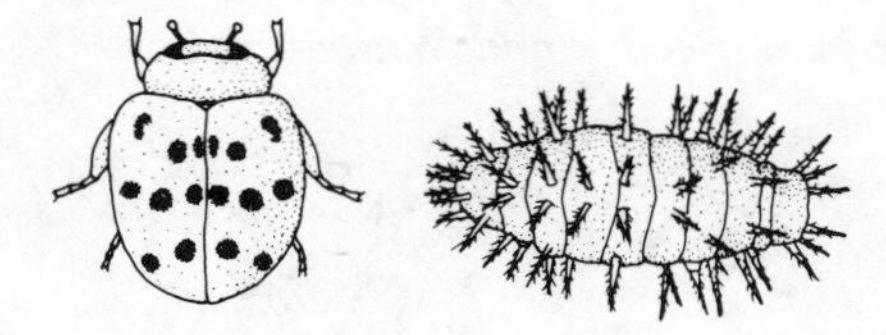

Mexican Bean Beetle — Adult and Larva
(Univ. of Arizona)

resemble their close kin, the lady beetles, but are larger, ¼ to ⅜″ long. They are yellow to coppery brown, with 8 black spots on each wing cover. The larvae are oval and about ⅓″ long when mature and have 6 rows of long, branched, black-tipped spines along their backs. It can sometimes become a serious pest in the home garden, because both the adults and larvae feed on lower surfaces of bean leaves, leaving only a network of leaf veins. The stripped, skeletonized leaves usually dry up and in severe injury the plants may be killed. Pods and stems are also attacked when heavy infestations occur. The bean leaf beetle can be found all through the United States.

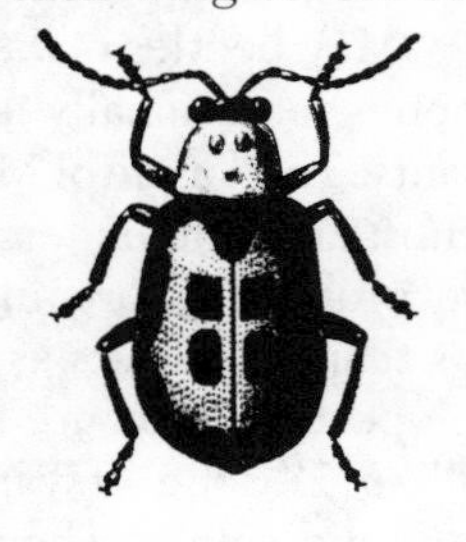

Bean Leaf Beetle
(Union Carbide)

Blister Beetles

There are several kinds of blister beetles, particularly *Epicauta* spp., that eat any of the above-ground parts of tomato, potato, pepper, eggplant and other garden vegetables. They frequently move in small herds and can quickly cause noticeable damage when they appear in large numbers. Blister beetles are identified by their long bodies and slender "shoulders", which are narrower than the head or abdomen. Body colorings vary as do markings and surface textures. Some are quite pretty, and most are from ½ to ⅝″ long. There is no reason to be concerned about the larvae since they do not feed on foliage.

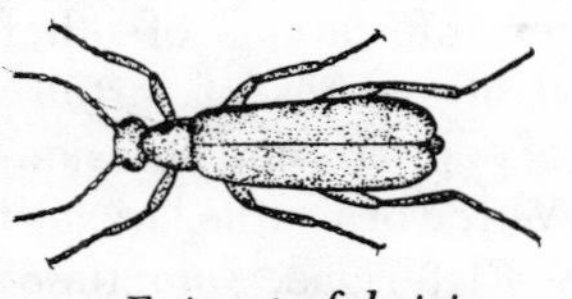

Epicauta fabrici
Ash-Gray Blister Beetle
(U.S. Public Health Service)

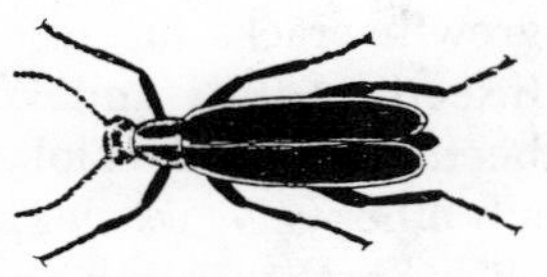

Epicauta pestifera
Margined Blister Beetle
(U.S. Pvblic Health Service)

Colorado Potato Beetle

Both adults and larvae of the Colorado potato beetle (*Leptinotarsa decim lineata*) or just plain potato beetles may be found feeding on potato leaves anywhere in the U.S. They also attack tomato and eggplant, and may completely consume infested plants when abundant. The adults are rather rounded, about ⅜″ long and ¼″ wide, with the head and thorax black-spotted and the wing covers bearing 10 black and white lengthwise stripes. It can be a serious garden pest in some parts of the nation.

Colorado Potato Beetle — adult and larva
(USDA)

Cucumber beetles

There are several species of cucumber beetles that can be potential problems on several plants, especially the cucurbits. The 12-spotted cucumber beetle (*Diabrotica undecimpunctata tenella*) is related to the western spotted cucumber beetle and to the spotted cucumber beetle known as the corn rootworm of mid-west and eastern states. It is recognized by its black head, yellow prothorax, and yellow or greenish-yellow wing covers bearing, naturally, 12 black spots. The western striped cucumber beetle (*Acalymma trivittata*) is about ⅕″ long with an orange-yellow prothorax and wing covers marked black and yellow lengthwise stripes. The banded cucumber beetle (*Diabrotica balteata*) has a green-yellow prothorax and yellow wing covers marked with 3 transverse green bands. Cucumber beetles feed on leaves, stems and fruit of cucurbits, and may produce numerous irregular shot-hole leaf perforations and scars on the rinds of developing fruit. The larvae feed on roots of many common plants and go under the broad identification of root worms. They may also stunt or kill the vines and attack fruit that touch the ground.

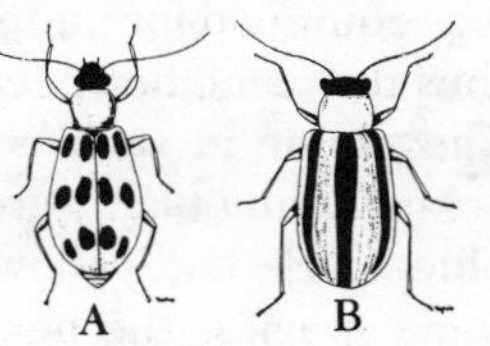

Cucumber Beetles:
A. Spotted; B. Striped.
(Univ. of Arizona)

Darkling Beetles

These are small, dark brown to gray-black beetles, common in western states, that sometimes attack seedlings of practically all garden plants by girdling or cutting off stems at or below ground level. Their activities are sometimes blamed on cutworms, since they also are most active at night. When abundant, they may become nuisances by accumulating around outdoor light and occasionally invading homes. The larvae are known as false wireworms and are seldom seen.

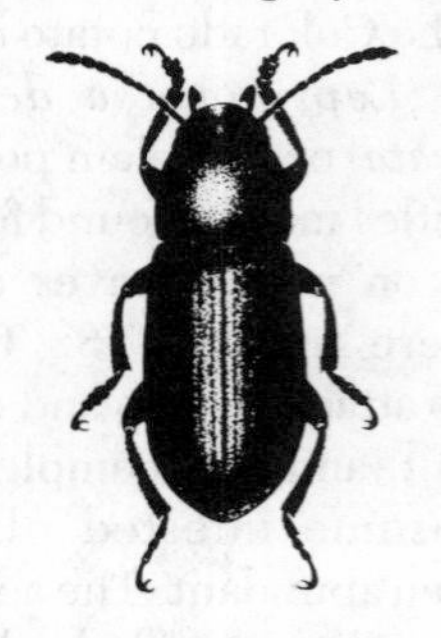

Darkling Beetle
(Univ. of Arizona)

Flea Beetles

Several species of flea beetles attack garden plants in all stages of growth. They range from tiny 1/16″ to medium 1/4″ long beetles, with strong hind legs equipped for jumping, thus the name flea beetles. They come in various colors including black, green, blue-black, and yellow. In some species, the head or prothorax may be red, brown or yellow and wing covers may be of a single color or with light and dark stripes. These jumping beetles attack potato, corn, carrot, cauliflower, bean and many other vegetables, eating round or irregular holes in the leaves. This gives the leaves a riddled appearance, and some plants, such as corn, may have a scalded appearance resulting from the feeding of adults on the upper leaf surfaces. When flea beetles are numerous, young plants may be killed in 1 or 2 days. The slender white larvae feed on roots and underground stems, or in some species, may feed on lower leaf surfaces or tunnel inside leaves. Some cause potato tubers to become pimply or silvery from small pin holes and burrows just below the surface.

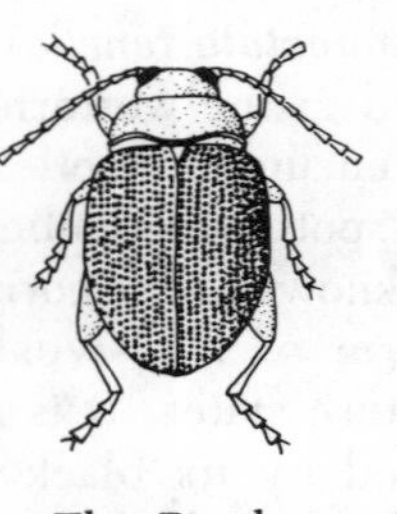

Flea Beetle
(Univ. of Arizona)

May or June Beetles and White Grubs

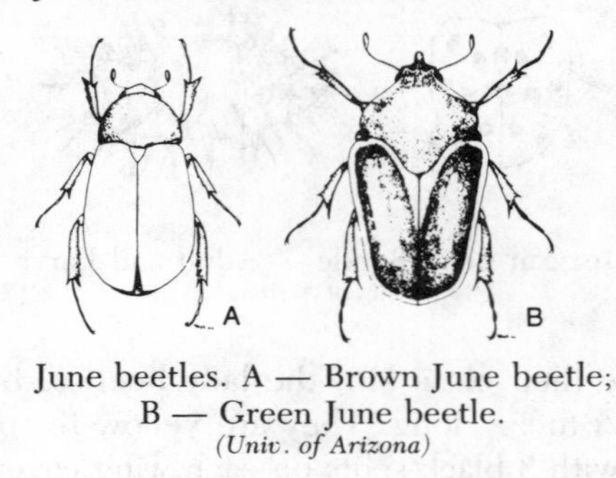

June beetles. A — Brown June beetle; B — Green June beetle.
(Univ. of Arizona)

May beetles of various species, also called June beetles, occasionally feed on foliage of home garden plants. The larvae of May beetles are known as white grubs, are root feeders and are probably more injurious than the adults as garden pests. Adults are from 1/2″ to 1″ long of various sizes and colors. They become active at dusk and are attracted to lights, flying awkwardly and crashing into objects with their heavy buzzing. The white grubs, or soil-infesting larvae, are whitish with a dark head and 3 pairs of short legs immediately behind the head. The grubs are always found in their typical "C"-shaped position. Underground damage occurs to practically all varieties of garden plants. The common green June beetle, or fig beetle, (*Cotinis mutabilis*) is not ordinarily a garden pest, although it may attack over-ripe tree fruit and melons. Their larvae feed on decaying organic matter and not on the living roots of garden plants.

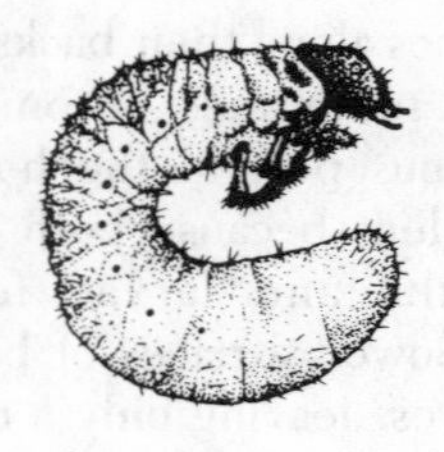

A White Grub
(Univ. of Arizona)

Miscellaneous Beetles

Several other beetles may occasionally become localized or minor pests of home gardens. Sap beetles, of the family Nitiduilidae, may invade maturing ears of sweet corn to feed on fermented kernels previously damaged by the corn earworm. They may also invade maturing melons and tomatoes injured by growth cracks or by earlier infestations of other insects. Adults and larvae of species of tortoise beetles may feed on foliage of peppers, and occasionally other vegetable leaves. Wireworms, the larvae of click beetles in the family Elateridae, sometimes tunnel into potato tubers and roots of other plants such as sweet potatoes.

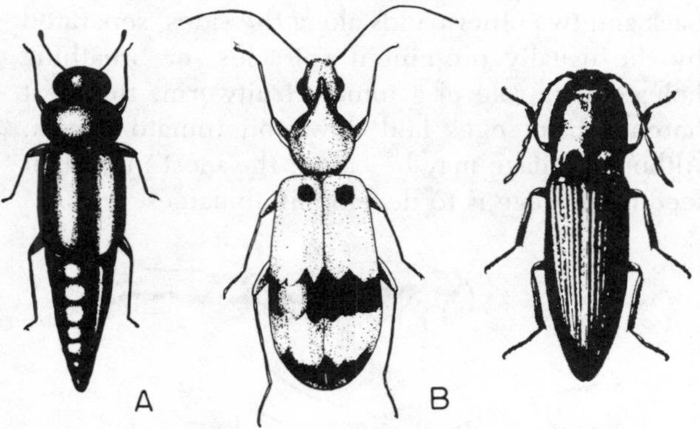

Non-injurious Beetles Found in Gardens. A — Fruit bud beetle; B — Notoxus beetle.
(Univ. of Arizona)

Click Beetle
(USDA)

Weevils

There are a large group of beetles whose head tapers into a snout with the mandibles at the tip, the weevils. Although the adults may be destructive pests, the grub-like, legless larvae, often found tunneling within plant tissues, also cause serious injury. Among these are the vegetable weevil, pepper weevil, and several related to the potato stalk borer.

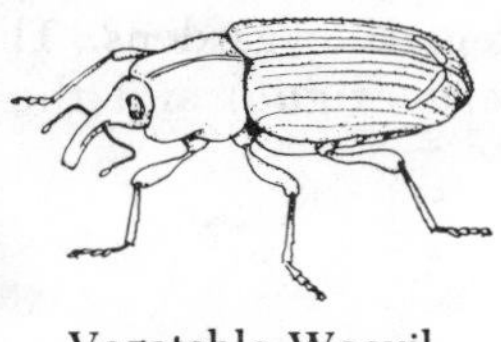

Vegetable Weevil
(Univ. of Arizona)

Vegetable weevil larvae (*Listroderes costirostris obliquus*) attack carrots, celery and other vegetables during spring and early summer months. These fleshy, cream-colored grubs feed on foliage and may gouge and tunnel the fleshy roots of carrots and other crops. Adult weevils are brownish-buff, ½″ long, with a "V" shaped, whitish marking on the wing covers. Adults feed on the foliage of various vegetables during the summer and fall months.

The stalk-boring weevils (*Trichobaris* spp.) related to the potato stalk borer, are occasional pests of potato, tomato and eggplant. The graying adults are ¼″ long. The legless larvae are white with darker heads and injure plants by drilling in the stems above and below ground level.

The pepper weevil (*Anthonomus eugenii*) is one of the smaller snout beetles, being about ⅛″ long, shiny black and covered with white fuzz. Adults feed on buds, tender pods and leaves of peppers and sometimes eggplants. Eggs are laid within the buds or pods where the larvae hatch and begin feeding. Injured peppers darken and decay, and the smaller peppers may drop in large numbers under heavy infestation.

Caterpillars

Several species of caterpillars commonly attack garden vegetables and may very well be the most serious category of insect pests. Some caterpillars, or worms, are general feeders on many cultivated and wild plants, while others are limited to a single or only a few hosts. Caterpillars are the young or larvae of moths and butterflies, and develop from eggs usually laid on or near the plants they attack. Most garden caterpillars, with a few exceptions, are larvae of dull-colored, night-flying moths. Caterpillars have chewing mouthparts and may damage leaves, buds, fruit, stems or roots. Some may fasten leaves together with silken webs to make protective shelters in which they feed, including the celery leaf tier (*Udea rubigalis*), the omnivorous leaf roller (*Platynota stultana*), and webworms (*Loxostege* spp.).

Cutworms of various species are general feeders on many garden vegetables. Newly transplanted succulent plants are particularly vulnerable, such as tomato and cabbage. Emerging seedlings are cut off at or below the soil surface, and roots and tubers may on occasion be attacked. Cutworms are active during the night and hide during the day beneath clods or in the soil near the plants attacked. A single cutworm may destroy several plants each night, progressing steadily down a row. These voracious larvae are the young of several inconspicuous moths, and when full grown are fleshy, dull in color, with or without markings, up to 2″ long, and commonly seen in their distinctive curled position in the soil. To protect transplanted garden seedlings, wrap stems of seedlings with newspaper or foil to prevent damage by cutworms.

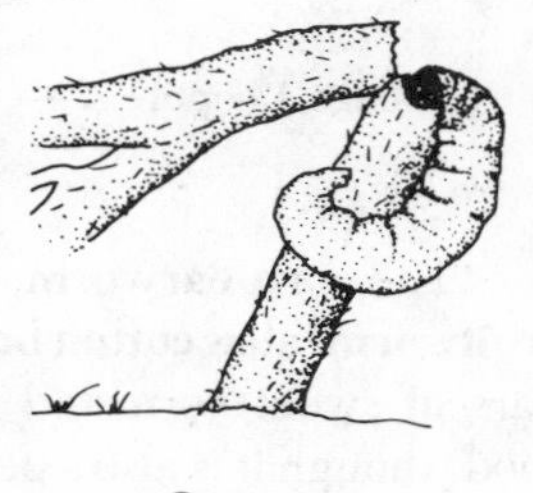

Cutworm
(Univ. of Arizona)

The **cabbage looper** (*Trichoplusia ni*) is perhaps the most common garden caterpillar since they feed on lettuce and several of the crucifers (cabbage family) found in all home gardens. With many generations each year, many kinds of garden plants may be attacked and young

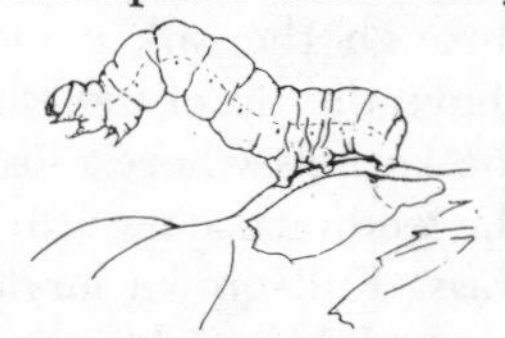

Cabbage Looper
(Univ. of Arizona)

plants may be consumed to the ground. Large holes may be eaten in older leaves, however when infestations are light or moderate this damage may be outgrown. The cabbage looper gets its name from its looping movement as it inches forward, and occasionally it is referred to as an "inch worm". Newly hatched loopers are pale green with a black head and are usually found feeding on the undersides of leaves. Older larvae, up to 1½″ long, are pale green or tan, with or without white lines along the sides. They may completely devour leaves and consume several times their body weight in plant tissue each day. Small plant-colored fecal pellets are easily identified evidence of their presence.

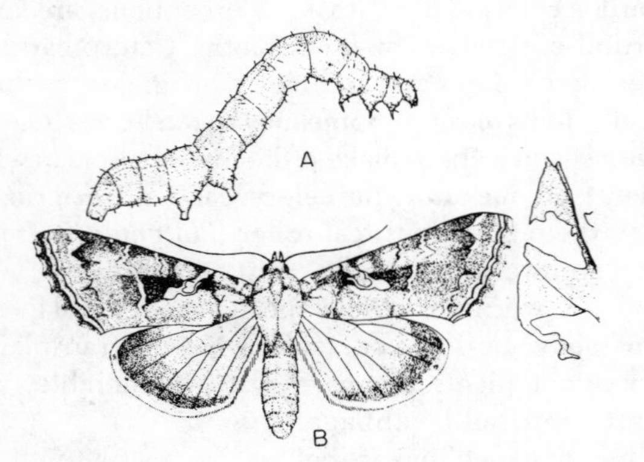

Cabbage Looper. A — Mature larva; B — Adult.
(Univ. of Arizona)

The **corn earworm**, (*Heliothis zea*) alias **tomato fruitworm**, alias **cotton bollworm**, prefers developing ears of sweet corn and tomatoes beyond all other food, though it is also a pest of other crops including lettuce. It is the major pest of home garden sweet corn; the principal damage is done when the developing kernels within the ears are eaten. Leaf whorls and developing tassels of younger corn plants may also be attacked. The larvae hatch from eggs laid on fresh corn silks and feed down through the silk channel above the tip of the ear to the kernels where most of the feeding occurs, leaving behind abundant feces or frass. Full-grown larvae are up to 1½″ long with brown heads and body coloration varying from green with touches of red or brown to almost black. All stages have a pair of dark lines extending along the back and two other bands along the sides, separated by the usually prominent spiracles, or breathing holes. In its role of a tomato fruitworm, this pest hatches from eggs laid down on tomato leaves. Although foliage may be eaten, the most important feeding damage is to developing tomatoes.

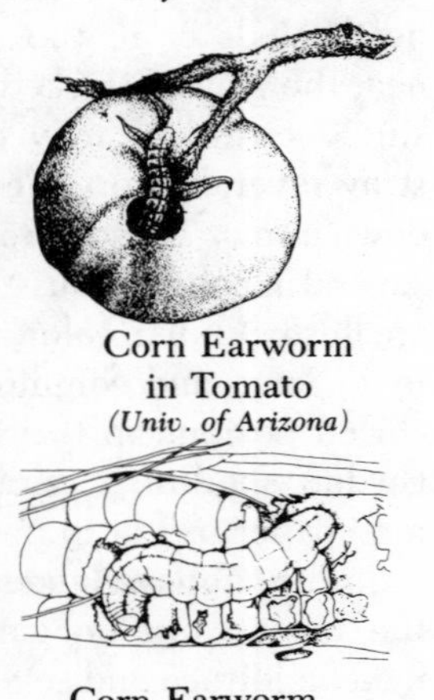

Corn Earworm in Tomato
(Univ. of Arizona)

Corn Earworm in Sweet Corn
(Univ. of Arizona)

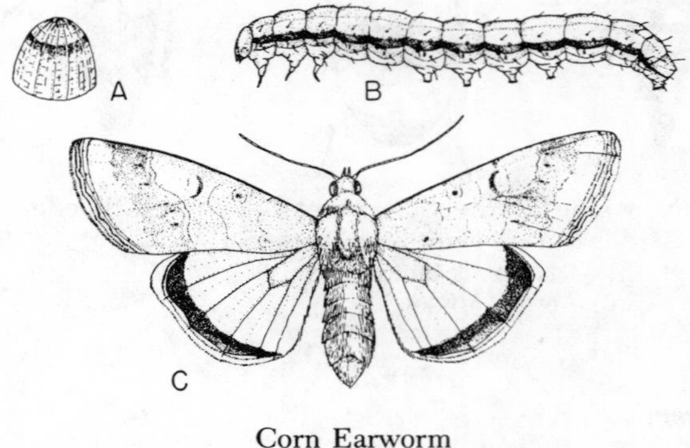

Corn Earworm
A — Egg; B — Mature larva; and C — Adult.
(Univ. of Arizona)

Armyworms may on rare occasion invade home vegetable gardens. The beet armyworm (*Spodoptera exigua*) and the yellow striped armyworm (*Spodoptera ornithogalli*) feed on beets and lettuce and some other leafy plants, particularly in the seedling stage. In large numbers they may consume entire plants. Mature beet armyworms are gray-green and up to 1¼″ long, with an irregular black spot on each side of the thorax above the sec-

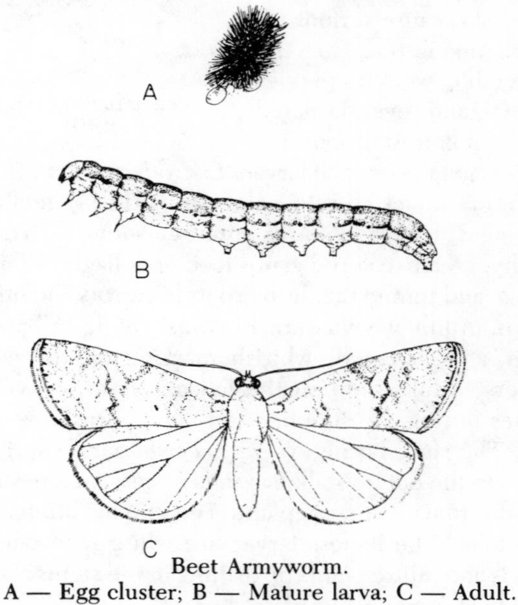

Beet Armyworm.
A — Egg cluster; B — Mature larva; C — Adult.
(Univ. of Arizona)

ond pair of legs. Mature yellow striped armyworms are up to 1¾″ long, generally purplish-brown, velvet-black on the back and have two outstanding yellow stripes on each side. Younger larvae of these two are variable in color, but similar in appearance and feeding habits. They can be distinguished from cabbage loopers by their 4, instead of 2, pairs of abdominal legs and their absence of looping movements.

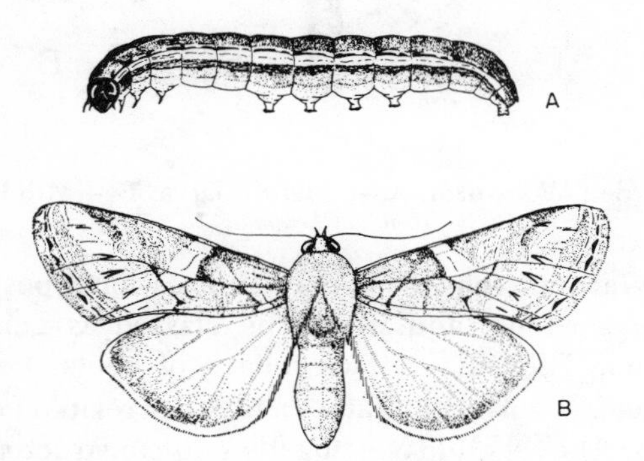

Yellow Striped Armyworm. A — Mature larva; B — Adult.
(Univ. of Arizona)

Fall armyworm larvae (*Spodoptera frugiperda*) feed inside the leaf whorls of developing sweet corn plants before tassels emerge and they may weaken or kill the growing tip. This damage is usually not observed until the ragged, shredded-looking leaves grow out. Fall armyworms resemble corn earworms in size and shape. When mature they have shiny, brown bodies with black spots. Their black head has a very distinguishing, white marking sesembling an inverted "Y".

Tomato and tobacco hornworms are occasional foliage pests on tomatoes. Hornworms are large, sphinx moth larvae, much larger than any other garden caterpillar. They derive their names from the spine-like horn at the end of the abdomen, which appears menacing to the untrained. The tomato hornworm (*Manduca quinquemaculata*) has 8 forward-pointing, V-shaped, white markings on each side, each enclosing a spiracle or breathing hole. Its horn is black and uncurved. The tobacco hornworm (*Manduca sexta*) has 7 diagonal white markings, extending upward and to the rear on each side of the body. Its horn is red and curved. Neither require chemical control in that they are readily removed by hand-picking.

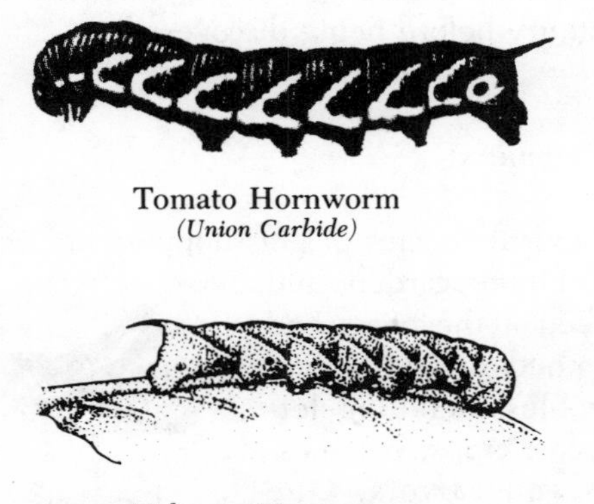

Tomato Hornworm
(Union Carbide)

Tobacco Hornworm
(Univ. of Arizona)

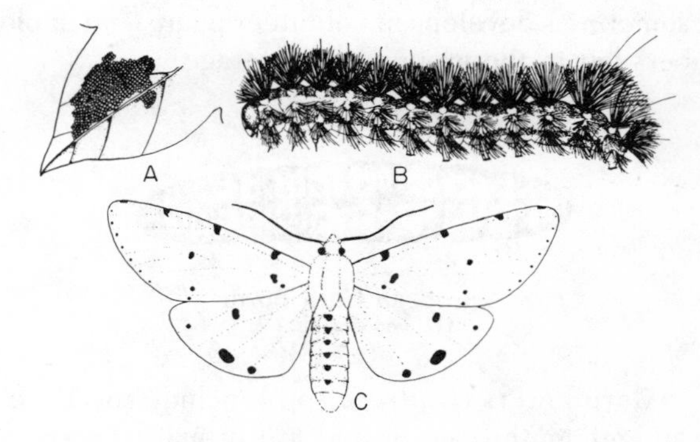

Saltmarsh Caterpillar. A — Egg cluster;
B — Mature larva; C — Adult.
(Univ. of Arizona)

Wooly worms are the saltmarsh caterpillar (*Estigmene acrea*), and may become a pest during the late summer and fall months. Since they usually migrate from commercial plantings, they are a problem only to gardeners adjacent to farming operations. The larvae develop in farm fields and then move out in hordes, feeding on nearly all kinds of vegetation in their path. Lettuce and other leafy vegetables are particularly vulnerable. Full-grown caterpillars are up to 2″ long, buff-colored, and covered with dense dark hairs, thus wooly worms.

Lesser cornstalk borers (*Elasmopalpus lignosellus*) kill or weaken seedling corn plants by entering at or below the ground line and tunneling inside, often destroying the growing tip. Beans, turnips, peanuts and other plants may also be attacked, and one larva is capable of injuring or killing several plants. The larvae are usually found outside the infested plants resting in silk tubes in the soil adjacent to the entrance holes. They move rapidly when disturbed, and reach ¾″ in length when mature. Johnson grass is a favored host and should be removed from the home garden area.

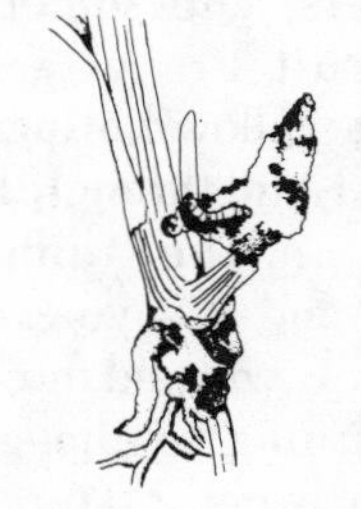

Lesser Cornstalk Borer
(Univ. of Arizona)

Potato tuberworms *(Phthorimaea operculella)* bore into potato tubers while growing and also in storage. They enter near the soil surface and make tunnels which become brown and corklike, as well as mining in growing potato foliage and stems. Other potato-related plants such as tomato and eggplant may also be attacked. There may be 2 or 3 generations per year, and the mature larvae are about ½″ long, pink or white, and a red-purple band down the back. It sometimes develops in volunteer plants and in old tubers left in the ground or in storage.

European Corn Borer
(Union Carbide)

Corn borers (*Diatraea* spp.) include the European and southwestern, and are primarily pests of field corn, but sweet corn is also attacked. The young caterpillars feed on leaf surfaces, producing translucent or skeletonized areas. They may also feed, several together, in leaf whorls of younger corn plants and may destroy the growing tip, causing a condition known as dead heart. Early in their development the larvae bore into the stalks, tunneling both upward and downward. Tunnels may extend from bases of the ears to the tap roots. Internal girdling, especially near the ground, may cause the stalks to break over. Mature larvae are white, about 1¼″ long, and may be "peppered" with conspicuous dark spots, depending on the season of the year.

The **squash vine borer** (*Melittia calabaza*) is one of the more aggravating of the caterpillar pests. It tunnels in vines of squash, pumpkins, melons, and cucumbers, and sometimes the fruit are attacked. Coarse yellow borings are pushed out through holes made in the tunneled areas, and the vines may suddenly wilt and die, with no warning to the green thumb owner. Mature borers are white with brown heads and up to 1¼″ long. The simplest control method is to slit each infested vine area with a blade and remove borers individually. Good preventive methods have not been worked out.

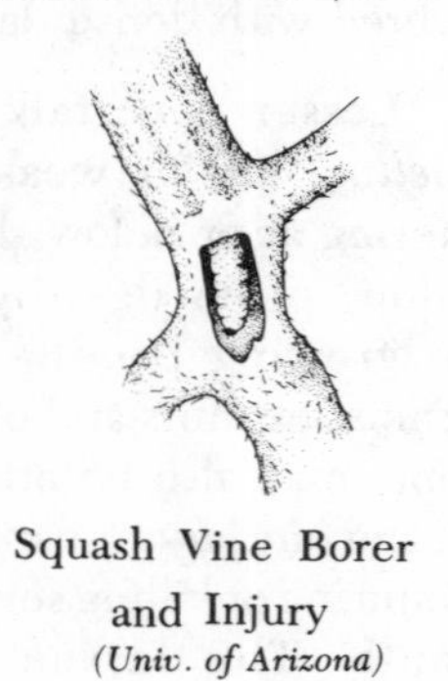

Squash Vine Borer and Injury
(Univ. of Arizona)

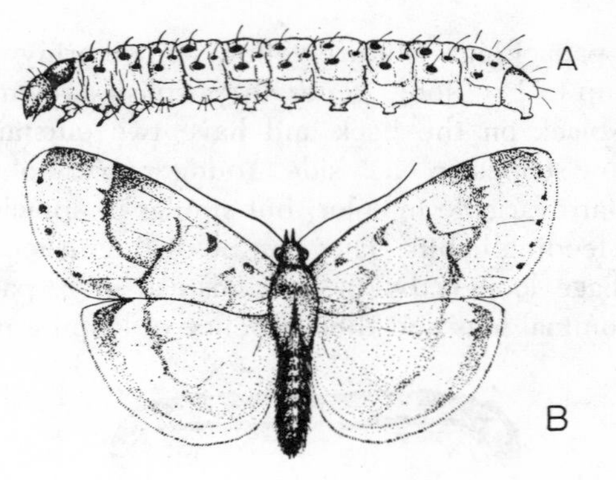

Garden Webworm. A — Mature larva; B — Adult.
(Univ. of Arizona)

Garden webworms may be occasional pests in home gardens, and would be classed as general feeders. They may attack foliage of beets, beans, cabbage, cucurbits, potatoes and other related vegetables. They web leaves together to construct shelters within which they feed until the enclosed areas are skeletonized, thus the name webworms. Younger plants may be killed. Several species occur across the United States.

Crickets

Field crickets, (*Gryllus assimilis*) can become pests feeding on the tender seedlings of most vegetables, during late summer and early fall. They are attracted to young lettuce plants, but also nibble on cole crops, beets, tomatoes and most other vegetables. This cricket may eat almost any part of any garden plant when abundant. It is nocturnal, feeding by night and hiding by day in soil cracks or under other cover, so can cause considerable injury before being discovered.

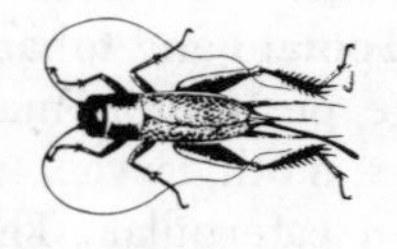

Field Cricket
(Univ. of Arizona)

Grasshoppers

Several species of grasshoppers are occasional pests of home gardens and may feed on the foliage and tender buds of the common vegetables including lettuce, cole crops, potatoes, beets and carrots. One large grasshopper, feeding

Grasshopper
(USDA)

on only one plant, may consume a surprising amount of foliage before being discovered. There are two common routes that grasshoppers enter gardens: Some develop from eggs laid in soil in or near gardens, and others may migrate by flying in from distant vegetation. In small gardens, on cool mornings when they are relatively inactive, grasshoppers may be caught and eliminated by hand. When chemical control is required treat infested areas while grasshoppers are still small.

Leafhoppers

Several species of leafhoppers may attack garden vegetables. Most commonly seen are the V-shaped hoppers, up to 1/8″ long, green to yellow in color and which are active fliers when disturbed. The nymphs resemble the adults, but are smaller, lack wings and consequently cannot fly. Leafhoppers, adults and nymphs, have the habit of walking sidewise, in cautious retreat. They have sucking mouthparts, and damage to foliage varies from unimportant to fatal, depending on the species, numbers and weather conditions. Vegetables commonly attacked by one or more species of leafhoppers are cantaloupes and other cucurbits, tomato, potato, pepper, eggplant, lettuce, beans, beet, chard, and radish. Damage to leaves is caused by their sucking sap, removing the green chlorophyll, and causing pale, circular spots or specks to appear. The potato leafhopper (*Empoasca fabae*) interferes with fluid movement inside leaves by plugging the phloem tubes, resulting in browning of leaf edges and later the entire leaf. Some species cause stunting and downward curling of leaves. Beet leafhoppers (*Circulifer tenellus*) sometimes transmit curly top virus, which infects various garden vegetables and may be particularly serious to tomatoes. Protection of tomato plants can be achieved by draping them with cheesecloth in the presence of dense populations.

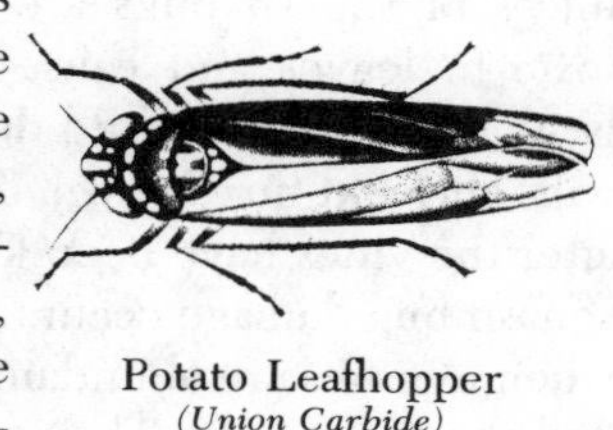

Potato Leafhopper
(Union Carbide)

Leaf Miners

There are several flies and moths whose larvae tunnel between leaf surfaces of vegetation, known as leaf miners. One common group is the serpentine leaf miners (*Liriomyza* spp.), tiny larvae of small flies. Adults are black, about 1/16″ long, with yellow markings on the face and back. Eggs are laid in leaf tissues, which hatch into pale larvae that mine gradually enlarged tunnels in random directions within the leaves, leaving dead tissue trails. Several larvae may mine the same leaf, resulting in yellowing and dropping of the leaf. Damage, however, is usually of a minor and sporadic nature, and occurs more commonly on seedling leaves than on larger leaves. Certainly their trails are not as noticeable on mature leaves as on recently emerged seedlings. Plants susceptible to attack are melons and lettuce. Other vegetables that may be infested by these general feeders are beans, eggplant, pepper and tomato. Generally, leaf miners are kept under control by several species of small parasitic wasps, and dead parasitized miners can occasionally be found in leaf tunnels.

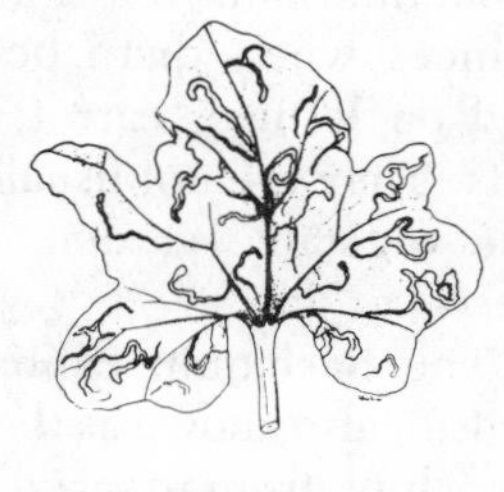

Leaf Miner Injury
(Univ. of Arizona)

Plant Bugs

Various species of plant bugs are sometimes seen in home gardens, including chinch bugs, false chinch bugs, lygus bugs, tarnished plant bugs, fleahoppers, squash bugs, and stink bugs. These all belong to the Order Hemiptera, which are the true bugs. They are usually of minor importance, although sporadic infestations may cause serious injury.

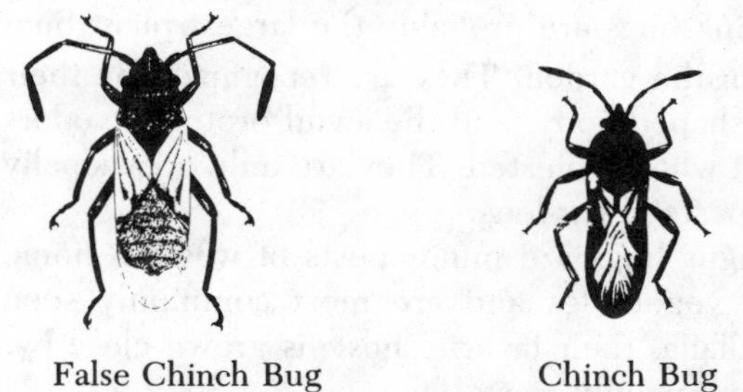

False Chinch Bug
(Univ. of Arizona)

Chinch Bug
(Union Carbide)

Chinch bugs and false chinch bugs normally develop on weeds and grasses until they become unpalatable or die then migrate over the ground in swarms to other sources of green food. Vegetable and flower gardens, lawns and young fruit trees are among the common targets. When numbers are heavy most garden vegetables may be seriously weakened or killed. Prevention centers on the control of weeds in neglected areas and on the borders of home gardens.

Fleahoppers may be found on foliage and tender fruit of melons in late spring and early summer. Fleahoppers are 1/8″ to 1/4″ long, green, black, yellow or sometimes mixed in color, and usually jump great distances when disturbed. Fleahoppers are also found on lettuce, carrot, chard and other garden plants. They are not usually considered pests in the home vegetable garden, but more as very active "guests".

The **harlequin cabbage bug** (*Murgantia histrionica*), also known as the calico bug from its red, black and white coloration, attacks cabbage and other crucifers including cauliflower, broccoli, turnip and radish. Occasionally it will feed on squash, sweet corn and garden beans.

Harlequin Bug
(Univ. of Arizona)

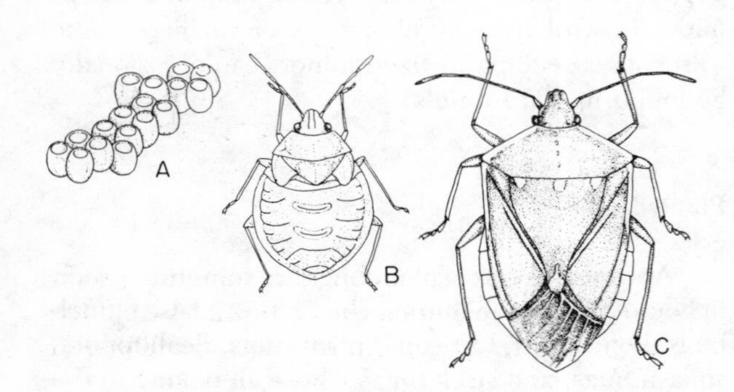

Stink Bugs. A — Egg cluster; B — Nymph; C — Adult.
(Univ. of Arizona)

Stink bugs are probably the largest plant bugs found in the garden. They are recognized by their shield-shaped body, and the awful protective odors emitted when molested. They are only occasionally minor pests in gardens.

Lygus bugs are minor pests of western home garden vegetables and are most commonly seen when alfalfa, their favorite host, is grown close by. They are mainly pests of buds, flowers and seeds and are found on many crops and weeds. They have on occasion caused minor injury to beans, cabbage and other crucifers, celery, corn and potato. Their feeding with piercing-sucking mouthparts through tender bean pods may cause the young beans within to become misshaped, with discolored areas around the points of penetration. Similar discolorations surround the punctures of other plants attacked. Lygus bugs vary in color as much as almost any other insect. They may be yellowish green, brown, or reddish brown. The antennae are whiplike and about 2/3 the length of the body. Control measures are seldom needed.

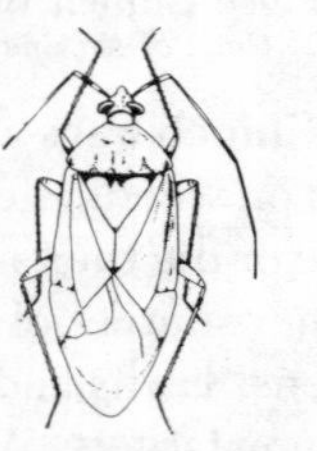

A Lygus Bug
(Univ. of Arizona)

Squash bugs (*Anasa tristis*) are generally found throughout the U.S. They are pests particularly of squash, pumpkin, cantaloupes, cucumbers and watermelons. Both adults and nymphs feed on plant juices. The nymphs hatch from brown eggs laid in clusters on the lower sides of leaves, and feed together in colonies. The salivary juices of squash bugs are toxic to leaves and cause large discolored areas to develop, causing leaves to wilt, curl and turn brown. The fruit may be attacked after the vines have been killed. The greater part of squash bug damage occurs in mid- and late summer when nymphs are abundant. Adults are flat-backed, and about 5/8″ long. They are blackish brown on top and mottled yellow beneath. Both nymphs and adults emit a disagreeable odor. The most direct control is to collect and destroy adults and egg clusters by hand. In late season bugs may be trapped and destroyed under pieces of board or burlap where they seek overnight shelter.

Squash Bug
(Union Carbide)

The **squash mirid** (*Pycnoderes quadrimaculatus*) is one of the smaller plant bugs related to lygus bugs and black fleahoppers. The adults are 1/8″ long, black mottled with gray and white and yellow legs. It is quite active and flies readily, being much smaller than the adult squash bug. It feeds on the foliage of squash and other cucurbits, lettuce, beans, and several weeds, usually in late summer and fall. Adults and nymphs leave feeding punctures in the lower leaf surface, which causes the upper surfaces to become gray. Attacked plants may be destroyed under heavy infestations.

Sowbugs and Pillbugs

Sowbugs and pillbugs are not insects, but belong to the crustaceans, somewhat related. They grow to about 1/2″ in length, are gray, with 7 pairs of short legs, and a hard, shell-like covering. When disturbed pillbugs frequently roll into a ball resembling an armadillo, thus the name. Sowbugs do not. Unlike

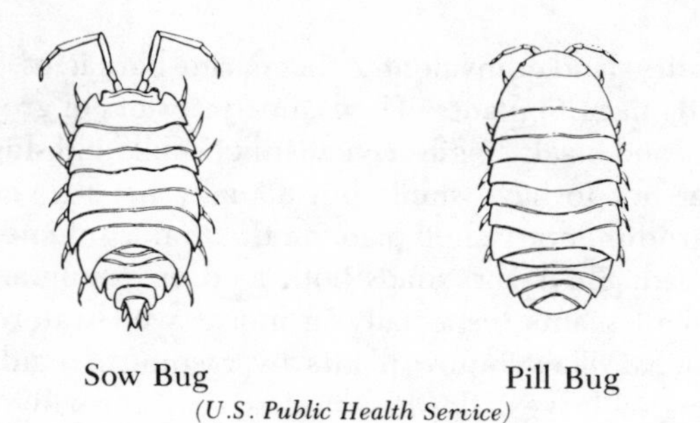

(U.S. Public Health Service)

insects they breathe by gills and live only in moist environments. Though they normally feed on decaying vegetable material, they may also attack seedlings, new roots and tender stems of growing plants. They are nocturnal and hide by day in damp environments such as under boards, flower pots, rocks, and in mulches and decaying plant material. Ventilation and dryness will help reduce their numbers, and control measures are seldom if ever needed.

Spider Mites

Several species of spider mites in the genus *Tetranychus* are among the plant feeding mites frequently found in varying numbers on garden vegetables. They remove plant juices from the leaves through piercing-sucking mouthparts, similar to certain insects, though they are not even distantly related to insects.

These small spider mites are almost invisible to the naked eye. Their color varies from a reddish yellow to light green, usually with some dark markings. The young resemble adults, newly hatched nymphs have only 6 legs while older nymphs and adults have eight. Spider mites live on the undersides of leaves, sometimes covering them with dusty webbing. Their feeding results in bronzed leaves, with yellow or brown discolorations. Heavily attacked leaves may become dry and brittle and drop from the plant. Mites are particularly a problem on cucurbits, beans and tomatoes.

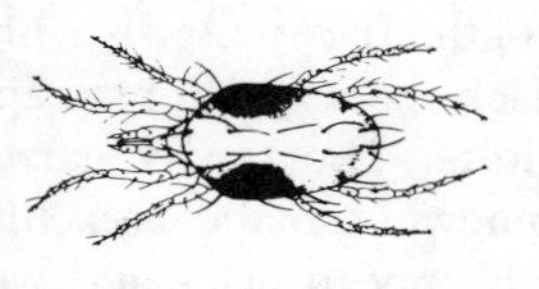
A Spider Mite
(Univ. of Arizona)

On tomatoes can occasionally be found the ultra-small tomato russet mite, which is only 1/125th of an inch in length, and virtually invisible to the naked eye. It is cream colored, elongated and has but two pairs of legs. It attacks only tomato, and feeds on the upper surfaces of leaves, causing the usual bronzing. It is rarely found in home gardens, and may not be recognized because of its small size. Injury normally appears first on the stalks at ground level and moves upward. In severe cases the fruit may be attacked.

Psyllids

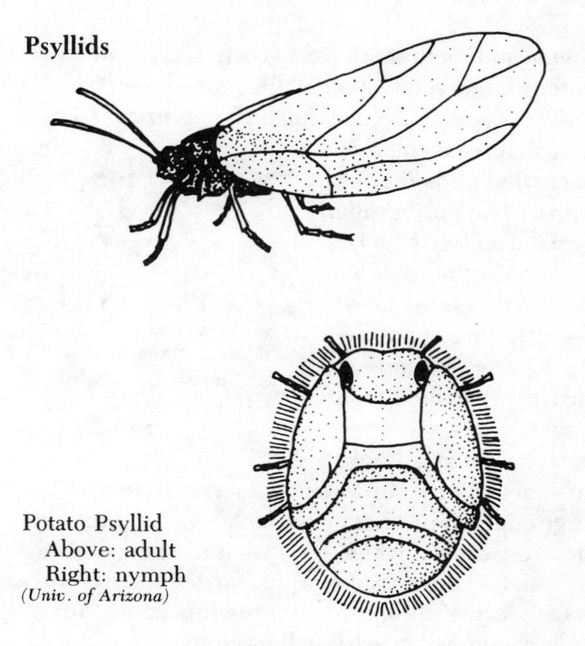
Potato Psyllid
Above: adult
Right: nymph
(Univ. of Arizona)

In the West, the potato (or tomato) psyllid (*Paratrioza cockerelli*) removes the plant juices from potato, tomato, pepper, eggplant, and related weeds of the family Solanaceae. The adults are small, about 1/8″ long, with dark gray-brown bodies. They are marked above with parallel whitish stripes on the thorax and an inverted white "Y" on the rear of the abdomen. Their wings are held rooflike and are longer than their bodies. When disturbed they jump easily with their strong hind legs, suggesting their rather close relationship to the leaf- or plant-hoppers. Sometimes psyllids are referred to as jumping plant lice. The young (nymphs) are flat, greenish-yellow, and resemble their other cousins, the scales. The older nymphs have bright red eyes. Psyllid nymphs are normally found attached to the under sides of leaves, feeding on plant juices through their piercing-sucking mouthparts. Their injury, however, is the result not of removing the plant juices, but rather from injecting their salivary fluid into the leaves while feeding during the nymphal stages, resulting in "psyllid yellows" on potato and tomato. The first symptoms of psyllid yellows are a slight yellowing of the growing tips, the midribs and leaf edges, and curling at the base of leaves. Progressively the entire plant becomes yellow to purple and growth ceases. Heavy infestations of nymphs early in the growing season invariably result in reduced yields, both in size and number of tubers.

Seedcorn Maggot

Seed and small seedlings of corn, beans, melons and other large-seeded vegetables may be attacked by the seedcorn maggot (*Hylemya platura*). Their feeding may weaken or destroy cotyledons and growing tips. The full grown maggots, which are the larvae of a gray fly, are ¼″ long, and resemble all other fly-maggots — white, legless, and tapered toward the headless front end which bears two destructive feeding hooks. Injury occurs in the spring as seeds begin to emerge, in damp, heavy soils heavy in organic matter, especially when cool weather has slowed emergence. (Beware, organic gardeners!) Little damage occurs under ideal growing conditions, as in late summer or early fall gardens.

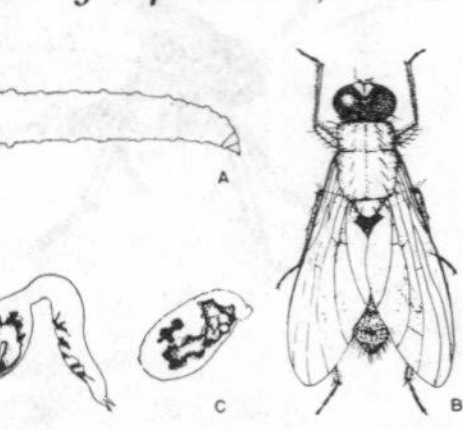

Seedcorn Maggot. A— Mature larva; B — Adult; C — Injured germinating seed. *(Univ. of Arizona)*

Slugs and Snails

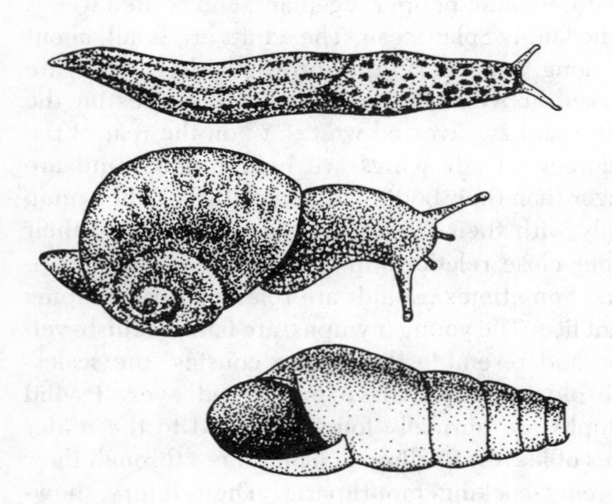

Top: A Slug
Center: Brown Garden Snail
Bottom: Decollate Snail (shell)
(Univ. of Arizona)

Snails and slugs belong to the mollusks, which include the shellfish oysters, clams, mussels, barnacles, and cuttlefishes. Slugs and snails are the mali-lcious dry-land equivalent to the marine barnacles, in their damage to plants. They come in colors of grey, orange and black. Snails have distinct shells and slugs appear not to have shells but actually have a very small rudimentary shell plate on the upper side near the head. Slugs and snails both feed on young and succulent plants, especially in moist, well-watered situations. They injure plants by rasping irregular designs in leaves, flowers, and stems, normally at night and during damp weather. In contrast, caterpillars feed on plants all the time, leaving large amounts of droppings instead of a slime trail. Often the silvery, slimy trails of snails and slugs are seen on walks, stepping stones, soil, grass and foliage before their damage is found Some crawl on houses or damage painted siding.

Slugs are more damaging than snails, and both are often present in greenhouses throughout the year and in many home gardens and border flower plantings for most of the summer. Eggs are found almost any time of the year, and most species overwinter in this stage. A few species overwinter as adults in concealed places.

The young resemble the adults and begin feeding as soon as they hatch. The life span is usually less than a year.

Two common garden snails are the brown garden snail or European brown snail (*Helix aspersa*), which has a roughly globular spiral shell up to an inch or more in diameter, and the decollate snail (*Rumina decollata*), which has a cone-shaped spiral shell about 1½ inches long and ½ inch wide at the open end.

Three species of slugs may be found in home yards and gardens. The spotted garden slug (*Limax maximus*) is the largest, and may reach 4 inches in length. It is yellowish-brown to gray and has large black spots which may appear as black bands in the adults. The tawny garden slug (*Limax flavus*) is somewhat smaller, reaching 3 to 3½ inches in length. It is gray to black and has yellow spots. Unlike the clear slime left by the spotted slug, the tawny slug leaves a trail of yellow. The gray garden slug (*Agriolimax reticulatus*) is smaller than the others, measuring up to 2 inches. It is flesh-colored, heavily spotted with gray or black, and it leaves a clear slime trail.

Slugs and snails require moist, shaded areas and can be discouraged by garden sanitation and foliage removal to improve ventilation and air movement.

Control by hand picking provides limited success, is tedious, but avoids the use of chemicals. Other methods include the use of molluscicides (snail and slug poisons) as poison baits, trapping under boards, and using beer pan traps. In the poison bait category are four compounds: Metaldehyde, carbaryl, mercaptodimethur, and mexacarbate, all sold as ready-to-use baits. Dusts or liquid formulations are recommended for greenhouse slug problems, while baits are more satisfactory for home garden and flower problems. Baits are more successful when distributed in infested areas just after a rain or after watering when these pests are most active.

Non-chemical control can be achieved by laying boards of any size one inch thick on moist garden soils. The next morning the underside should be heavily laden with these slimy culprets which can be removed and destroyed.

Beer pan traps are the most interesting of all control methods. To make these fascinating traps, place small cans in the ground with the lip at soil level, spaced at 3 to 10-foot intervals. Then fill to about one-half with beer. Empty the traps and refill twice weekly.

Thrips

One thrips or ten thrips, the name is both singular and plural. These are slender, tiny, active insects about 1/20″ long. Their colors vary from light yellow to dark black, depending on the species. They can readily be found in the blossoms of most garden flowers by thumping the blossom over a sheet of light-colored paper. Their wings are narrow and fringed, but may be absent. Thrips are somewhat unusual in that they possess rasping-sucking mouthparts. Most species are plant feeders, but intermingled among them may be predatory species which feed on mites and other insects. The plant feeders rasp surfaces of young leaves and growing tips to expose the plant juices within, causing the leaves to become distorted. Older tissues become blotched appearing silvery or leathery in affected areas. Two common thrips are the onion thrips (*Thrips tabaci*) and the western flower thrips (*Frankliniella occidentalis*). The latter may be found in both vegetable and flower gardens on practically all species. The onion thrips, when abundant, cause onion leaf blades to become spotted and wither at the tips, giving the plants an aged, grayish appearance. The flower thrips cause occasional damage, particularly in the slow-growing period of early spring and late fall, by feeding on the tender growing tips of developing garden plants. Because adults are strong fliers, there is really no way to avoid their emigration.

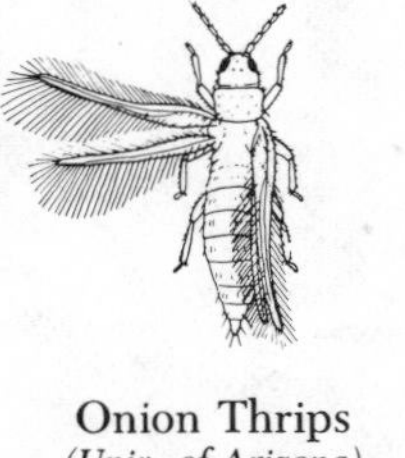

Onion Thrips
(Univ. of Arizona)

Whiteflies

Whiteflies first become noticeable as swarms of tiny 1/16th inch, white, mothlike insects that fly up from plants when they are disturbed. These are the adults. The nymphs are very different, tiny and scale-like, flat and fringed with white waxy filaments. They have sucking mouthparts, are confined to the undersides of leaves, and secrete sticky honeydew.

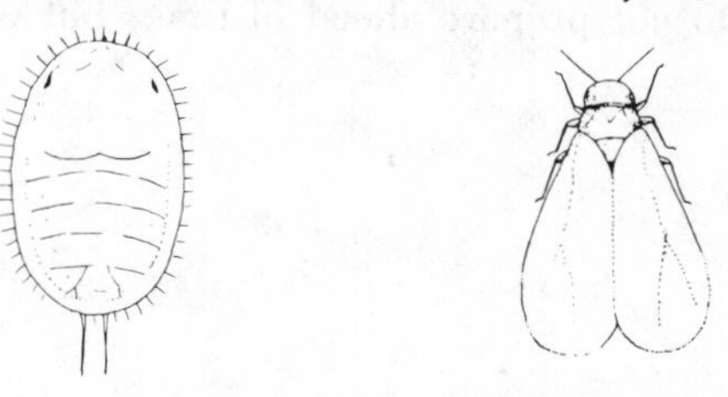

Whitefly: Left, Nymph; Right, Adult.
(Univ. of Arizona)

Whiteflies are very common pests in greenhouses, and are much more noticeable because they fly around the bench plants at eye level. In the garden they are usually a pest on tomatoes, though they may be found on other bushy plants such as beans. Whiteflies are most frequent early in the season, but may develop later also, moving from weed hosts in the area. Damage is usually slight to otherwise healthy plants, so insecticidal control is seldom suggested.

Applying Insecticides Properly in Vegetable Gardens

Table 10 shows generally which of the natural or organic controls is preferable for insect groupings. In Table 11, Vegetable Insect Control, insecticidal controls are emphasized only because there simply are not enough non-chemical methods available to match the efficacy of currently available insecticides. In the application of insecticides to garden plants, it is extremely important to observe the DAYS-WAITING-TIME to harvest. This is the waiting period required by federal law between the last application of a pesticide to a food crop and its harvest. These waiting periods are expressed as the number of days from the last application of the insecticide until the vegetable can be harvested. The number of days that you must wait before harvest varies depending upon the insecticide and the crop; therefore, you must look on the insecticide label for this information. If you observe the days-to-harvest waiting time, there will be no reason to fear that your vegetables will contain harmful insecticide residues as your family sits down to enjoy the home-grown evening repast.

Occasionally it is convenient to have a multi-purpose spray mix that will control insects, mites, and disease in one fell swoop. For the vegetable garden mixture, add 2 tablespoons of carbaryl (50% WP), 4 tablespoons of malathion (25% WP) and 2 tablespoons of maneb (80% WP) to one gallon of water. Do not prepare ahead of time, but only as needed.

In every case, it is very important to read and observe all directions and precautionary statements on the label. And in the application of all pesticides, cover plants thoroughly with the suggested materials for maximum protection and benefit.

Because Table 11 carries mostly insecticidal recommendations, the reader will assume that this is where the emphasis is placed. It is not. Instead, it should be strongly emphasized that an insecticide, or any other pesticide for that matter, is to be used *only* when needed. They are not to be used as a routine gardening procedure or scheduled, but only when and as needed. Any other system is environmentally unsound and counter to good judgement.

TABLE 10. General Selection Chart For Pest Control in the Garden and on Ornamentals Using Natural Controls.

	Bacillus thuringiensis	Dusting Sulfur*	Hand Removal	Lime Sulfur	Milky Spore Disease	Nicotine Sulfate	Petroleum Oils	Pyrethrins	Rotenone (cubé)	Ryania	Sabadilla	Soap Solution
Sucking												
Aphids						x	x	x	x	x	x	x
Leafhoppers						x		x	x	x		
Mealybugs						x	x					x
Scales			x	x		x	x					x
Spider mites		x		x		x	x			x		x
Spittlebugs						x			x			
Thrips						x	x	x	x	x	x	x
Whiteflies						x	x	x	x			x
Chewing												
Beetles								x	x	x		
Japanese beetles			x		x					x		
Weevils			x					x				
Caterpillars	x		x					x	x			
Grasshoppers			x								x	
Snails & slugs			x									
Burrowing												
Codling moths			x							x	x	
Leaf miners			x			x			x			
Corn earworms			x							x		
Borers			x									
Soil Insects												
Cutworms			x									
Grubs			x									
Lawn moths			x					x	x			
Nuisance												
Ants			x					x				x

* Do not use sulfur on vegetables to be canned.

TABLE 11. Insect Control Suggestions for the Vegetable Garden, With and Without Chemicals[1]

Plant	Pest	Chemical Control	Non-Chemical Control
Asparagus	Asparagus beetle	Carbaryl, Rotenone, or Malathion. Treat spears and vines when beetles appear and repeat as needed. Do not harvest until 1 day after treatment, nor repeat applications within 3 days.	Remove beetles by hand
Beans	Aphids	Dimethoate, Nicotine Sulfate, or Malathion. Apply to foliage when aphids appear. Repeat as needed. Wait 1 day before harvest, 3 days for Nicotine.	Knock off with hard stream of water.
	Bean beetles Flea beetles	Carbaryl or Malathion. Apply to underside of foliage at first sign of leaf-feeding and repeat as needed. Wait 1 day before harvest.	Remove beetles by hand.
	Spider mites	Dicofol, Sulfur, or Ethion. Apply at first sign of off-color stippling of foliage. Repeat in 2 weeks if needed. Wait 7 days before harvest for Dicofol and 2 days for Ethion.	
	Potato leaf-hopper	Carbaryl, Nicotine Sulfate, or Malathion. Treat when leafhoppers appear and repeat as needed. Wait 1 day before harvest, 3 days for Nicotine.	
	Seedcorn maggot	Plant only insecticide-treated seed.	Avoid planting in soil containing too much humus. Don't overwater.
	Leafhoppers, Spider mites, Bean beetles and Thrips	Disulfoton, 1% granules, will control these pests if applied near seed at planting time. CAUTION: Use only as directed on label. Do not apply within 60 days of harvest.	
Beets	Flea beetles	Carbaryl. Treat at first sign of small circular holes in leaves and repeat as needed. Wait 3 days before harvest, and 14 days if tops are to be eaten.	
	Leaf miner and Aphids	Malathion. Spray when mines appear at 7 day intervals as needed. For aphids treat on first appearance and as needed. Wait 7 days before harvest.	Wash aphids off with hard stream of water.

[1] Refer to TABLE 35 for trade or proprietary names of recommended pesticides.

Plant	Pest	Chemical Control	Non-Chemical Control
Cabbage Cauliflower Collards Broccoli Brussels sprouts	Aphids	Malathion, or Diazinon. Spray on appearance and weekly as needed. Wait 7 days before harvest.	Remove with hard stream of water.
	Cabbage worms	Carbaryl, or Diazinon. Apply when worms are very small and repeat every 10 days until harvest. Wait 7 days before harvest for Diazinon and 3 for Carbaryl.	*B. thuringiensis* pathogenic spores. Repeat weekly as needed. Begin treatment when worms are small.
	Flea beetles	Carbaryl or Malathion. Treat when tiny holes are found in foliage and repeat as needed. Wait 7 days for Malathion and 3 for Carbaryl before harvest.	
	Cabbage maggot	Diazinon. Use as a transplant water treatment. CAUTION: follow label directions exactly. Apply 1 cupful of the mixture into each transplanting hole.	
	Aphids, Flea beetles, Leafhoppers, Spider mites and Thrips	Disulfoton, 1% granules, will control these pests if applied as a band on each side of the seed furrow at planting time, or as a side dressing after plants become established, or mix in with the soil in transplant. CAUTION: use only as directed on label.	
Carrots	Six-spotted leafhopper	Carbaryl, Rotenone, or Malathion. Leafhopper transmits carrot yellows disease. Apply when leafhoppers are first seen and repeat as needed. Wait 0 days for Carbaryl and 7 days for Malathion before harvest.	
Corn, sweet	Corn earworm Corn borer	Carbaryl, or Ryania. Spray or dust foliage and silks. Apply when tassels begin to emerge and at 2-3 day intervals through silking. There is no waiting period when used as directed.	Examine silks daily and remove newly laid white eggs. Inject ½ medicine dropperful of mineral oil into silk channel as silks start to dry. Follow planting dates prescribed for your area. This helps avoid egglaying of moths.
	Seedcorn maggot	Purchase insecticide-treated seed.	Avoid planting in soil containing too much humus, and don't overwater.

Plant	Pest	Chemical Control	Non-Chemical Control
Corn, Sweet (Continued)	Flea beetles	Carbaryl, or Diazinon. Flea beetles transmit Stewart's disease in the North Central U.S. Apply when plants emerge and repeat 2 or 3 times at 5-day intervals. Early applications are necessary if infection is to be avoided.	
	Sap beetles	Carbaryl, Rotenone, or Diazinon. Apply to damaged ears of corn as described under corn earworm.	Beetles feed on overripe and cracked fruit including corn ears. Harvest corn when ripe and remove all fallen or rotten fruit from vicinity of garden.
Cucumbers	Aphids	Malathion or Nicotine Sulfate. Spray as needed when vines are dry. Wait 1 day before harvest for Malathion and 3 days for Nicotine.	Remove with hard stream of water.
	Cucumber beetles	Carbaryl, Rotenone or Methoxychlor. Apply when seedlings emerge, and at 5-day intervals. Repeat after a rain.	
	Spider mites	Dicofol or Ethion. Apply when mites appear and as needed. Wait 0 days for Ethion and 2 days for Dicofol before harvest.	
	Squash vine borer	Carbaryl, Rotenone, or Methoxychlor. Begin a 3-treatment series, at 7-day intervals, when first borer entrance signs are found. Wait 1 day before harvest.	Break growing tips to encourage branching. Cover main stems with soil or mulch to prevent egg laying by moths.
Eggplant	Aphids	Malathion or Nicotine Sulfate. Apply when aphids appear and repeat as needed. Wait 3 days defore harvest.	Remove with hard stream of water.
	Flea beetles	Carbaryl. Treat at first sign of tiny holes in leaves and repeat as needed. 0 days waiting period.	
	Potato beetle See Potatoes		
	Spider mites	Ethion or Sulfur. Apply when mites appear and repeat as needed.	
Lettuce	Leafhoppers	Carbaryl or Rotenone. Spray or dust when they appear and weekly as needed. Wait 14 days on leaf and 3 days on head lettuce for Carbaryl before harvest.	

Plant	Pest	Chemical Control	Non-Chemical Control
Lettuce (Continued)			
	Aphids	Malathion or Rotenone. Treat when aphids appear. Wait 14 days on leaf and 7 days on head lettuce before harvest for Malathion.	Carefully remove with hard stream of water.
Muskmelons	Aphids	Dimethoate, Rotenone, Diazinon, or Malathion. Aphids may transmit a mosaic disease. Treat when aphids appear and repeat as needed. Wait 3 days for Dimethoate and Diazinon before harvest.	Remove with hard stream of water.
	Cucumber beetles	Follow procedures as prescribed under Cucumbers.	
	Spider mites	Follow procedures as prescribed under Cucumbers.	See "Aphids".
Onions	Onion maggot	Diazinon. Spray very dilute mixture in the row at planting time.	Avoid intensive use of humus or overwatering.
	Thrips	Malathion, Rotenone, or Diazinon. Treat when thrips appear and at 5-10 day intervals as needed. Wait 3 days for Malathion and 10 days for Diazinon on green onions before eating.	
Peas	Aphids	Malathion, Nicotine Sulfate, or Diazinon. Apply when aphids first appear and repeat as needed. Wait 0 days for Diazinon and 3 days for Malathion or Nicotine before harvest.	Remove with hard stream of water.
Peppers	Aphids	Malathion, Nicotine Sulfate, or Diazinon. Apply when aphids first appear and repeat as needed. Wait 3 days for Malathion and Nicotine and 5 days for Diazinon before harvest.	Remove with hard stream of water.
	Flea beetles	Carbaryl. Spray or dust as needed.	
	Hornworms	Carbaryl or Rotenone. Spray or dust as needed.	Remove larvae by hand.
	Pepper weevil	Carbaryl or Rotenone. Treat as needed when pods begin to set.	
	European corn borer	Carbaryl or Ryania. In North Central states apply as preventive treatment every 3 days after blossoms appear and fruit forms. Damage occurs in late July and August.	Plant resistant varieties.
Potatoes	Aphids	Malathion, Nicotine Sulfate, or Dimethoate. Apply to foliage when aphids appear and repeat as needed.	Remove with hard stream of water.

Plant	Pest	Chemical Control	Non-Chemical Control
Potatoes (Continued)			
	Flea beetles	Carbaryl, Rotenone, or Methoxychlor. Treat when beetles appear and weekly as needed.	
	Leafhoppers	Carbaryl or Nicotine Sulfate. Treat when they are first seen and repeat as needed.	
	Potato beetle	Carbaryl, Rotenone or Methoxychlor. Spray or dust on appearance of larvae (slugs) and weekly as needed.	Remove larvae by hand.
	Soil insects (cutworms, white grubs, wireworms)	Diazinon. Apply to soil as a preplanting treatment to spaded ground and work into a 3-5′′, then plant potatoes. Lasts only one season.	Don't plant potatoes in same ground two successive years. Rotate crops.
Pumpkins and Squash	Aphids	Malathion, Rotenone, or Diazinon. Apply when aphids first appear and repeat as needed. Wait 3 days for Malathion on pumpkin and 1 for squash before harvest. Wait 3 days for Diazinon on squash.	Remove with hard stream of water.
	Cucumber beetles	Carbaryl, Rotenone, Malathion, or Methoxychlor. Follow procedures as prescribed under Cucumbers. Wait 1 day for all materials except 3 days for Malathion on pumpkin before harvest.	
	Squash bugs	Carbaryl or Sabadilla. Treat when bugs appear and repeat as needed.	Hand-pick adults and brown egg-masses from plants. Trap adults under boards laid beneath plants.
	Squash vine borer	Malathion, Rotenone or Methoxychlor. Apply 5 times at 7-day intervals when first borer entrance signs are found. Wait 1 day for Malathion and Methoxychlor on squash and 3 days for Malathion on pumpkin before harvest.	See Cucumbers. Butternut squash is resistant to vine borer.
	Spider mites	Dicofol or Pyrethrins. Treat as needed. Wait 2 days before harvest for Dicofol.	
Radishes	Aphids	Malathion, Rotenone or Diazinon. Treat when aphids first appear and repeat as needed. Wait 7 days for Malathion and 10 days for Diazinon before harvest.	Carefully remove with hard stream of water.

Plant	Pest	Chemical Control	Non-Chemical Control
Radishes (Continued)			
	Flea beetles	Carbaryl, or Diazinon. Treat when small holes are seen in leaves and repeat as needed. Wait 3 days for Carbaryl and 10 days for Diazinon before harvest.	
	Radish maggot	Diazinon. Spray or dust soil at 7 day intervals after planting. Wait 10 days before harvest.	
Spinach	Leaf miner	Malathion, or Nicotine Sulfate. Spray when mines appear at 7-day intervals as needed. Wait 7 days before harvest.	
Tomatoes	Aphids	Malathion, Rotenone or Diazinon. Treat when aphids first appear and repeat as needed. Wait 1 day before harvest.	Usually do not require control.
	Cutworms	Carbaryl or Rotenone. Apply when plants are first set out and repeat twice at weekly intervals.	Search beneath clods during day for cutworms. Look for plant parts in or near hiding place. Place a paper collar in the soil 1″ deep and 2″ high, 1″ from plant. This serves as a barrier that cutworms will not climb.
	Flea beetles	Carbaryl. Treat after plants are set and weekly for two weeks.	
	Blister beetles	Carbaryl, Rotenone, Diazinon or Methoxychlor. Treatment is needed later in season when first seen on plants. Repeat as needed. Wait 1 day for Diazinon and Methoxychlor before harvest.	"Herd" out of garden with small limb or broom to weedy margin where they originated.
	Hornworms and Fruitworms	Carbaryl. Spray or dust 3 to 4 times at 10-day intervals. Begin when first fruits are small.	Hand-pick worms when nibbled foliage is noticed. Apply Biotrol, Thuricide or Dipel HG pathogenic spores.
	Whitefly	Malathion, Rotenone, or Endosulfan. Treat when whiteflies can be shaken from foliage, at 15-day intervals for 3 applications and repeat as needed. Wait 1 day before harvest.	Inspect undersides of leaves at place of purchase. Do not buy if tiny, oblong, white, motionless insects are found. These are whitefly nymphs.
	Sap beetles	Carbaryl or Rotenone. Treat when beetles appear and repeat as needed.	Buy crack-resistant varieties. Harvest fruit when ripe and remove all fallen or rotten fruit from vicinity of garden.

Plant	Pest	Chemical Control	Non-Chemical Control
Tomatoes (Continued)			
	Aphids, Flea beetles Leafhoppers, Leaf miners, and Mites	Disulfoton 1% granules. Apply at planting time only. Mix in with the soil in the transplant hole before setting out plant. CAUTION: use only as directed on label.	
Turnips	Aphids	Malathion, Rotenone or Diazinon Treat when aphids are seen and repeat as needed. Wait 3 days for Malathion and 10 days for Diazinon before harvest.	Remove with hard stream of water.
	Flea beetles	Carbaryl or Diazinon. Treat when small holes are first seen in leaves and repeat as needed. Wait 3 days for Carbaryl (14 days if tops are eaten) and 10 days for Diazinon before harvest.	
	Turnip maggot	Diazinon. Spray soil at 7-day intervals after planting. Wait 10 days before harvest.	
Watermelons	Aphids	Malathion or Nicotine Sulfate. Treat when aphids appear and repeat as needed.	Remove with hard stream of water.
	Cucumbers beetles	Follow procedures as prescribed under Cucumbers.	
	Spider mites	Follow procedures as prescribed under Cucumbers.	

APPLE, PLUM, CHERRY, PEACH AND GRAPE

What greater pleasure can man derive from his domestic garden activities than offer a friend fresh fruit, preserves and jellies, or a glass of wine, with the casual statement, ". . . from this year's crop." There is many a slip from the cup to the lip in the production of the wine, the preserves, or that bowl of fresh fruit, since diseases and insects are also fond of your fares, from the time of blossom to the rich colors of ripening.

In the case of plum, cherry and peach pest control, you can improve your overall pest control program by keeping trees pruned and the areas around the trees and nearby fence rows free of brush and weeds. In the fall or winter collect and bury peach and plum mummies (rotted and dried fruit) to reduce brown rot and the survival of certain insects. If brown rot is a problem, when the fruit rots before it ripens, spray with captan at the full bloom stage. Prune and burn all infected branches if black knot is prevalent on plums. Spray sour cherries right after harvest with Dodine or Benomyl to control cherry leaf spot, which causes defoliation.

Grapes must be pruned every year. Do not use 2,4-D anywhere near grapevines. These chemicals will damage foliage and reduce production more than you can imagine. Symptoms of this damage are elongated terminal growth, downward cupping of old leaves and fan-shaped growth of new leaves.

As for apples, you have been spoiled by those beauties brought from the grocer. It's extremely difficult to grow fruit as high in quality as that produced commercially, but you can produce healthy apple trees and quality fruit through tree and orchard sanitation, pruning and training trees properly and fertilizing properly. Of the five fruits listed, apples are the most difficult to carry to perfection.

The universal multi-purpose spray mix including disease control that can be used on fruit trees and grape vines is as follows: to one gallon of water add 3 tablespoons of dicofol (18% WP), 2 tablespoons of malathion (25% WP), 3 tablespoons of methoxychlor (25% WP) and 2 tablespoons of captan (50% WP). Do not prepare ahead of time, but only when needed.

In all cases, it is very important to read and observe all directions and precautionary statements on the label. And in the application of pesticides cover plants thoroughly with the suggested materials for maximum protection and benefit.

Apple Maggot

Apple Maggot — Adult
(Union Carbide)

Of the orchard fruits attacked by the apple maggot, apples are the most seriously damaged. Other hosts are hawthorn, plum, pear, and cherries. The adult is a small, dark brown fly, with light and dark markings on the body and wings. Damage is caused by the larvae, which are white tapered maggots slightly smaller than those of house flies. The adults emerge from the overwintering puparia beginning in mid-June and continue for about a month. A week or more after emergence the females begin laying their tiny white eggs beneath the skin of the young fruit. Susceptible varieties of apple are the early maturing, such as Delicious, Cortland and Wealthy, those that are sweet. The eggs hatch in a few days and the maggots mine the fruit with brown tunnels, which may cause premature fruit drop. The larvae will mature in as little as 2 weeks in early varieties and as much as 3 months in hard winter apples. On maturing the maggots drop to the ground, enter the soil, and become puparia. Usually there is only one generation per year.

Cultural control consists of collecting all infested fallen apples and the elimination of hawthorn in the immediate vicinity. The fruit of ornamental flowering crabapple are a prime source of developing populations and should be sprayed or have all fruit collected periodically. Chemical control depends on using insecticide sprays to kill the adults before they lay their eggs. These can be included in the third through fifth or sixth cover sprays.

Cherry Fruitworm

This is the larva of a very small gray moth related to the oriental fruit moth. Its original or native host was probably wild cherry, but it is also known to attack blueberry. Mature larvae hibernate in galleries in the bark and pupate in the spring. The moths emerge from early June through mid-July and lay small, flat eggs on the fruit. After one week hatching takes place and the white larvae with black heads bore into the green cherries and feed around the pit. Their development is completed in about 3 weeks, and they leave the cherries to begin hibernation. There is usually only one generation each year.

Cherry Fruit Flies

These are the larvae of the cherry fruit fly, the black cherry fruit fly and the western cherry fruit fly. These maggots all feed in the flesh of cherries around the pit, often causing deformed fruit. Both sweet and sour cherries are attacked. The adults are about one-half the size of the house fly and have dark bands across their wings. Their bodies are dark with yellow markings.

Typically winter is passed as puparia in the soil. The adults emerge in late spring and lay eggs in the fruit. The maggots complete their development in the cherries usually at harvest time, and drop to the ground to change to puparia. There is only one generation each season.

Cultural control consists of destroying infested fruit and cultivation to destroy the puparia in the soil. Insecticidal control is directed at the adults just as they emerge and before they begin egg laying, requiring 2 to 4 applications.

Codling Moth

The codling moth occurs wherever apples are grown and is generally considered one of its most important pests. Pear, quince, English walnut, and sometimes other fruit are also injured. Damage is caused by the worm or larval stage which tunnels into fruit, usually to the core.

The moth is less than one inch in wingspread, with the front wings a gray brown, crossed with lines of light gray and bronzed areas near the tips. The larva is white, occasionally slightly pink, with a brown

Coding Moth
(Union Carbide)

head. They survive the winter as fully developed larvae, hibernating in cocoons on or near apple trees. In the spring the larvae pupate, and emerge as active moths the same time that apples are in bloom, and live 2 to 3 weeks. Eggs are white, about the size of a pin head, and are deposited on leaves, twigs and fruit. The eggs hatch in a few days and the small larvae bore into the fruit, often through the calyx. After 2 to 3 weeks the larvae spin cocoons and pupate, to emerge in 2 weeks as the next generation of moths. In the southern apple-growing areas, there may be three and a partial fourth generations.

Weather and several parasitic wasps play important roles in the numbers of codling moths that appear each year, but the most practical control is achieved with a spray schedule. This will usually combine a fungicide with an insecticide. Proper timing of sprays is essential and moth activity can be determined by using pheromone traps baited with the synthetic sex lure codlelure. The first spray is referred to as the petal-fall and is applied when nearly all petals have fallen and before the calyx closes. It is important to follow this with several cover sprays, and a second-generation spray applied about 10 weeks after petal-fall, usually late July or early August for most fruit-growing areas.

European Red Mite

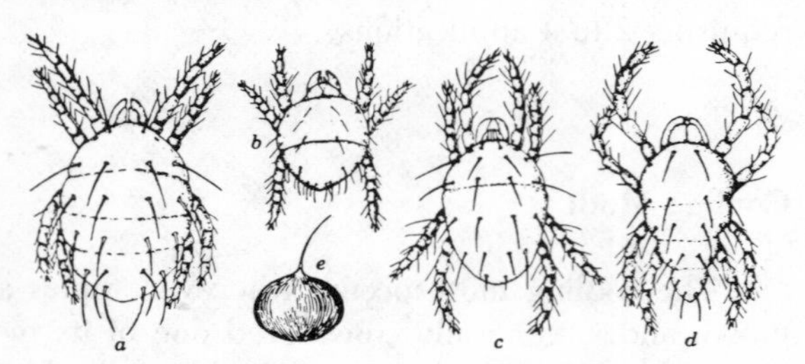

The European Red Mite, *Panonychus ulmi* (Koch): *a*, adult female; *b*,first instar; *c*, deutonymph; *d*, adult male; *e*, egg. *(Newcomer & Yothers, USDA)*

This spider mite is one of the most important fruit tree pests in the northern United States and adjacent areas of Canada. It attacks apple, pear, peach, plum, prune, and may be found on other deciduous trees and shrubs. Its damage results from the removal of plant juices with its piercing-sucking mouthparts, causing bronzed and off-colored foliage. Under heavy infestations, defoliation and undersized poorly colored fruit are produced.

The mites are rust-colored, though colors range from dark green to bright red. Winter is passed as the egg, usually on branches and twigs of the host trees. Hatching takes place during the pink bud stage, and damage may be seen shortly after young foliage appears. There may be 6 or 8 overlapping generations per season.

Considerable natural control by predators occurs in the home orchard, however control with sprays is usually necessary. Good control can be had by killing the over-wintering eggs using a delayed dormant (½ inch green tip stage) application of a superior-type oil emulsion. The eggs become increasingly susceptible to control as hatching time approaches. If mite control is required in mid-summer use a miticide least damaging to the predators, such as dicofol, where resistance is not a problem.

Eyespotted Bud Moth

Also referred to as the bud moth, this pest is distributed in all principal apple producing areas, but has been a serious pest only in the northeastern and northwestern states. The moth is a bit smaller than the codling moth, is dark brown with a light-colored band. The mature larva has legs, is ½ ich long, brown with a light stripe down its back, and has a black head.

There is only one generation a year, beginning when the adults appear in midsummer and begin egg laying on either side of leaves. After several days the eggs hatch and the larvae begin feeding on leaves, where their damage is most important. Occasional damage occurs to fruit that are in contact with infested leaves. Early in the fall the immature larvae hibernate in protected portions of the tree. In the spring they begin their feeding again as the buds swell and open, thus the name, bud moth. In early summer they mature, pupate in silk chambers usually in crumpled leaves, and emerge as adults beginning in mid-June. The life cycle is started again with egg laying. Very low winter temperatures (−21°F) and numerous parasites and predators play important roles in its natural control. Chemical control is for the larvae feeding in the buds and is applied at petal fall and the first cover spray.

Leaf Rollers

There are three leaf-rolling caterpillars in this group: the fruit-tree roller, oblique-banded leaf roller, and redbanded leaf roller. Their life cycles vary considerably and each will be discussed briefly.

The fruit-tree leafroller attacks apples and all other orchard fruits, and may require special control measures because larval feeding on blossom buds may prevent fruit setting. Damage to fruit results where leaves are attached to fruit with silk, the larvae feeding within. Foliage injury is not as important. The insect overwinters in the egg stage and hatching begins when buds open. The larvae feed on buds, blossoms, leaves and fruit, reaching maturity in June. They pupate inside rolled leaves and emerge as moths in about 2 weeks. Eggs are layed shortly after and the adults die. There is only one generation each year.

The oblique-banded leaf roller is of lesser importance. The caterpillars attack foliage in the spring, and fruit and foliage in the summer and fall. They feed on many different hosts including greenhouse crops and roses. After hatching the young act as leaf miners, then feed inside rolled leaves tied with silk. They overwinter as immature larvae in silk cases, and renew their feeding with the appearance of young leaves. The moths emerge in June and begin egg laying shortly afterward. There may be two generations per year, depending on latitude.

Redbanded Leafroller
(Union Carbide)

The redbanded leaf roller is most abundant in the Cumberland-Shenandoah apple region, and attacks apple, cherry, plum, peach, grape, other fruit, vegetable crops, ornamentals and weeds. The caterpillars feed on rolled or folded leaves held together with silk. Fruit are blemished when they are attached to infested leaves with silk. This pest overwinters in the pupal stage in protected cases on the ground. Moths emerge in the spring and lay their eggs in clusters on the tree bark. The eggs hatch near petal-fall, and the larvae feed, develop then pupate and emerge as adults in July. The cycle is repeated and there may be 3 to 4 overlapping generations, again depending on latitude.

There are several parasites and predators of the eggs and larvae of all three leafrollers which serve to some extent as natural controls. Chemical control is for the larvae and is included in the petal-fall and first through fifth cover sprays.

Oriental Fruit Moth

Oriental Fruit Moth
(Union Carbide)

Peach and quince are the most commonly attacked hosts, however apple, pear, plum and other fruit are occasionally attacked, particularly if they are near infested peach trees. Injury to trees occurs when the larvae tunnel into the tips of growing twigs, which prevents normal tree development. After attack new lateral shoots develop just below the damage, giving the tree a brushy appearance. Larvae also attack the fruit at the end of the growing season when twigs harden.

This pest is closely related to the codling moth, and both the larval and adult stages resemble each other. The larvae hibernate in cocoons on the tree, on the ground or on near-by objects. Pupation takes place in the spring and moths emerge at peach tree blooming. The eggs are laid on twigs and foliage, and later on the fruit. The larvae mature in two weeks when they leave the twigs to spin a cocoon and pupate. After 10 days the adults emerge and the cycle begins again, requiring about one month. There may be 4 to 5 generations a year.

There are several parasites and predators that attack eggs and larvae, thus exerting some level of natural control. However, chemical control is usually necessary, and is achieved by including the appropriate insecticide in the second, third, and remaining cover sprays.

Peach Tree Borer

The peach tree borer attacks not only peach, but plum, prune, cherry, almond, apricot, and nectarine. The injury is caused by the larvae boring beneath the bark near ground level, sometimes girdling the trunk. It occurs all over the United States where peaches are grown. The adult is truly beautiful. It is a clearwing moth with blue, yellow and orange markings. The moths fly by day and resemble wasps. The

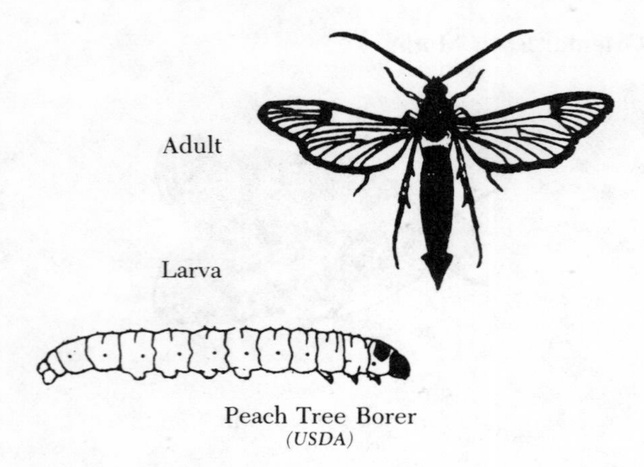

Peach Tree Borer
(USDA)

larvae overwinter in their tunnels at the bases of trees. They complete their development and leave their tunnels to pupate in the soil near the tree base. Pupae and adults can be found most of the growing season. Eggs are laid usually on tree trunks, requiring 8-9 days to hatch. The young bore into the trunk causing gum to seep from the holes along with their borings. There is one generation per year.

Natural control is not complete, and must be aided by chemical control. This consists of fall treatment of the soil around each tree trunk with paradichlorobenzene which acts as a fumigant. Spraying the tree trunks with an insecticide to kill adults, eggs and newly hatched larvae before they enter the bark is also effective. Pheromone, or sex attractant, baited traps will help determine the best timing for sprays.

Peach Twig Borer

As implied by its name, the peach twig borer is primarily a pest of peach, but it also attacks plum, apricot and almond. Its injury is to the twigs and fruit similar to the damage of the oriental fruit moth. The immature hibernating larvae emerge in the spring when growth of the twigs begins and bore into twigs and buds. Their tunneling stops growth or kills the shot. The larvae leave the twigs when development is completed and pupate on branches or trunk. Some 2 weeks later the adults emerge and lay eggs on leaves and fruit. As the season progresses each new generation tends to feed more and more on the fruit. There may be up to 4 generations per year.

Peach Twig Borer
(Union Carbide)

The twig borer is not a problem, in trees receiving dormant sprays of lime sulfur or oil emulsion each year. Otherwise, insecticidal control is achieved beginning with the petal fall and first through third cover spray.

Plum Curculio

The plum curculio is a hard-bodied snout weevil that attacks stone fruits, and is broadly distributed east of the Rocky Mountains. It also attacks apple, pear, quince and related fruits. Injury to fruit is extensive and begins in the spring from feeding of the adults, followed by punctures in the fruit from females laying their eggs, then feeding of the larvae within the fruit, and last from fall feeding by adults.

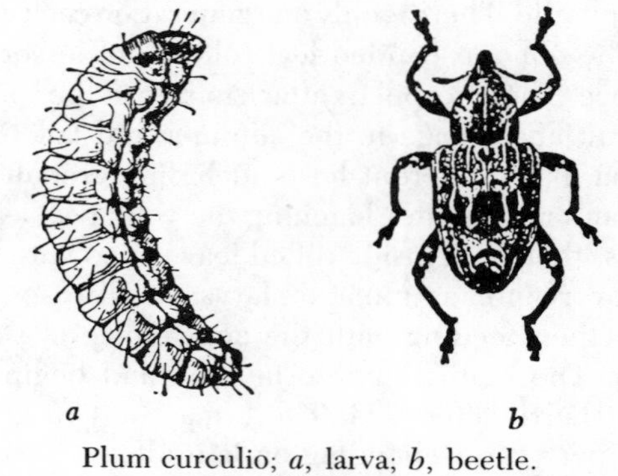

Plum curculio; *a*, larva; *b*, beetle.
(USDA)

The adults are brown with gray markings and have 4 small bumps on their backs. The larvae are curved, legless, and white with a brown head. The plum curculio overwinters as the adult in protected places near the fruit trees, and appears in the spring about the time trees begin to leaf. The adults feed for a month or more during which time they lay their white eggs in the holes where they have fed. These holes are cresent-shaped and remain with the fruit until harvest marring them and lowering their quality. Eggs hatch in 4-5 days and the grubs or larvae feed for 2-3 weeks when they mature. The larvae then bore their way out of the fruit and drop to the ground to pupate in the soil. The new generation of beetles emerge in a month and usually feed on the fruit before going into hibernation to await spring.

Many of the hibernating beetles die during the winter, while birds and other predators take their toll. There are several species of wasps that attack the eggs or larvae, and a fungus is known to infect both the larvae and weevils. An old method of control consists of jarring the weevils from the tree onto a sheet in the early morning and destroying them.

Destruction of infested, fallen fruit and cultivation beneath the trees to kill the pupae is quite helpful. Chemical control is for the adults and is usually applied at the petal-fall, and first and second sprays.

Scale Insects

The two most important scale pests of orchard fruits are the San Jose and the oystershell scales.

The San Jose scale attacks most cultivated fruits and a large number of ornamental shrubs and trees. The osage orange is often a reservoir for reinfestation. Tree health declines and fruit are blemished by the nymphs and adults removing plant juices from any part of the plant but particularly the wood. Heavy infestations may kill trees.

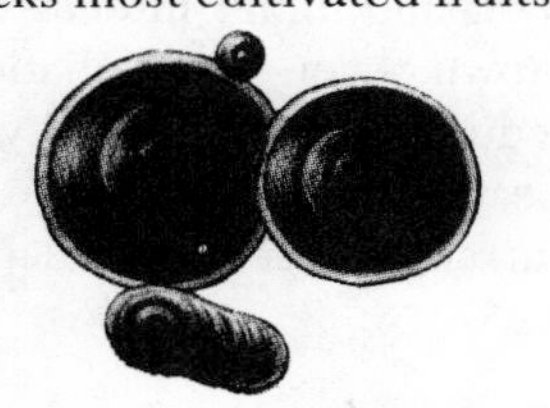

San Jose Scale
(Union Carbide)

The mature female scale is circular, the size of a pinhead, dark brown with a yellow center. Young scales are lighter, becoming dark with age. The males are smaller and oval in shape. Winter is passed in the immature scale stage beneath the scale coverings attached to the trees. In the spring development continues and winged males appear and mate with the females under the edge of their scale coverings. The female produces young or "crawlers" which move from under her covering to sites on the tree. The young are bright yellow, soon settle down to insert their mouthparts into the bark, leaves or fruit to feed, and lose their legs and antennae during the first molt. Their growth continues and is completed in about 6 weeks. Two or more generations may occur each season.

Spread within the tree is by crawlers, and from tree to tree on the feet of birds and on other insects. Some perhaps are carried by wind. Several parasitic wasps and predacious lady beetles and mites exercise a fairly high degree of natural control. Dormant sprays of lime-sulfur or superior-type oil emulsions are more effective than mid-summer efforts with the broad-spectrum insecticides.

The oystershell scale is a pest of apple, pear, occasionally other fruit trees, and many shade and ornamental plants. Lilac and ash trees are particularly susceptible. The scales are brown to gray, and considerably larger than the San Jose scale. Winter is passed as tiny white eggs beneath the scale of the female. In the spring the eggs hatch and the young crawlers migrate over the fruit tree to settle down and remove plant juices with their piercing mouthparts. The remainder of the process and cycle is identical to the San Jose scale, as are the controls.

Woolly Apple Aphid

This aphid occurs in most of the apple-producing areas of the world. Not only is it a serious pest of apple, it attacks, elm, mountain ash, and species of hawthorn. It feeds on plant juices from the usual above-ground plant parts and roots, making it a complex pest. The most serious injury is the below-ground, gall-forming feeding. These aphids are recognized by the woolly covering at the rear end of their blue-black bodies.

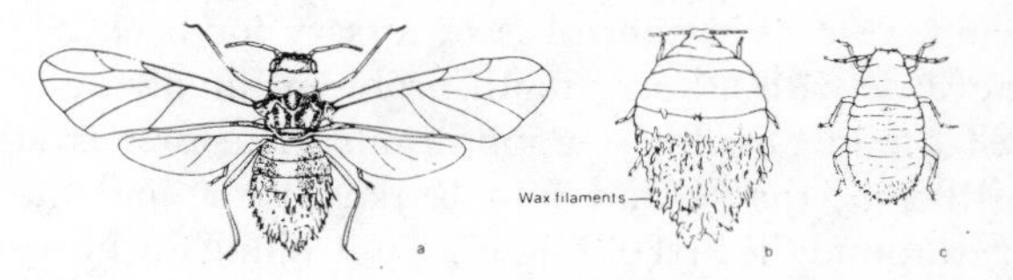

Eriosoma lanigerum, the Woolly Apple Aphid.
a, Winged female; *b*, apterous female; *c*, male
(From Baker, 1915.)

The alternate host, is the elm tree, and they move back and forth depending on the season, partially explaining their somewhat complex life cycle. Syrphid fly larvae, lady beetles and a chalcid wasp all are important in keeping the aphid checked. Dormant sprays listed in the table of spray schedules are quite effective in controlling this pest.

Grape Berry Moth

The berry moth, which occurs east of the Rockies, is the only common insect that does heavy damage to grape berries. The small brown moths begin to emerge from overwintering pupae when grape foliage has unfolded, usually the early part of June. The small scalelike eggs are laid on grape stems, blossom clusters or the small berries. The larvae feed on blossoms and small berries, usually leaving a silken thread as they move, resulting in webbed clusters.

Grape Berry Moth Larva
(Union Carbide)

One larva may practically destroy an entire cluster. After 3 to 4 weeks feeding they each cut a semicircular slit in the leaf, fold it over and tie it with silk, thus forming a cocoon where they pupate.

Second-generation moths emerge in 10 to 14 days and begin to lay their eggs on the berries. It is this second generation of larvae that cause the greatest damage, usually in late August, by eating their way into the berries and feeding on the pulp and seeds.

There are 2 generations per year, and the mature larvae of the second generation pupate inside the cocoons made in the leaves, which fall to the ground in the fall. Cultural control is quite valuable, especially cultivation or plowing under and burying the cocoons containing the over-wintering pupae, thus preventing adult emergence the following spring. Chemical control is necessary for backyard vineyards with a berry moth problem, involving at least 3 and possibly 4 applications. The first is at postbloom, the second 7 to 10 days later, and the third around the first of August. Controlling the berry moth with insecticides will also control the grape flea beetle.

Grape Flea Beetle

The grape flea beetle occurs only in the eastern United States, from the Mississippi valley eastward. It damages also plum, apple, pear, quince, beech and elm. Neglected vineyards, however are its favorite host. Their damage is caused by the adults eating the buds and unfolding leaves of grape, and the larvae skeletonizing the leaves.

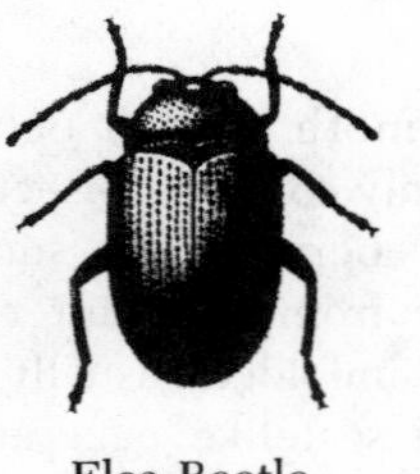

Flea Beetle
(Union Carbide)

The adults are about ¼ to ⅙ inch long, robust, with a metallic blue-green color. They emerge from hibernation and begin egg laying in the bark and on the leaves and buds. The brown larvae marked with black spots hatch from the eggs in a few days, feed on the undersides of the leaves for 3 to 4 weeks, then drop to the ground to pupate in the soil. They emerge as adults in 10 to 14 days, feed the rest of the summer and go into hibernation in the fall. There is only one generation each season. Control of the grape berry moth also controls the grape flea beetle.

Grape Leafhoppers

Wherever grapes are grown in the United States and Canada several species of leafhoppers can be found sucking plant juices from the lower leaf surfaces. The foliage becomes blotched with small white spots and under intense damage will turn yellow or brown, even to defoliation. The result of this is a reduction in the quantity and quality of the grapes. Leafhoppers may also attack apple, plum, cherry, currant, gooseberry, blackberry and raspberry.

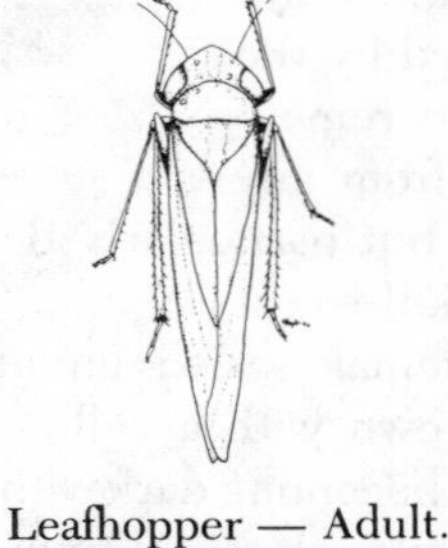

Leafhopper — Adult.
(Univ. of Arizona)

There may be 8 or more species involved in the heading of leafhoppers. They are all small, usually no longer than ⅛ inch, pale yellow, with red, yellow, or black markings on the wings.

Leafhoppers pass the winter as adults in protected areas under plant debris. As the spring days warm they become active and feed on any green plant before grape leaves develop. The eggs are laid in the leaf tissue, usually from the underside. Hatching occurs within 2 weeks, and the tiny nymphs move about and feed on the lower leaf surfaces, molting 5 times before becoming winged adults. This development requires 3 to 5 weeks and there are 2 to 3 generations each season.

Partial natural control of leafhoppers results from several parasites and predators and one fungus. In dense populations, however, chemical control is necessary and is included in prebloom and postbloom sprays, and later, when needed. The heavy use of insecticides on grapes may result in mite problems.

Grape Leaf Skeletonizer

These small yellow caterpillars are a pest all over the United States, but usually only on vines grown in home gardens. They attract attention because of their feeding only on the upper surfaces of the leaves, sometimes side by side in rows across the entire leaf. When fully developed the caterpillars are a little less than ½ inch long. The adult is a small, black, narrow-winged moth. Winter is passed as pupae in cocoons on leaves in ground trash. The adults appear late in spring and lay their bright yellow eggs in clusters on the lower leaf surfaces. There are 2 and sometimes 3 generations a year. Vines receiving regular sprays for other pests are not usually infested with skeletonizers.

Rose Chafer

Actually the rose chafer is a general feeder, and attacks not only grapes and roses but many tree fruits, raspberry, blackberry, strawberry and several garden flowers. The chafer occurs generally east of the Rockies, and is more abundant in areas having light sandy soil. The beetles feed on leaves, flowers, and fruit, while the larvae, which resemble small white grubs, feed in the soil on roots of various weeds and grasses. The larvae pass the winter in the soil and pupate near the surface in May. The tan, long-legged beetles, ½ inch long, appear in late May and June. The adults may cause serious damage in the years when they reach abundance.

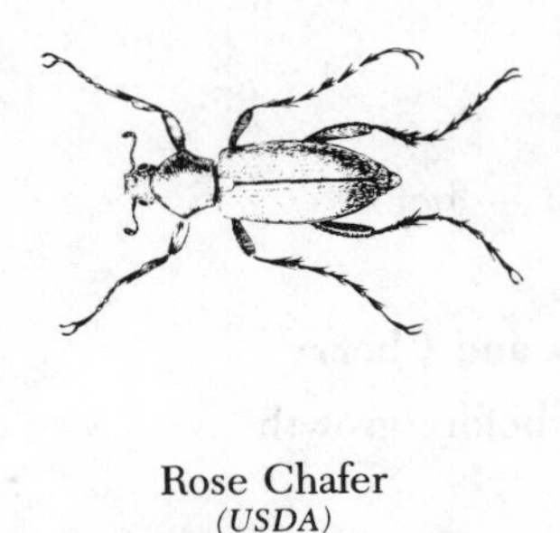

Rose Chafer
(USDA)

Feeding, mating, and egg-laying occur into early July. The eggs are deposited one at a time below the soil surface near the larval host plant. After hatching, the grubs feed until cold weather, becoming nearly full grown. There is one generation each year. The petal-fall spray for berry moth on grape also controls rose chafers, however, control is seldom needed on grape or other fruits.

TABLE 12. Pest Control Spray Schedules for Apple, Peach, Plum, Cherry and Grape in the Home Garden.[1]

Schedule (When to Spray)	Pesticides to Use	Pest to Control
Apple		
Dormant — before any green starts to show	Dormant oil or lime sulfur	Aphids, mites, and scales
Prebloom — leaves ¼ inch long until flowers start to open (3 sprays, 7 days apart)	Fungicides: Captan, Benomyl, Dodine, or Zineb	Apple scab
Bloom — at peak of blossom	Fungicide: same as prebloom. Do not use insecticides during bloom.	Apple scab
Petal Fall — When ¾ or more of blossom petals have dropped.	Fungicide: same as prebloom	Apple scab
	Insecticide: Ryania, Diazinon, or Malathion + Methoxychlor	Aphids, codling moth, leaf roller plum curculio
First Cover — 7 to 10 days after petal fall	Fungicide: same as prebloom	Apple scab
	Insecticide: same as petal fall	Codling moth, oystershell scale, plum curculio
Additional Covers — apply regularly at 10 to 14 day intervals	Fungicide: same as prebloom	Apple scab

[1] Refer to TABLE 35 for trade or proprietary names of recommended pesticides.

Schedule (When to Spray)	Pesticides to Use	Pest to Control
Additional Covers (Continued)		
	Insecticide: Ryania, Diazinon, Carbaryl, or Malathion + Methoxychlor	Aphids, codling moth, leaf roller, apple maggots, fruit moth
Post-Harvest — before leaf drop	Fungicide: Benomyl	Apple scab (over-wintering stage)
Peach, Plum and Cherry		
Dormant — before growth in spring	Lime sulfur	Aphids, mites, peach leaf curl, black knot
Petal Fall — when 90% of blossoms have fallen	Fungicide: Benomyl, Captan or Ferbam, plus	Brown rot
	Insecticide: Carbaryl, Pyrethrins or Methoxychlor	Plum curculio, eyespotted bud moth, prune moth, leaf rollers, twig borers
First Cover — 10 days after petal fall	Fungicide: Captan, Sulfur, Benomyl, or Ferbam, plus	Brown rot, leaf spot
	Insecticide: Carbaryl or Rotenone	Plum curculio, cherry fruitworm
Additional Covers — continue to spray at 10 to 14 day intervals until fruit begins to ripen	Fungicide: Captan, Sulfur, Benomyl, or Ferbam, plus	Brown rot, leaf spot
	Insecticide: Carbaryl or Rotenone	Cherry fruit fly, leaf roller, fruit moth
Grape		
When new shoots are 6 to 8 inches long	Fungicide: Zineb, Folpet, Captan, Ferbam, or Benomyl	Black rot
Prebloom — just before blossoms open	Fungicide: same as above	Black rot
	Insecticide: Carbaryl, Rotenone or Malathion + Methoxychlor	Leafhoppers, rose chafer, flea beetles
Post-bloom — just after blossoms have fallen	Fungicide: Zineb, Folpet, Captan, or fixed copper + hydrated lime	Downy mildew
	Insecticide: Rotenone, Methoxychlor, or Diazinon	Berry moth, rose chafer, leafhoppers, skeletonizers
When berries begin to touch in the cluster or are about pea-size	Insecticide: same as above	Berry moth, leafhoppers, skeletonizers
	Fungicide: same as above	Downy mildew

CITRUS

If you are one of the rare "northerners" who manage to produce citrus in your backyard by the skillful use of winter protection in the form of tents, plastic tarps, and temporary or permanent greenhouses, then you probably have few if any of the pest control problems plaguing your southern citrus growing neighbors. However, for those gardeners who raise citrus as easily as they do bermudagrass, they may indeed have pest problems (Table 13).

As a general practice, citrus trees should receive a moderate pruning every year, removing especially those limbs infested with scales. The trees should be fertilized twice, once in the spring and again in the fall. In the spring, fallen and shriveled fruit still on the tree should be removed and destroyed, since they invite secondary pests such as flies, wasps and bees, and ground beetles. At all times, weeds and grass should be removed or maintained at lawn mower height, since they too harbor pests, including gnats, flies and mosquitoes.

Citrus grown strictly for ornamental purposes, such as sour orange requires little or no pest control attention. Scale insects, however, can eventually produce partial defoliation and should be given attention when present. And, please, pick up your fallen oranges. They ruin the desired effect.

Whether your citrus are grown for gustatory or visual purposes, in the use of pesticides, cover trees thoroughly with the suggested materials for maximum benefit and read and observe all direction and precautionary statements on pesticide containers.

Citrus Thrips

Thrips are found on all varieties of citrus in all citrus-growing areas. Their damage is caused by both the nymphs and adults by their rasping the surfaces of leaves and fruit. Growth is stunted and the leaves and fruit take on a leathery silver appearance. Buds and blossoms are sometimes killed under dense populations, and new leaves may be gnarled and dwarfed.

Thrips overwinter as eggs deposited in leaves and stems, which hatch in early March. The young are pale and wingless with red eyes. As they mature they grow larger and become orange in color. The adults are only about 1⁄10 inch in length, and have 4 narrow, fringed wings. The life cycle from egg to adult requires from 2 to 4 weeks, depending on the temperature, and there may be as many as 12 generations a year.

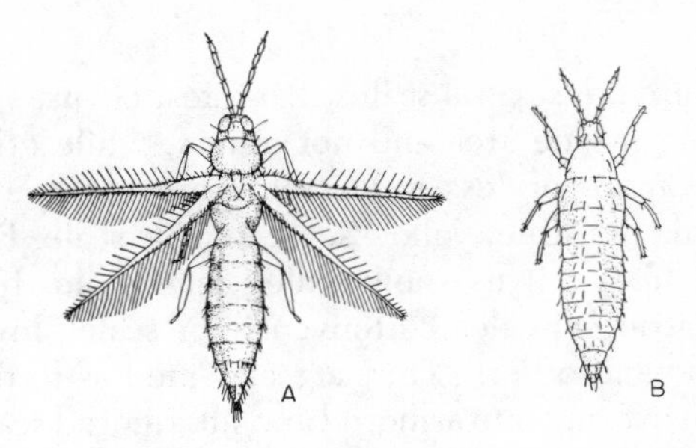

Thrips. A — Adult; B — Mature nymph
(Univ. of Arizona)

There are other species which may be found on citrus, but the home gardener cannot distinguish between them. They are similar in life cycle, and are more often found in blossoms where their feeding is considered unimportant.

Thrips are controlled to some extent by predator thrips, mites, spiders, and lady beetles. Insecticides are usually not necessary in that damage is primarily only of cosmetic or appearance value and does not actually alter the quality or quantity of fruit. Only rarely will chemical control be of value. (See also Thrips under Garden Pests).

Orange Dog

The orange dog caterpillar is a master of disguise in a citrus tree, resembling more a large bird dropping than it does a worm. These unusual larvae will pupate and develop into beautiful swallowtail butterflies. They overwinter as pupae attached to the tree or more commonly in ground trash. In the spring, the adults emerge, mate, and the females lay their eggs singly and randomly throughout the plant during the months of May through August. There may be up to 3 generations a year.

California Orange Dog
(Union Carbide)

Surprisingly, the orange dog is usually a pest only in the home yard, and never a problem in commercial groves. Even in the backyard it is seldom a problem, since the larvae normally eat only foliage. Because they occur in small numbers they can be easily removed by hand.

Scale Insects

There are several scales that infest citrus, some occurring in one area and not others, while others may be more or less universal. Among these are California red scale, yellow scale, purple scale, Florida red scale, citrus snow scale, chaff scale, black scale, citricola scale, cottonycushion scale, brown soft scale and others. They are classified as to their scale or covering into armored and unarmored scales, but in general their life cycles are similar as are their controls.

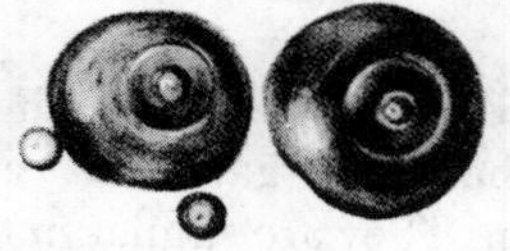

California Red Scale
(Union Carbide)

The scales overwinter either as eggs or young beneath the female scale, or the female gives birth to living young in the spring. The young are "crawlers" and move about the plant until they settle down and insert their piercing mouthparts to remove plant juices. They lose their legs and antenna and begin developing the scale, or grow larger without a scale depending on their classification. There are usually several generations a year, and natural controls by predators such as syrphid fly larvae and lady beetles sometimes are adequate to hold their numbers to unimportant levels. Chemical control is usually best achieved through the use of summer oil emulsions or dormant oils, though several of the organo-phosphate insecticides can be used in emergency conditions. (See also Scale Insects under deciduous fruits)

TABLE 13. Insect Control Suggestions for Citrus, With and Without Chemicals (Includes grapefruit, kumquat, lemon, lime, mandarin, orange, tangerine, and tangeloe)[1]

Pest	Chemical Control	Non-Chemical Control
Orange dog	Malathion. Apply when worms appear.	Remove larvae by hand or spray with Dipel
Scales	Malathion, Nicotine, or Diazinon. Treat when crawlers appear in spring and as needed. Summer oil emulsion (2%) or in fall.	Allow adequate time for biological control by vedalia and other lady beetles, before applying insecticides.
Spider mites	Dicofol, Tetradifon or Oxythioquinox. Apply when mites become numerous and as needed. Sulfur dust is also effective and the least disruptive to non-pests.	Mites may be hosed off with strong stream of water.
Whitefly	Malathion, Nicotine, or Dimethoate. Apply as spray when young begin to increase on leaf undersides.	
Thrips	Dimethoate, Ryania, or Malathion. Apply when 75% of blossoms have fallen, to protect foraging honeybees. Sulfur dust is also effective and the least disruptive with regard to non-pests.	
Aphids	Malathion or Dimethoate. Apply when aphids appear.	Remove with strong stream of water.
Leafhoppers	Malathion, Rotenone, Ryania, or Diazinon. Apply only if damaging leafhopper populations appear.	A light spray of whitewash, though unattractive, is repellent to leafhoppers.
Leaf roller	Carbaryl, Rotenone, or Diazinon. Apply when leaves are closed.	Larvae can be removed by hand or controlled with *B. thuringiensis* pathogenic spores.

[1] Refer to TABLE 35 for trade or proprietary names of recommended pesticides.

STRAWBERRIES

Strawberries, at one time available only to the affluent, can be produced by any interested person in the garden, barrel planters, window box, or even pots. Generally, there are few pests that specifically attack strawberries, but several seem to thrive on them simply because nothing else is available at the time.

To keep the protection of strawberries simple, insect and disease control have been combined into a spray schedule (Table 14). Chemical weed control is not suggested for home gardeners or amateurs. However, you should keep the plant rows narrow and free of weeds, for they not only inhibit growth and are unsightly, but they harbor pests. Heavy mulches are the best weed control.

As the best safeguard against disease, choose plants grown from virus-free stock, and select those known to be disease-resistant. When using sprays, mix all materials together and apply at one time, and apply enough to wet the foliage thoroughly. When using dusts, apply the material under leaves, but avoid overuse. Do not use dusts after berries begin to ripen.

If you intend to plant strawberries in areas that were previously in sod or weeds, you should control resident soil insects by first treating the soil with diazinon spray or granules and stirring the soil thoroughly. Follow the label instructions. Combine the insecticide with the upper 4-6 inches of soil immediately after application.

As fruit is produced, it is best to remove ripe and overripe or damaged berries from the garden, because they attract sap beetles, birds, wasps and hornets, bees, sowbugs or pill bugs, and a host of other hungry pests. Don't however, throw away such damaged fruit. Instead, wash them, cut out the damaged area and make strawberry preserves for next winter's hot biscuits.

Root Weevils

There are several species known as root weevils which, in addition to strawberries, attack raspberries, loganberries, blueberries, grapes, nursery stock and flower garden plants. Damage is caused by the larvae feeding on the roots and the adults feeding on the leaves. The adults are light brown to black snout beetles with rows of small pits on their backs. The larvae are white legless grubs, with pink tinge and brown heads. Most overwinter as nearly grown larvae in the soil among the roots of the host plants. In the spring they pupate and emerge as adults in June. Egg laying begins in about 2 weeks, in the crown of the plants. After hatching the larvae burrow into the soil and begin feeding on the roots.

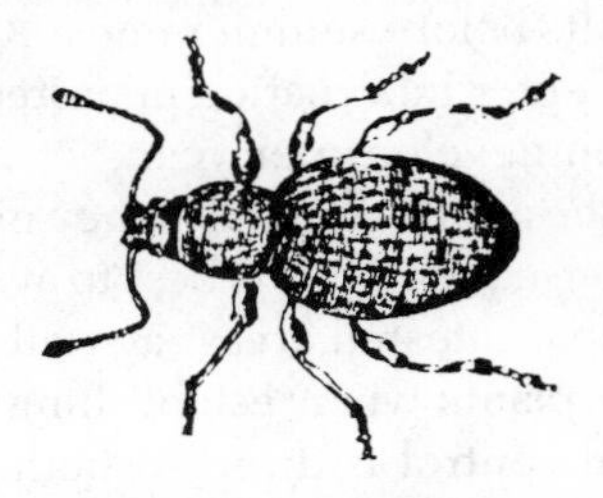

Strawberry Root Weevil
(USDA)

Cultural control consists mainly of making new plantings in uninfested soil. Control with insecticides is directed at the adult stage since the larvae are protected beneath the soil.

Spittlebugs

Spittlebugs are pests of perennial plants as strawberries, nursery stock, and legume forage crops. They injure plants both as nymphs and adults by removing plant juices with their piercing-sucking mouthparts. Some of the effects are stunting of growth, shortening of internodes, dwarfing, rosetting, and a general loss of vitality. The adults resemble robust leafhoppers, with many varied color patterns. Eggs are deposited in rows between sheaths and stems of plants near the soil surface. The yellow nymphs become green as they reach full development. Nymphs form the spittle mass by mixing air with the excretion from their alimentary canal, thus giving it the appearance of spittle.

Control with chemicals is best when applied before the adults begin egg laying. In small strawberry patches the nymphs can be removed by hand as easily as the plants can be sprayed.

Strawberry Crown Borer

Crown borer damage is caused by the larvae boring into the crown of the strawberry plant and feeding on the interior, sometimes so completely that growth is stopped or the plant killed. Some injury results from the feeding cavities made by adults in which they lay their eggs, as well as small holes they eat in the leaves.

The dark brown snout beetles cannot fly, and hibernate in or near the host plant in the winter. They become active when strawberries bloom, and begin laying eggs and continue into August. The small white, legless grubs feed in the crowns, pupate, and emerge as adults before summer ends. After feeding for a time they enter hibernation in protected debris. One generation develops per year.

Because the adults do not fly, they must crawl to their hosts. Consequently, it is best to avoid planting strawberries near infested areas, as well as purchasing borer-free plants when establishing new plantings. Chemical control is directed both toward the adults and larvae exposed in the crown.

Strawberry Leafroller

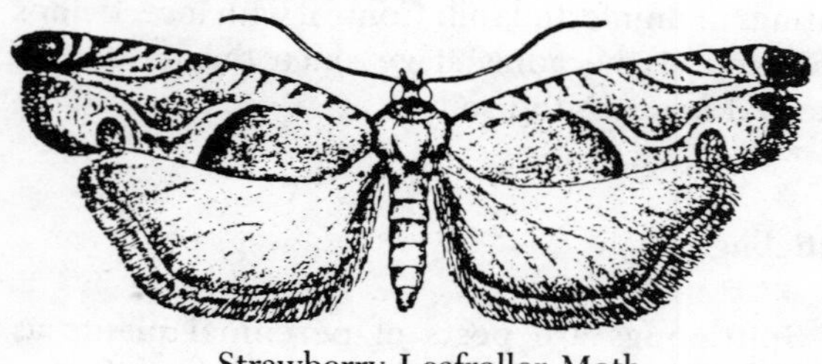

Strawberry Leafroller Moth
(J. B. Smith)

Damage caused by this leafroller, or leaf folder, results from the caterpillars feeding within the folded, rolled, or webbed leaves causing them to turn brown and die. Adults are rusty brown moths with light yellow markings, while the larvae are pale green to brown and up to ½ inch in length when fully developed. It overwinters as larvae or pupae in folded leaves or in other nearby shelters. Adults emerge in April or May and lay their eggs on foliage. The eggs hatch in a week and the larvae complete their feeding, pupate, and emerge as adults in 40 to 50 days. There are 2 generations each season.

Several parasites act to suppress the leafroller, though it may be necessary to use insecticides under heavy population pressure.

Strawberry Weevil

This small, brown snout beetle with the black patches on its wings is a native of North America and is a pest not only of strawberry, wild and domestic, but also of dewberry, brambles, and redbud. The larvae are small white curved grubs. The hibernating weevils emerge in early spring, and lay eggs in their feeding punctures in strawberry blossom buds. They then nibble away part of the bud stem, causing it to wilt, fall over, and usually drop to the ground. The larvae complete their development in these buds, pupate and emerge as adults. After a short feeding period they enter hibernation, thus making it a 1-generation per year pest. Only staminate flowers provide proper nutrition and developmental conditions making the pistillate varieties of strawberries relatively immune from attack. If insecticide applications are made in the spring at the first sign of weevil activity there should be very little damage.

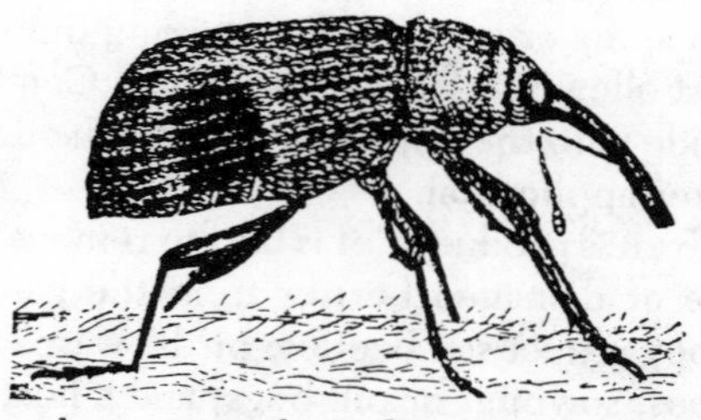

Strawberry Weevil, *Authonomus signatus*.
(Baerg, Ark. Agr. Exp. Sta.)

TABLE 14. Pest Control Spray Schedule for Strawberries[1]

Schedule (When to spray)	Pesticides to Use	Pests to Control
Prebloom — just as first blossom buds appear in spring	Insecticide: Methoxychlor + Malathion, or Nicotine	Aphids, spittlebugs, plant bugs, spider mites, leaf roller, weevils
	Fungicide: Dodine, Basic Copper, Captan, Thiram, Anilazine or Folpet.	Leaf scorch, leaf spot, leather rot
Every 10 days following first spray until blossoms appear	Insecticide: Same as above	Aphids, spider mites, plant bugs, leaf roller

[1] Refer to TABLE 35 for trade or proprietary names of recommended pesticides.

Schedule (When to Spray)	Pesticides to Use	Pest to Control
Every 10 days following first spray (Continued)		
	Fungicide: Same as above. Do not use Dodine within 14 or Anilazine within 5 days of harvest	Gray mold, leaf spot
Every 10 days starting in early bloom until 3 days before first picking	Fungicide: Captan, Thiram or Benomyl	Berry mold and rot, leaf spot
After bloom and on Sept. 1 and Oct. 1	Fungicide: Any listed in Prebloom spray Insecticide: Malathion + Methoxychlor, or Rotenone	Fungus diseases Aphids, leaf roller

LAWN AND TURF

Several insects, mites and related arthropod pests damage lawns and other turf. They cause the grass to turn brown and die, retard its growth and spread, or they build unsightly mounds that may eventually smother the grass. Some of these pests infest the soil and attack the roots, while others feed on the blades and stems, while yet others suck juice from the plants.

Other insects and insect-like pests inhabit lawns but do no damage. They may merely annoy by their appearance, or they may attack man and his pets.

Both of these categories can be controlled with insecticides, and a few can be managed with non-chemical methods. The suggestions in Table 15 are applicable not only to home lawns, but also to athletic fields, golf courses, parks, cemeteries, and to the areas along roadsides.

Not every pest, not every type of turf, and not every control method are listed. However, a general scope is presented, representing most of the common lawn situations. The solutions were derived from the latest research and recommendations of state agricultural experiment stations. If your problem isn't answered, or you want more detailed information, contact your County Extension Agent in your county.

Billbugs

Injury to turf and lawns by billbugs, or snout beetles, is caused by larvae eating the roots and crowns of grasses and by the adults feeding on stems and foliage. Their easily identified feeding punctures in grasses show up as a series of transverse holes of the same size and shape in the leaf. They result from a single puncture through the leaf in the bud stage before it has unfolded. Punctures in stems cause more damage but are less noticeable.

Billbugs are likely to be more numerous in well watered grass than lawns given haphazard treatment. The adults vary in color from light olive-yellow to brown and black, and they have the characteristic elongated mouthparts or snout. The larvae are white, short, thick-bodied, curved, legless grubs. There is only one generation a year. The overwintered adults are produced in late summer and may be active and feed for a while, or remain in the pupal cells until spring. In the spring the adults appear and feed almost continuously, laying eggs in feeding punctures near the base of the grass stems. The larvae develop quickly feeding in crowns and larger roots. In midsummer pupation occurs in the soil or in feeding cavities near the base of plants. In 10 to 14 days the adults emerge completing the one-year cycle.

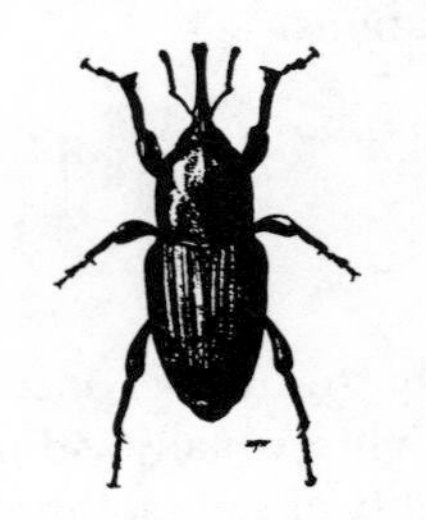

Maize Billbug, *Sphenophorus maidis*. (*USDA*)

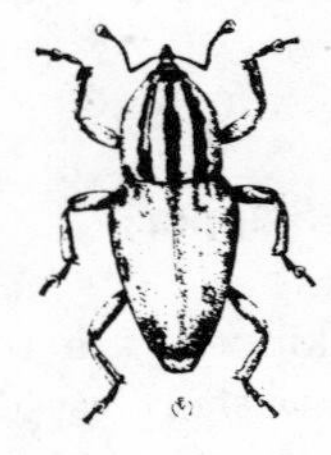

Clay-colored Billbug, *Sphenophorus aequalis*. (*USDA*)

Control by natural enemies is ineffective, and requires the use of one or more applications of insecticide when adults first appear in the spring.

Chiggers

Sometimes called red bugs, chiggers are actually the larval, parasitic stage of a small mite. They are about 1/150 inch long, and bright red. Young chiggers attach to the skin of humans, domestic animals, wild animals including reptiles, poultry and birds. More commonly than not they will enter hair follicles or pores and insert their mouthparts to suck up cellular fluids. They become engorged in about 4 days then drop off to change to nymphs.

Unlike the larvae, nymphs and adults feed on insect eggs, small insects, and other small creatures found near woody decaying material.

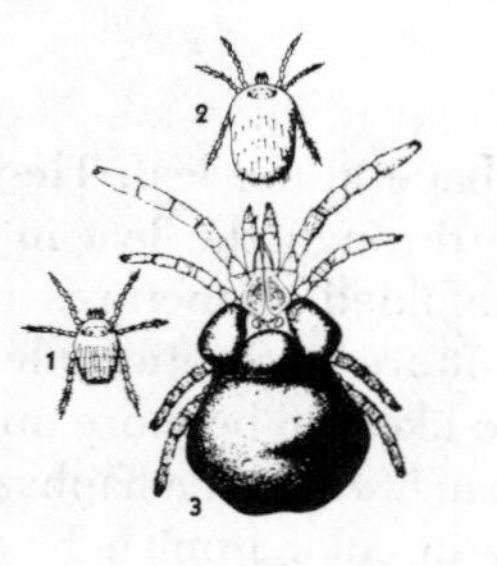

A Species of Chigger
(Eutrombicula batatas)

1. Unengorged larva
2. Engorged larva
3. Adult

From Michener. Ann. Ent. Soc. Amer. 1946

Once the lawn is established and the area becomes "civilized", the chigger problem gradually disappears. Insecticide dusts or sprays will aid immensely in resolving the problem. For personal protection use one of the two repellents, Deet or Rutgers 6-12, on the skin, socks and trousers. Repeat applications several times if exposure is continuous. Sulfur dust applied to the lawn at 7-10 day intervals is quite helpful in holding down populations.

Fiery Skipper

Skippers are the small, butterfly-like insects that have short, rapid, darting flights, usually at dusk, and not the long, sustained flights of the moths and other butterflies. They are small, yellowish-brown butterflies. Larvae of the fiery skipper feed on the leaves of common lawn grasses, but attack bentgrass most severely. Early infestation is indicated by isolated, round bare spots, 1 to 2 inches in diameter. The spots may become numerous enough to destroy most of the grass on the lawn. The fiery skipper is occasionally a pest of lawns in California.

Frit Flies

The frit flies, grass stem maggots, and eye gnats are all closely related and may be found in the same general environments. Frit flies are true flies, and the immature stages are maggots. The larvae feed in the lower portion of grass stems of healthy, well watered lawns that are left rank. Close mowing and usually one treatment of some short-lived insecticide will dispel the problem of hovering flies.

Japanese Beetles

The Japanese beetle is destructive in both the larval and adult stages. The larvae feed in the soil, devouring the roots of a large number of plants, and are especially injurious to turf in lawns, parks, golf courses, and pastures, and are sometimes a pest in nurseries and gardens. The adults feed on foliage, flowers, and fruits during their period of activity, which is only on warm sunny days. More than 275 kinds of plants, including fruit and shade trees, ornamental shrubs, flowers, small fruit, and garden crops are often damaged.

The beetles are ½ inch in length, shiny metallic green with coppery brown wing covers, making it one of the prettier beetles. The larvae resemble all of the other white grubs, being C-shaped and having 6 legs.

Japanese Beetle
(Union Carbide)

Larva or Grub
(USDA)

It overwinters in the soil in the half-grown larval stage. In the spring they crawl near the surface and feed. In late May and June they pupate, and adults appear in late June and are active through September. Eggs are laid in the soil at 1-inch depth, with hatching 2 weeks later. The young feed until cold weather sets in then retreat below the frost line. There is only 1 generation per year.

Larvae are controlled by incorporating milky disease spores in the soil of the bacteria, *Bacillus popilliae*. Moles, skunks, and birds also aid in the reduction of the grubs. Adults are difficult to control because they feed on so many different hosts. The only effective method of adult control is with the use of insecticides which is less than satisfactory.

Leaf Bugs

Leaf bugs are those tiny, 1/10 to 1/4 inch long true bugs that are easily seen crawling through a lawn. Most everyone who sits down in a grassy, weedy spot in early summer has seen these green, black, or red, often flecked, spotted, or striped insects. Both the nymphs and adults cause damage to grass by removing the plant juices from tender grass leaves. Their injury is recognized at first by small yellow dots on the leaves where their first scattered punctures are made. Later, with increased activity, leaves will yellow and die, leading to small areas and eventually to spots as much as a yard wide that die, appearing as if left unwatered. Prevalent among these leaf bugs are lygus bugs, tarnished plant bugs and apple redbugs.

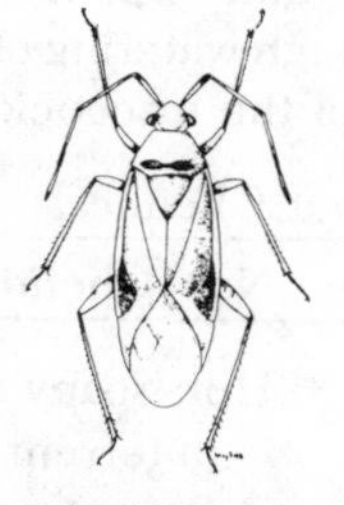

Tarnished Plant Bug
(Univ. of Arizona)

Cultural control consists of keeping the area free of weeds and mowing regularly. Chemical control is the only effective route when dead spots are observed and plant bugs identified as the cause.

Lucerne Moth

The larvae of this insect prefer clover and other legumes that may be found in your lawn, but they also infest grass. The adult is a grayish-brown moth, and has 2 pairs of dark spots on each forewing. The larvae resemble the sod webworms, but are slightly larger. Their damage occurs mostly in late summer.

Mole Crickets

Mole crickets are voracious feeders, and damage is known to occur to garden vegetables, and strawberries as well as grasses. Their damage is the result of underground feeding by the nymphs and adults at or near the soil surface, where they chew roots, tubers, underground stems and most fruit that touch the ground. Of the four common species, the southern mole cricket is the most abundant. They all resemble each other except for size. Adults are about 1½ inches in length, brown, tiny eyes, and short mole-like legs equipped for digging. The winter is passed as nymphs or adults in the soil. In spring the females lay their eggs in cells constructed in the soil, several to a cell. After hatching the young develop rapidly, with most of them becoming adults by early fall.

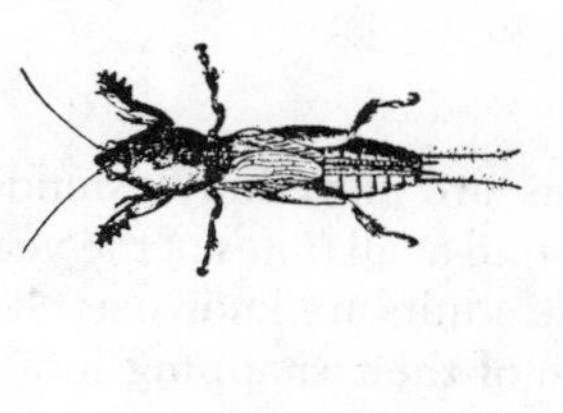

Mole Cricket.
(USDA)

There is very little in the way of natural controls. Chemical control can be aided by watering lawns heavily prior to applying the insecticide. When the soil becomes flooded they leave their burrows and swim about in an effort to locate dry quarters, and may increase their exposure to the insecticide.

Sod Webworm

Sod webworms, or grass moths, are not only pests of lawns, but also of golf courses, and occasionally cultivated crops planted in soil which was previously grassy sod. The yellow-to-white larvae spin threads as they move about and feed, which webs soil and leaves together near the surface, often forming tubes in which they live. They feed at the soil surface and slightly below. The light brown moths are less than an inch in length.

There are several species, but their life cycles are all similar. They overwinter as larvae in silk cases covered with soil near ground level. They continue feeding in the spring and pupate close to their feeding tunnels. The adults appear in early to mid-summer,

Sod Webworm
(Union Carbide)

and lay their eggs on grass at dusk. The eggs hatch in 5 to 7 days and the larvae begin the cycle anew. One to 3 generations may occur each year, depending on the latitude and species.

Natural control depends on insect and vertebrate predators, especially ants and birds, several insect parasites, as well as the use of *Bacillus thuringiensis*. Chemical control is quite simple and always effective.

Wireworms

Wireworms are the shiny, slender yellow or brown larvae found at all times of the year in most any type of soil. The adults are known as click beetles, so named because of their snapping into the air when turned upside down. Wireworms feed on many food crops as well as grass and turf lawns. They injure plants by eating seeds in the soil, by cutting off small underground stems and roots, and by boring in the larger stems, roots and tubers.

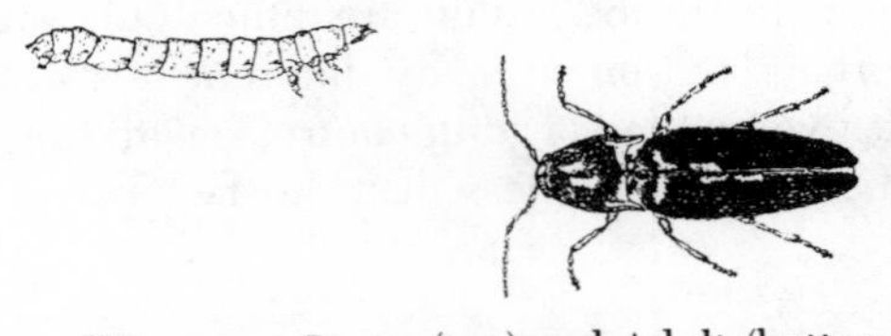

Wireworm Larva (top) and Adult (bottom)
(USDA)

Wireworms usually overwinter in the soil as larvae and complete their life cycles in one year, while others may take more than one year. The eggs are laid one at a time beneath the soil surface during the spring or early summer by the female click beetles. After hatching the young begin moving through the soil in search of food. All sizes of the larvae may be found feeding on the roots of grass during the spring and summer.

Cultural control is not very effective, since the most direct method is to deprive them of their food source. Either flooding or drying of the soil during the summer will kill many wireworms, but this too is impractical. This leaves us only with chemical control which is best achieved with the use of the granular formulations of several effective insecticides. After application a thorough watering of the lawn improves the effectiveness of the insecticide.

TABLE 15. Insect Control Suggestions for Lawn and Turf With and Without Chemicals[1]

Pests	Chemical Control	Non-Chemical Control
Ants	Diazinon, Pyrethrins, Chlorpyrifos or Carbaryl. Apply directly to nest as drench and lightly along trails.	Hot, soapy water will act as a deterrent for several days. Repeat as needed.
Armyworms	Diazinon, Rotenone, Chlorpyrifos or Trichlorfon. Apply at first sign of feeding damage.	
Billbug	Diazinon or Propoxur. Adults can be controlled by treating in April-May and larvae with May-June application.	
Chinch bugs	Carbaryl, Chlorpyrifos or Diazinon. Water lawn before treatment. Apply preferably as granular formulation in early June with second application 2-3 weeks later. If second generation occurs in August repeat applications.	Plant resistant varieties.
Clover mite	Dicofol, Diazinon or Chlorpyrifos. Apply treatment at first sign of an infestation.	

[1] Refer to TABLE 35 for trade or proprietary names of recommended pesticides.

Pests	Chemical Control	Non-Chemical Control
Cutworms	Carbaryl, Diazinon, or Chlorpyrifos. Apply at first sign of feeding damage.	Mow closely and roll lawn at dusk with heaviest roller manageable.
Earthworms	Not really a pest. No chemical control.	Probably indicates over-watering. Rake or roll worm casting mounds into surrounding lawn.
Earwigs	Chemical control not recommended since they do no damage. If enough of a pest Diazinon or Chlorpyrifos is effective.	Probable indication of over-watering. Trap in flower beds with boards, folded newspaper or dark plastic and destroy.
Fiery skipper	Carbaryl, Diazinon or Chlorpyrifos. Apply when bentgrass lawns show isolated, round ½″ diameter bare spots.	
Fleas and Chiggers	Diazinon, Rotenone, Chlorpyrifos, or Lindane. Spray all areas thoroughly, especially pet sleeping areas. Repeat in 10 days.	Fleas — open up bedding and expose for long periods to sun. Chiggers — keep lawn mowed to minimum height. Make light application of dusting sulfur.
Frit Fly	Diazinon or Chlorpyrifos. Apply to lawns on first appearance of hovering flies.	
Grasshoppers	No chemical control recommended.	Mow lawn frequently enough to qualify as well-kept.
Grubworms (White grubs)	Diazinon or Chlorpyrifos. Apply uniformly to lawn when grubs are small, late spring.	Apply milky disease spore dust and water into soil.
Japanese beetle grubs	Diazinon, or Chlorpyrifos. Apply uniformly to lawn when grubs small, early spring.	Apply milky disease spore dust and water into soil.
Leaf bugs	Diazinon or Malathion. Apply when first dead spots are observed in lawn.	
Leafhoppers	Carbaryl or Diazinon. Apply when leafhoppers are numerous, but before grass shows yellowing.	
Lucerne moth	Carbaryl or Diazinon. Usually a pest on clovers, this caterpillar can be controlled with one application.	Remove clovers from lawn by hand as soon as they appear.
Millipedes	Carbaryl or Diazinon. Treat when millipedes become abundant.	
Meadow spittlebug	No chemical control recommended.	Regular mowing.
Mole crickets	Diazinon or Chlorpyrifos. Apply to areas where burrowing and root feeding occur.	

Pests	Chemical Control	Non-Chemical Control
Scales	Disulfoton 15% G. Apply granules and water into the soil as near the time of crawler stage as can be determined.	
Sod webworm	Carbaryl, Diazinon, or Chlorpyrifos. Spray area for both generations, June and August.	
Thrips	Not a lawn pest but may annoy humans. Diazinon or Chlorpyrifos spray is effective.	Keep down nearby rank grass, flowers, and weeds.
Ticks	Diazinon, Chlorpyrifos, or Carbaryl. Spray entire lawn giving special attention to dog runs and sleeping areas.	Check dogs and remove all ticks every 3 days during tick season.
Wild bees	Malathion, Rotenone or Diazinon. Spray individual burrows as they appear.	Discourage by filling burrows with sand or coffee grounds.
Wireworms	Diazinon. Apply as granules in early spring to control these larvae of various click beetles.	

PESTS OF TREES, SHRUBS, WOODY ORNAMENTALS, ANNUALS AND PERENNIALS

Millions of dollars are spent annually by homeowners to beautify their homes and surroundings through the artistic and utilitarian planting of trees and shrubs. This beautification is not only pleasing to the neighbor and the homeowner, but it also increases the value of the property.

This increasing interest in and growing demand for more ornamental plants have been the incentives for nurserymen to experiment and grow an even greater selection of plants, including many exotics. The results are that never before have we had such a variety of plant materials and accessories from which to choose.

Usually, however, each of these species is the favorite food of one or more species of insect or mite. Most of the time natural enemies and other factors keep these pests at levels low enough that plants are not damaged. However, a season seldom passes that conditions are not favorable for the multiplication of at least one pest, and control measures must be applied to prevent serious injury to the plant.

We can never anticipate just which pests will become a problem, nor how severe the infestation will become. But there will always be potentially injurious pests in the area from the first warm spring days until frost and freezing in the fall and winter.

Tables 16 and 17 have been prepared with the intent that this information will help you to identify the pest, and know when and how to safely and effectively control the more troublesome pests of ornamental plants. As a result you can derive more enjoyment from your landscaping efforts.

Before closing, let me leave with you a universal insect and mite control spray mixture recipe that can be used on practically all ornamental trees, shrubs, and flowers. It is a multi-purpose spray mix, including disease control. To one gallon of water, add 2 tablespoons of carbaryl (50% WP), 1 tablespoon of dicofol (18% WP), 4 tablespoons of malathion (25% WP) and 1 tablespoon of zineb (65% WP). Do not prepare ahead of time, but only when needed.

In every case, it is very important to read and observe all directions and precautionary statements on the label. And in the application of all pesticides cover plants thoroughly with the suggested materials for maximum protection and benefit.

TABLE 16. Insect and Mite Control Suggestions for Trees, Shrubs, and Woody Ornamentals, With and Without Chemicals.[1]

Plant	Pest	Chemical Control	Non-Chemical Control
Andromeda (Pieris)	Lace bug	Carbaryl, Nicotine Sulfate, or Malathion. Treat as needed, underside of leaves.	
	Spider mites	Dicofol or Tetradifon. Repeated applications may be necessary.	
Alder	Leaf miner	Nothing labeled.	
	Woolly aphid	Malathion, Endosulfan or Lindane. Treat when aphids appear and as needed.	
Arborvitae	Bagworms	Carbaryl, Rotenone, Diazinon, Acephate or Malathion. Treat when bagworms are small, usually early to mid-June.	*B. thuringiensis* spores. Remove bags by hand in the fall and destroy.
	Leaf miner	Malathion, Nicotine Sulfate, or Methoxychlor. Spray twice at 10-day intervals in June.	
	Scales	Dormant oil, Carbaryl, Malathion or Diazinon. Use oil as dormant treatment in spring, or apply others against crawlers from late June to early July. May have to repeat in early September.	
	Spider mites	Dicofol or Tetradifon. Apply after new growth begins and in fall. Disulfoton 15% G. CAUTION: Follow label directions exactly.	A light dusting with sulfur each month reduces populations.
Ash	Aphids	Malathion, Endosulfan, Disulfoton 15% G, or Acephate. Treat when aphids appear and repeat when needed.	Remove with hard stream of water on smaller trees and lower limbs of large trees.
	Borer	Lindane or Chlorpyrifos. Apply to trunk and large branches in late May or early June, again in August-September.	Use adult sex-lure traps 3 weeks earlier than usual treatment time.
	Ash flower gall mite	Demeton Methyl or Disulfoton 15% G. Apply to soil when first blossoms form.	
	Fall webworm	Carbaryl or Acephate. Apply when webs appear.	*B. thuringiensis* spores.
	May beetles	Carbaryl or Methoxychlor. Treat when adults are seen and as needed.	

[1] Refer to TABLE 35 for trade or proprietary names of recommended pesticides. Many of the pests mentioned in this table are illustrated on pages 118-122.

Plant	Pest	Chemical Control	Non-Chemical Control
Ash (Continued)			
	Scale	Malathion, Nicotine Sulfate, or Carbaryl. Treat against crawlers in May-June. Repeat in 10 days.	
	Leafhoppers	Carbaryl or Methoxychlor. Treat when first seen, repeat as needed.	
	Leaf roller	Carbaryl. Apply on first appearance of leaves rolled together.	*B. thuringiensis* spores.
	Plant bugs	Carbaryl or Malathion. Apply when nymphs first appear and as needed.	
Azalea	Bark scale	Malathion, Diazinon, Carbaryl or Acephate. Treat for crawlers in June-July.	
	Leaf miner	Malathion or Nicotine Sulfate. Make 2 applications, mid-June and mid-July.	
	Spider mites	Dicofol or Oxythioquinox. Treat when mites appear and repeat as needed.	Remove with hard stream of water.
	Whitefly	Disulfoton 15% G., Endosulfan, Acephate or Diazinon. Treat when flies appear and repeat at 5-day intervals. Repeat as needed.	
	Lace bug	Malathion, Nicotine Sulfate, Carbaryl or Acephate. Treat when first bugs are seen, repeat as needed.	
	Borer	Lindane. Treat trunk and large branches in May, repeat twice at 3-week intervals.	
Bald-Cypress	Bagworms	Carbaryl, Diazinon, Acephate, or Malathion. Treat when bags are small, usually in June.	*B. thuringiensis* spores.
Barberry	Aphids	Malathion, or Disulfoton 15% G. Treat when aphids become numerous.	Remove with hard stream of water.
	Scale	Carbaryl or Nicotine Sulfate. Treat when crawlers appear.	
	Webworms	Carbaryl. Treat when larvae appear, usually July.	
Birch	Aphids	Malathion, Diazinon, or Acephate. Treat when winged aphids appear and as needed.	Remove with hard stream of water on smaller trees.
	Bagworms	Carbaryl, Diazinon, Malathion or Acephate. Apply when bags are small, usually June.	*B. thuringiensis* spores.

Plant	Pest	Chemical Control	Non-Chemical Control
Birch (Continued)			
	Fall webworm	Carbaryl, Rotenone, Diazinon, or Methoxychlor. Treat when webs appear.	*B. thuringiensis* spores.
	Japanese beetle	Carbaryl. Apply when beetles or riddled foliage are seen and as needed.	
	Scale	Malathion, Nicotine Sulfate, or Carbaryl. Apply for crawlers, usually May, and repeat in 10 days.	
	Borer	Lindane. Difficult to do, but entire tree should be treated twice, May-June.	
	Leaf miner	Carbaryl, Nicotine Sulfate, Malathion, or Acephate. Apply when adults are present, usually mid-May and again in late June.	
	Leafhopper	Carbaryl, Methoxychlor, or Malathion. Treat when leafhoppers appear and repeat as needed.	
Bittersweet	Scale	Nicotine Sulfate, Malathion, Carbaryl or Acephate. Apply in May-June when crawlers are present, repeat three times at 10-day intervals.	
	Skeletonizer	Carbaryl. Apply when observed, usually mid-July.	
Box elder	Box elder bug	Carbaryl or Endosulfan. Spray bugs in early summer.	Trap under boards or folded paper and destroy.
Boxwood	Leaf miner	Malathion, Diazinon, Carbaryl or Lindane. Apply when first mines appear, April-May.	
	Psyllid	Lindane or Nicotine Sulfate. Treat when young appear, usually May, repeat as needed.	Remove with hard stream of water.
	Scale	Acephate or Nicotine Sulfate. Treat when crawlers appear, usually June.	
	Spider mites	Dicofol, Oxythioquinox, or Disulfoton 15% G. Treat when mites appear and again in 10 days. Repeat if needed.	Remove with hard stream of water.
Buckthorn	Bagworms	Carbaryl, Diazinon, Malathion, or Acephate. Apply when bags are small, usually June.	*B. thuringiensis* spores.

Plant	Pest	Chemical Control	Non-Chemical Control
Buckthorn (Continued)			
	Japanese beetle	Carbaryl. Apply when beetles or riddled foliage is seen and as needed.	
Catalpa	Catalpa sphinx	Carbaryl or Acephate. Treat at first sign of small larvae.	Shake larvae off and use for fish bait.
Cercis (see Red bud)			
Chestnut	Leafhoppers	Malathion or Disulfoton 15% G. Treat when hoppers appear and as needed.	
Clematis	Blister beetle	Methoxychlor. Spray or dust beetles as needed.	Remove by hand. Wear gloves and long-sleeve shirt.
	Borer	Lindane. Spray 3 times at 10-day intervals, beginning usually in May.	Wrap trunk with heavy tree wrap.
Cotoneaster	Aphids	Malathion, Endosulfan, or Lindane. Treat when aphids appear and as needed.	Remove with hard stream of water. Wait for predator control.
	Lace bug	Malathion, Carbaryl, or Acephate. Treat when young are seen and repeat as needed to protect young foliage.	
	Leafhoppers	Disulfoton 15% G., Carbaryl or Methoxychlor. Treat on first appearance and as needed.	
	Pear slug	Carbaryl or Diazinon. Treat foliage 2 weeks after petal-fall and again in 2 weeks.	Remove by hand where possible.
	San Jose scale	Carbaryl or Acephate. Treat against crawlers in late June, repeat twice at 10-day intervals.	Import Vedalia beetles, their natural enemies.
	Webworm	Trichlorfon or Carbaryl. Treat when webs first appear.	*B. thuringiensis* spores.
	Spider mites	Dicofol, Tetradifon or Oxythioquinox. Treat twice at 10-day intervals and as needed.	Remove with hard stream of water.
Crabapple	Aphids	Malathion. Treat as needed.	Knock off with strong stream of water.
	Borers	Lindane. Spray trunk twice at 10-day intervals beginning usually in June.	Wrap trunk with heavy tree wrap.
	Leafhoppers	Carbaryl, Methoxychlor, or Malathion. Treat when leafhoppers appear and as needed.	

Plant	Pest	Chemical Control	Non-Chemical Control
Crabapple (Continued)			
	Scale	Acephate or Nicotine Sulfate. Treat when crawlers appear, usually late May.	
Deutzia	Aphids	Malathion, Lindane or Disulfoton 15% G. Apply when aphids appear and as needed.	Remove with hard stream of water.
	Leaf miner	Nicotine Sulfate. Treat when first mines appear and as needed.	
Dogwood	Borer	Endosulfan or Lindane. Treat trunk and bases of low branches in May and repeat 3 times at 3-week intervals.	Wrap trunk with heavy tree wrap.
	Leafhopper	Carbaryl, Malathion or Methoxychlor. Treat when leafhoppers appear and as needed to protect new foliage.	
	Scale	Malathion, Nicotine Sulfate, or Carbaryl. Apply against crawlers in late May and repeat in 10 days.	Remove with hard stream of water.
Dutchman's Pipe	Mealybug	Malathion. Spray 3 times at 14-day intervals beginning mid-May.	
	Scale	Malathion or Carbaryl. Apply against crawlers in early June and repeat in 10 days.	
Douglas Fir	Gall aphid	Lindane, Carbaryl or Disulfoton 15% G. Apply in early spring before budbreak or in October.	
	Bagworms	Carbaryl, Rotenone, Diazinon, or Acephate. Apply when bags are small, usually June.	*B. thuringiensis* spores. Remove by hand in fall.
Elm	Woolly aphid	Endosulfan, Diazinon, Lindane, or Carbaryl. Treat in spring when leaves are expanding.	Remove with hard stream of water on smaller trees.
	Bark beetles	Call your local County Agent or U.S. Forest Service Office.	
	Elm leaf beetle	Carbaryl, Rotenone, Methoxychlor, or Acephate. Apply when first larvae appear, usually when leaves are fully opened, and again in July.	
	Scale	Carbaryl, Acephate or Disulfoton 15% G. Apply against crawlers in June-July.	
	Fall webworm	Carbaryl, Diazinon, or Methoxychlor. Treat when webs appear.	*B. thuringiensis* spores.

Plant	Pest	Chemical Control	Non-Chemical Control
Elm (Continued)			
	Leafhoppers	Carbaryl, Malathion, or Diazinon. Treat when hoppers appear and as needed to protect new foliage.	
	Japanese beetle	Carbaryl. Apply when beetles or riddled foliage is seen and as needed.	
Euonymus	Aphids	Malathion, Nicotine Sulfate, Acephate or Disulfoton 15% G. Treat when aphids appear and as needed.	Knock off with strong stream of water.
	Scale	Malathion, Nicotine Sulfate, Carbaryl or Acephate. Use against crawlers in May-June and repeat twice at monthly intervals.	
	Bagworms	Carbaryl, Rotenone, Acephate or Malathion. Treat when bags are small, usually mid-June.	*B. thuringiensis* spores. Remove by hand in fall.
	Leafhoppers	Carbaryl, Methoxychlor, or Disulfoton 15% G. Apply when hoppers are seen and as needed.	
	Spider mites	Oxythioquinox, Dicofol or Disulfoton 15% G. Treat when mites appear and again in 10 days or as needed.	Remove with hard stream of water.
Fir	Bagworms	Rotenone, Diazinon or Malathion. Apply when bags are small, usually late June.	*B. thuringiensis* spores. Remove by hand in fall.
	Balsam twig aphid	Lindane, Diazinon, or Disulfoton 15% G. Treat when aphids are first noticed, usually early May.	Knock off with heavy stream of water.
	Pales weevil	Lindane. Treat young trees in May and again in July.	Pull and burn stumps before July.
	Pine needle scale	Diazinon, Malathion, Carbaryl or Acephate. Apply against crawlers.	
	Spider mites	Oxythioquinox, Dicofol, or Disulfoton 15% G. Apply twice, 10 days apart, and as needed.	
Firethorn (Pyracantha)	Aphids	Malathion, Lindane, or Disulfoton 15% G. Apply when aphids appear and as needed.	Remove with hard stream of water.
	Lace bug	Lindane, Malathion or Acephate. Apply when bugs first appear and as needed to protect new foliage.	
	Spider mites	Disulfoton 15% G. Apply twice, two weeks apart.	Remove with hard stream of water.

Plant	Pest	Chemical Control	Non-Chemical Control
Flowering Fruit Trees (Ornamental apricot, cherry, crabapple, peach, pear, plum, quince)			
	Aphids	Malathion, Endosulfan, or Disulfoton 15% G. Treat as needed.	Knock off with heavy stream of water.
	Borers	Lindane. Apply to trunks in May-June and repeat 4 times at 3-week intervals.	Wrap trunks during this period with tight trunk wrap.
	Fall webworm	Carbaryl, Diazinon, or Methoxychlor. Treat when webs appear.	*B. thuringiensis* spores.
	Leafhoppers	Carbaryl, Malathion, Methoxychlor, or Disulfoton 15% G. Treat when leafhoppers appear and as needed.	
	Peach tree borer	Endosulfan or Lindane. Make first application to trunks June-July, then two more spaced 3 weeks apart. Or use paradichlorobenzene crystals around base of trunk in the fall.	See Borers.
	Lesser peach tree borer	Endosulfan. Follow instructions for peach tree borer only beginning one month earlier.	See Borers.
	Pear slug and Psylla	Ethion + dormant oil. Apply during dormant season only. For psylla use Carbaryl or Diazinon 2 weeks after petal fall and repeat in 2 weeks.	Remove slugs by hand where possible.
	Japanese beetle	Carbaryl. Apply when beetles or riddled foliage is observed and as needed.	
	Bagworms	Carbaryl, Diazinon or Acephate. Apply when bags are small, usually early to mid-June.	*B. thuringiensis* spores. Remove by hand in the fall.
	Scales	Dormant Oil, Diazinon or Nicotine Sulfate. Use oil only in dormant stage. Apply others when crawlers appear.	
	Spider mites	Dicofol, Tetradifon or Disulfoton 15% G. Apply when mites appear and again in 10 days, and as needed.	Knock off with hard stream of water.
	Spring cankerworm	Methoxychlor, Rotenone, or Carbaryl. Apply when worms appear, early to mid-May.	*B. thuringiensis* spores.

Plant	Pest	Chemical Control	Non-Chemical Control
Flowering Fruit Trees (Continued)			
	Tent caterpillars	Carbaryl, Rotenone, Diazinon or Acephate. Eggs usually hatch when buds break in spring. Treat then or when webs appear.	*B. thuringiensis* spores.
	Woolly aphid	Malathion, Carbaryl, Diazinon or Chlorpyrifos. Treat when bluish-white threads from aphids appear, May-June.	
Forsythia	Spider mites	Oxythioquinox, Dicofol, or Disulfoton 15% G. Begin applications when mites appear and repeat in 10 days, and as needed.	Knock off with hard stream of water.
Golden Rain-tree	Leafhoppers	Carbaryl, Methoxychlor or Disulfoton 15% G. Apply when leafhoppers appear and as needed.	
Hackberry	Lace bugs	Malathion, Carbaryl or Disulfoton 15% G. Treat when first small nymphs appear, usually in May.	
	Nipple gall psyllid	Lindane. Apply in April-May.	
	Scale	Dormant Oil, Carbaryl, or Acephate. Use oil only as dormant application, and the others against crawlers in May.	
Hawthorn	Aphids	Lindane, Malathion, Diazinon or Acephate. Treat when aphids appear and as needed.	Knock off with strong stream of water.
	Bagworms	Rotenone, Diazinon or Acephate. Treat when bags are small, usually June.	*B. thuringiensis* spores.
	Fall webworm	Chlorpyrifos. Treat when webs appear.	
	European red mite	Dicofol, Diazinon, Oxythioquinox, or Disulfoton 15% G. Apply just prior to beginning of growth.	Remove with hard stream of water.
	Japanese beetle	Carbaryl. Apply when beetles or riddled foliage appears and as needed.	
	Tent caterpillars	Malathion, Rotenone or Acephate. Apply at first sign of leaf feeding while larvae are small.	*B. thuringiensis* spores.
	Lace bugs	Lindane, Carbaryl, or Disulfoton 15% G. Treat when first small nymphs appear, usually in May.	
	Leafhoppers	Carbaryl, Methoxychlor or Disulfoton 15% G. Treat when hoppers appear and as needed.	

Plant	Pest	Chemical Control	Non-Chemical Control
Hawthorn (Continued)			
	Leaf miner	Diazinon or Acephate. Apply when leaves are fully expanded or at sign of browning early to mid-May.	
	Scurfy scale	Dormant Oil, Malathion or Carbaryl. Use oil only as dormant treatment, and other materials against crawlers in May.	
	Terrapin scale	Dormant Oil, Carbaryl, Nicotine Sulfate, or Diazinon. Use oil only as dormant application, and others when crawlers appear in June.	
	Oystershell scale	Dormant Oil, Carbaryl, or Diazinon. Use oil only as dormant application and others against crawlers in May and again in 10 days.	
	Pear slug	Carbaryl or Diazinon. Make foliage application 2 weeks after petal fall and repeat in 2 weeks.	Remove by hand where possible.
Hemlock	Bagworms	Rotenone, Carbaryl, or Chlorpyrifos. Treat when bags are small, usually in June.	*B. thuringiensis* spores.
	Looper	Naled or Rotenone. Apply when worms appear, July-August.	
	Scale	Dormant Oil, Carbaryl, or Acephate. Use oil only as dormant application, and others when crawlers appear, usually July.	
	Strawberry root weevil	Lindane. Apply to foliage and soil around infested plants in June.	
	Spider mites	Dicofol, Tetradifon or Disulfoton 15% G. Treat when mites appear, and again in 10 days or as needed.	Knock off with hard stream of water.
Holly	Southern red mite	Dicofol, Oxythioquinox or Disulfoton 15% G. Treat when mites appear and again in 10 days or as needed.	Remove with hard stream of water.
	Leaf miner	Diazinon or Disulfoton 15% G. Apply non-systemic material to foliage in May for adult control. Apply systemic in June to control larvae in mines.	
Honey Locust	Bagworms	Carbaryl, Acephate or Diazinon. Treat when bags are small, usually in June.	*B. thuringiensis* spores. Remove by hand in fall.
	Locust borer	Lindane. Spray trunks thoroughly in early September.	Wrap trunks with heavy wrap.

Plant	Pest	Chemical Control	Non-Chemical Control
Honey Locust (Continued)			
	Spider mites	Dicofol, Diazinon or Disulfoton 15% G. Treat when mites appear and again in 10 days and as needed.	
	Plant bugs	Carbaryl. Apply when bugs appear and as needed.	
	Pod gall midge	Rotenone or Nicotine Sulfate. Apply to growing tips in spring and again at 10-day intervals until controlled.	
	Honey locust scale	Dormant Oil or Diazinon. Use oil only in dormant season, and Diazinon when crawlers appear.	
	Leafhoppers	Malathion, Nicotine Sulfate, or Disulfoton 15% G. Apply when leafhoppers appear and as needed.	
	Mimosa webworm	Carbaryl, Acephate or Rotenone. Apply at first signs of foliage browning from first generation in July and second in August.	*B. thuringiensis* spores. Remove by hand.
	Oystershell scale	Dormant Oil, Malathion, or Diazinon. Use oil only in dormant season, and other materials when crawlers appear in May-June.	
Honeysuckle	Aphids	Lindane, Endosulfan or Disulfoton 15% G. Apply when aphids appear and as needed.	Knock off with strong stream of water.
	Leaf miner	Nicotine Sulfate or Rotenone. Treat foliage at first signs of mines, May-June.	
	Spider mites	Dicofol, Oxythioquinox, or Disulfoton 15% G. Apply when mites appear and as needed.	Remove with hard stream of water.
Hornbean	Bagworms	Carbaryl, Diazinon or Acephate. Treat when bags are small, usually June.	*B. thuringiensis* spores. Remove by hand in fall.
	Leafhoppers	Disulfoton 15% G, Carbaryl or Methoxychlor. Apply when leafhoppers appear and as needed.	
Inkberry	Leaf miner	Nicotine Sulfate or Rotenone. Apply in April and again in 10 days.	
Ivy (Boston and English)	Scale	Malathion or Nicotine Sulfate. Spray at 2-week intervals during June-July.	
	Spider mites	Dicofol, Oxythioquinox, or Disulfoton 15% G. Apply when mites appear and repeat as needed.	Remove with hard stream of water.

Plant	Pest	Chemical Control	Non-Chemical Control
Ivy (Continued)			
	Leafhoppers	Rotenone, Disulfoton 15% G, or Acephate. Apply when leafhoppers appear.	
	Japanese beetle	Carbaryl or Malathion. Treat when beetles first appear, usually mid- to late June, and as needed.	Remove by hand.
Juniper	Bagworms	Rotenone, Chlorpyrifos, or Acephate. Apply when bags are small, usually June.	*B. thuringiensis* spores. Remove by hand in fall.
	Midges	Dimethoate. Drench soil around plants in April and treat foliage in May. Repeat foliage treatment at 40 day intervals if needed.	
	Scale	Dormant Oil, Carbaryl or Malathion. Use oil only in dormant season and other materials beginning in May and repeat at 10 day intervals until controlled.	
	Webworm	Diazinon, Naled or Carbaryl. Apply in April-May and again in September if needed.	Remove by hand.
	Spruce spider mite	Dicofol, Oxythioquinox, or Disulfoton 15% G. Apply when mites appear in April and as necessary.	Remove with hard stream of water.
	Tip dwarf mite	Oxythioquinox, Dicofol or Chlorpyrifos. Apply May-June as needed.	Make light application of sulfur dust.
Larch	Bagworms	Malathion, Rotenone, or Chlorpyrifos. Apply when bags are small, usually early to mid-June, and again in 10 days if needed.	*B. thuringiensis* spores. Remove by hand in fall.
	Woolly aphid	Lindane, Malathion or Disulfoton 15% G. Apply in early May when aphids appear.	
Laurel	Borer	Lindane. Spray main stem and branches in mid-May and again in June.	Wrap trunk with heavy fabric.
	Lace bug	Carbaryl or Malathion. Make two applications, May and July, to undersides of leaves.	
	Taxus weevil	Methoxychlor or Malathion. Spray foliage and soil surface in late June.	
	Whitefly	Malathion or Nicotine Sulfate. Make two applications, June and July.	

Plant	Pest	Chemical Control	Non-Chemical Control
Lilac	Borer	Endosulfan, Lindane, or Chlorpyrifos. Treat stems in May-June.	
	Leaf miner	Nicotine Sulfate or Rotenone. Apply at first indication of mining.	
	Japanese beetle	Carbaryl. Make 2 applications, July and August.	
	Oystershell scale	Ethion + Dormant Oil, Malathion or Acephate. Use Ethion and oil only as dormant spray, and other materials against crawlers in May-June.	
Linden	Aphids	Lindane, Acephate, or Disulfoton 15% G. Apply when aphids appear.	Remove with hard stream of water.
	Cankerworms	Methoxychlor, Acephate, or Carbaryl. Treat at first sign of infestation, usually May.	*B. thuringiensis* spores.
	Fall webworm	Rotenone, Acephate, or Chlorpyrifos. Apply when webs appear, usually in June.	*B. thuringiensis* spores.
	Japanese beetle	Carbaryl, Methoxychlor or Malathion. Apply when beetles appear in June and as needed.	
	Leaf beetles	Rotenone or Malathion. Spray at first sign of beetles or feeding injury, June-July.	
	Scurfy scale	Dormant Oil, Malathion or Carbaryl. Use oil only as dormant treatment, and others against crawlers in May.	
	Bagworms	Carbaryl, Chlorpyrifos or Acephate. Apply when bags are small in June.	*B. thuringiensis* spores. Remove by hand in fall.
	Cottony maple scale	Dormant Oil, Diazinon or Acephate. Apply oil only in dormant season, and other materials in late June-July, concentrating on leaf undersides.	
Locust	Pod gall	Methoxychlor. Begin in May, and repeat at 2-week intervals.	
	Borers	Lindane. Treat trunks in early September.	Wrap trunk with heavy fabric.
	Leaf miner	Lindane, Diazinon or Acephate. Apply as foliage is developing and again in July.	
	Spider mites	Dicofol, Oxythioquinox or Tetradifon. Apply when mites appear and as needed.	Remove with hard stream of water.

Plant	Pest	Chemical Control	Non-Chemical Control
London Plane	Lace bug	Carbaryl or Malathion. Spray in May and again in June.	
	Plum borer	Methoxychlor or Lindane. Apply 3-6 times to trunk and limbs, beginning in June.	Wrap trunk with heavy fabric.
	Scale	Dormant Oil, Diazinon or Acephate. Apply oil only in dormant season, and other materials against crawlers in June.	
Magnolia	Scale	Dormant Oil, Malathion, or Acephate. Use oil only as dormant treatment, and other materials against crawlers in August-September, and as needed.	
	Yellow poplar weevil	Carbaryl. Treat foliage when adults appear, usually June-July.	
	Leaf miner	Naled. Apply when mines first appear and as needed.	
Mahonia	Barberry aphid	Malathion, Chlorpyrifos, or Disulfoton 15% G. Treat when aphids appear.	Knock off with strong stream of water.
	Scale	Carbaryl or Nicotine Sulfate. Apply when crawlers appear.	
	Webworm	Trichlorfon or Carbaryl. Apply when worms appear.	*B. thuringiensis* spores.
Maple	Aphids	Malathion, Diazinon, Acephate or Disulfoton 15% G. Treat when aphids appear and as needed.	Remove with hard stream of water.
	Bagworms	Rotenone, Diazinon or Acephate. Treat when bags appear, usually June.	*B. thuringiensis* spores. Remove by hand in fall.
	Borers	Lindane. Apply to trunk and lower branches in May-June-July.	Wrap trunk with heavy fabric.
	Cottony scale	Dormant Oil, Diazinon or Carbaryl. Use oil in dormant season only, or one of the other materials in July and repeat in 10 days.	
	Eriophyid mite	Carbaryl or Chlorpyrifos. Apply in early spring when buds start to open or when red or yellow patches appear on the undersides of leaves.	
	Leafhoppers	Carbaryl, Diazinon or Disulfoton 15% G. Apply when leafhoppers appear, and every 3 weeks as needed.	
	Bladder gall mite	Carbaryl, Diazinon or Disulfoton 15% G. Apply when leaves are ¼ expanded and again in 10 days, with special attention to leaf undersides.	

Plant	Pest	Chemical Control	Non-Chemical
Maple (Continued)			
	Oystershell scale	Dormant Oil, Carbaryl, or Ethion + Oil. Apply oil or Ethion and oil late in dormant season, or Carbaryl in May and repeat in 10 days.	
	Spider mites	Disulfoton 15% G, Oxythioquinox or Chlorpyrifos. Treat when mites appear and again in 2 weeks. Repeat as needed.	Remove with hard stream of water.
	Webworm	Carbaryl, Diazinon, or Chlorpyrifos. Treat when webs appear.	*B. thuringiensis* spores.
	Japanese beetle	Carbaryl or Rotenone. Apply when beetles or riddled foliage appear and as needed.	
	Cankerworm	Acephate, Rotenone, or Carbaryl. Treat at first sign of infestation, usually May.	*B. thuringiensis* spores.
Mimosa	Webworm	Carbaryl, Disulfoton 15% G, or Acephate. Apply late June-July and in August if foliage begins to brown.	*B. thuringiensis* spores.
Mountain Ash	Aphids and Woolly aphids	Diazinon, Chlorpyrifos, or Disulfoton 15% G. Treat in early May and as needed.	Remove with hard stream of water on smaller trees.
	European red mite	Dicofol, Chlorpyrifos or Disulfoton 15% G. Treat in early June and again 10 days later.	See Aphids.
	Japanese beetle	Carbaryl or Rotenone. Apply when beetles or riddled foliage appear and as needed.	
	Sawfly	Carbaryl or Diazinon. Apply 2 weeks after petal fall and again in 2 weeks.	Remove by hand in smaller trees.
	Lace bug	Malathion, Disulfoton 15% G, or Acephate. Apply in May and repeat in 10 days.	
Mountain Laurel	Leaf miner	Malathion or Disulfoton 15% G. Apply when mines appear and as needed to protect new growth.	
	Scale	Malathion, Nicotine Sulfate, or Carbaryl. Apply when crawlers appear in June and again in 10 days.	
	Lace bug	Lindane, Disulfoton 15% G, or Carbaryl. Apply in early June.	
	Borer	Lindane. Apply to trunk and large branches in May and twice again at 3-week intervals.	Wrap trunk with heavy fabric.

Plant	Pest	Chemical Control	Non-Chemical Control
Oak	Aphids	Disulfoton 15% G, Diazinon or Acephate. Apply when aphids appear and as needed.	Remove with hard stream of water.
	Bagworms	Carbaryl, Chlorpyrifos or Acephate. Treat when bags are small, usually June.	*B. thuringiensis* spores. Remove by hand in fall.
	Borers	Lindane. Spray trunks in May, June, July and August.	Wrap trunk with heavy fabric.
	Webworms	Acephate, Diazinon, or Chlorpyrifos. Apply when webs appear, usually June.	*B. thuringiensis* spores.
	Galls	Lindane. Treat in early June when first galls appear.	Prune and destroy stem and twig galls while still green to reduce further infestations.
	Golden scale	Dormant Oil, Nicotine Sulfate or Diazinon. Use oil only as dormant spray in spring, and other materials against crawlers beginning in May, and repeat twice 10 days apart.	
	Japanese beetle	Carbaryl or Rotenone. Apply when beetles or riddled foliage appear and as needed.	
	Leaf miners	Carbaryl, Diazinon or Acephate. Apply against adults, when leaves are ½ expanded, covering upper leaf surfaces.	
	Leafhoppers	Chlorpyrifos, Carbaryl or Disulfoton 15% G. Apply when leafhoppers appear and as needed.	
	May beetles	Carbaryl or Rotenone. Treat when leaves are being eaten, usually June. (Beetles are night feeders).	Remove by hand at night.
	Lace bug	Lindane, Acephate or Disulfoton 15% G. Apply when bugs appear, usually May, and as needed.	
	Spider mites	Disulfoton, Oxythioquinox, or Chlorpyrifos. Apply when mites appear and as needed.	
	Pin oak sawfly	Carbaryl or Diazinon. Apply when larval feeding is seen.	
	Skeletonizers	Lindane, Acephate or Chlorpyrifos. Treat when riddled leaves are seen, usually June and again in August.	
	Cankerworms	Methoxychlor, Acephate or Carbaryl. Apply when worms are small, usually May.	*B. thuringiensis* spores.

Plant	Pest	Chemical Control	Non-Chemical Control
Oak (Continued)			
	Tent caterpillars	Acephate, Chlorpyrifos or Carbaryl. Apply when webs appear, usually April.	*B. thuringiensis* spores.
	Twig pruner	Chemical control usually ineffective.	Rake and destroy fallen twigs by last of April.
Pachysandra	Euonymus and Oystershell scales	Carbaryl, Nicotine Sulfate, or Malathion. Apply when crawlers appear, May-June and again in 10 days.	
	Spider mites	Dicofol, Disulfoton 15% G, or Chlorpyrifos. Apply when mites appear and as needed.	Remove with hard stream of water.
Pieris (Andromeda)	Lace bug	Lindane, Acephate, or Disulfoton 15% G. Apply in May, and again in 10 days.	
	Spider mites	Dicofol, Tetradifon or Disulfoton 15% G. Apply when mites appear and again in 10 days, and as needed.	Remove with hard stream of water.
Pine	Aphids	Diazinon or Disulfoton 15% G. Apply once in May, and possibly again in August if aphids persist.	Remove with hard stream of water on smaller trees.
	Bagworms	Carbaryl, Acephate or Chlorpyrifos. Treat when bags are small, usually June.	*B. thuringiensis* spores. Remove by hand in fall.
	European shoot moth	Diazinon or Malathion. Apply to terminal growth in April and again in June.	Prune off infested terminals in May.
	Nantucket tip moth	Disulfoton 15% G or Lindane. Begin treatment in early spring when growth starts and as needed.	
	Northern pine and Pales weevil	Lindane. Spray seedlings and young twigs in April, May, July and September.	Pull and destroy stumps by June.
	Pine bark aphid	Dormant Oil, Lindane, Diazinon or Disulfoton 15% G. Use oil as dormant treatment, or other materials when aphids appear, usually May.	Remove with hard stream of water.
	Needle scale	Ethion + Oil, Acephate, Diazinon or Chlorpyrifos. Apply Ethion + oil as dormant treatment, or others when crawlers appear in May, July and August.	
	Tortoise scale	Dormant Oil, Diazinon or Acephate. Apply oil as dormant treatment in spring or other materials when crawlers appear in June.	

Plant	Pest	Chemical Control	Non-Chemical Control
Pine (Continued)			
	Webworm	Carbaryl or Trichlorfon. Apply when larvae appear, usually in July and August.	*B. thuringiensis* spores.
	Sawflies	Carbaryl, Rotenone, or Acephate. Apply when larvae appear and begin feeding on needles, usually early May.	Remove by hand on smaller trees.
	Spruce spider mite	Oxythioquinox, Dicofol, Diazinon, or Disulfoton 15% G. Apply when mites appear, again in 10 days, and as needed.	
	White pine weevil	Lindane. Apply to leaders when weevils appear, usually April.	
	Zimmerman pine moth	Endosulfan, Trichlorfon, or Lindane. Apply in early April or September to control larvae.	
	Eriophyid mite	Carbaryl, Dicofol, or Tetradifon. Apply when mite activity begins, usually May.	
	Root collar weevil	Lindane. Treat 3 times, May, August, and September.	
Poplar	Oystershell scale	Dormant Oil, Carbaryl or Malathion. Apply oil before growth begins, or other materials against crawlers in June as needed.	
	Tentmaker	Carbaryl or Malathion. Apply when first feeding or webbing appears.	*B. thuringiensis* spores.
Privet	Rust mite	Dicofol, Oxythioquinox, or Disulfoton 15% G. Apply when mites appear and as needed.	Remove with hard stream of water.
	Thrips	Carbaryl, Diazinon, or Disulfoton 15% G. Apply when thrips become active. Beat growing tips on dark paper to count.	Control seldom needed.
	White peach scale	Carbaryl. Apply when crawlers appear, usually May-June.	
Pyracantha (see Firethorn)			
Redbud (Cercis)	Leafhoppers	Malathion, Disulfoton 15% G, or Chlorpyrifos. Apply when leafhoppers appear and as needed.	
Rhododendron	Azalea bark scale	Malathion, Diazinon or Carbaryl. Apply when crawlers appear in June.	
	Black vine weevil	Lindane or Diazinon. Treat plants and soil around plants.	

Plant	Pest	Chemical Control	Non-Chemical Control
Rhododendron (Continued)			
	Borer	Lindane or Endosulfan. Apply to trunk and branches in May and again twice at 3-week intervals.	
	Lace bug	Malathion, Acephate, or Disulfoton 15% G. Treat in May when bugs appear, and as needed to protect new growth.	
Roses	Aphids	Malathion, Chlorpyrifos, Endosulfan or Disulfoton 15% G. Apply when aphids appear and as needed.	Knock off with strong stream of water.
	Chafer	Carbaryl or Rotenone. Spray or dust foliage when beetles or damage appear and as needed.	Protect with gauze netting.
	Japanese beetle	Carbaryl or Rotenone. Spray when first beetles appear and at weekly intervals until beetle season is over.	Protect with gauze netting.
	Spider mites	Dicofol, Oxythioquinox, Tetradifon, Disulfoton 15% G, or Chlorpyrifos. Apply when mites appear and as needed.	Knock off with strong stream of water.
	Leafhoppers	Carbaryl or Malathion. Treat 3 times at 3-4 week intervals beginning May-June.	
Spirea	Aphids	Malathion, Acephate, Nicotine Sulfate, or Disulfoton 15% G. Apply when aphids appear and as needed.	Knock off with strong stream of water.
	Leaftier	Carbaryl or Rotenone. Apply when leaves are observed folded or tied.	Remove tied leaves by hand
Spruce	Aphids	Disulfoton 15% G., Malathion, or Chlorpyrifos. Apply when aphids appear and as needed.	Remove with hard stream of water.
	Bagworms	Carbaryl Rotenone, or Diazinon. Treat when bags are small in June.	*B. thuringiensis* spores. Remove by hand in fall.
	Balsam twig aphid	Malathion or Diazinon. Apply when aphids appear in April-May.	Remove with hard stream of water.
	Gall aphids	Malathion, Diazinon or Disulfoton 15% G. Apply just before buds break in the spring or after galls open in late summer or fall.	
	Needle scale	Diazinon or Malathion. Apply against crawlers in May and again in July and August.	
	Sawflies	Carbaryl or Rotenone. Apply when larvae appear in early spring.	Remove by hand where possible.
	Bud scale	Carbaryl or Diazinon. Apply June-July when crawlers appear.	

Plant	Pest	Chemical Control	Non-Chemical Control
Spruce (Continued)			
	Needle miner	Carbaryl, Diazinon or Acephate. Treat foliage in mid- and late June.	
	Spider mite	Dicofol, Tetradifon, Disulfoton 15% G., or Chlorpyrifos. Apply when mites appear in April and a second spray 10 days later, and as needed.	Remove with hard stream of water.
	Pine weevil	Lindane. Treat leaders in spring when beetles appear in mid-April.	
Sweet gum	Bagworms	Carbaryl, Diazinon or Acephate. Apply when bags appear in June.	*B. thuringiensis* spores. Remove by hand in fall.
	Fall webworm	Trichlorfon or Carbaryl. Apply when worms and webs first appear.	*B. thuringiensis* spores.
	Leaf miner	Chlorpyrifos, Diazinon or Acephate. Apply when first mines appear and as needed.	
	Scale	Dormant Oil or Carbaryl. Apply oil only as dormant treatment in spring, and Carbaryl against crawlers on leaves in June or September when young scales appear on twigs and buds.	
Sycamore	Aphids	Malathion, Acephate, or Disulfoton 15% G. Apply when aphids appear and as needed.	Remove with hard stream of water.
	Bagworms	Carbaryl, Rotenone, Diazinon, or Acephate. Apply when bags appear in June.	*B. thuringiensis* spores. Remove by hand in fall.
	Fall webworm	Acephate, Carbaryl or Chlorpyrifos. Apply when first webs or worms are seen in June.	*B. thuringiensis* spores.
	Japanese beetle	Carbaryl or Rotenone. Apply when beetles or riddled foliage appears and as needed.	
	Leaf folder	Carbaryl. Apply when leaves are folded together.	Remove folded leaves by hand.
	Leafhoppers	Malathion, Acephate or Disulfoton 15% G. Apply when leafhoppers appear and as needed.	
	Lace bug	Malathion, Acephate or Disulfoton 15% G. Make 2 applications 10 days apart beginning in May.	
	Terrapin scale	Carbaryl. Apply when crawlers appear on leaves in June.	

Plant	Pest	Chemical Control	Non-Chemical Control
Taxus (see Yew)			
Tulip Tree	Aphids	Malathion, Acephate, or Disulfoton 15% G. Apply when aphids appear and as needed.	Remove with hard stream of water.
	Leaf miner	Chlorpyrifos, Nicotine Sulfate, Diazinon or Acephate. Apply when first mines appear and as needed.	
	Scale	Dormant Oil or Carbaryl. Apply oil only as dormant treatment, and Carbaryl when crawlers appear in August.	
	Yellow poplar weevil	Carbaryl. Apply for adult control in June-July.	
Viburnum	Aphids	Malathion, Acephate, or Disulfoton 15% G. Apply when aphids appear and as needed.	Remove with hard stream of water.
	Spider mites	Dicofol, Diazinon or Disulfoton 15% G. Apply when mites appear in spring and again in 10 days, and as needed.	See Aphids.
Walnut (If walnuts are to be eaten check insecticide labels for days-waiting-time from last application to harvest.)	Aphids	Malathion, Oxythioquinox, or Diazinon. Apply when aphids appear and as needed.	Remove with hard stream of water on smaller trees.
	Fall webworm	Carbaryl or Rotenone. Apply when webs first appear, mid-June thru July.	*B. thuringiensis* spores.
	Leafhoppers	Malathion or Carbaryl. Apply when leafhoppers appear and as needed.	
	Gall mite	Carbaryl, Diazinon or Chlorpyrifos. Apply when leaves are half open or when mites appear on leaf petioles.	
	Walnut caterpillar	Methoxychlor. Apply when caterpillars appear, usually in June.	*B. thuringiensis* spores.
Willow	Aphids	Diazinon, Acephate or Disulfoton 15% G. Apply when aphids appear and as needed.	Remove with hard stream of water.
	Borers	Lindane. Treat trunk thoroughly monthly from May thru August.	Wrap trunk with heavy fabric.
	Fall webworm	Diazinon, Acephate or Chlorpyrifos. Apply when webs appear, usually June.	
	Leaf beetles	Carbaryl, Rotenone, or Acephate. Apply on appearance of leaf feeding and as needed.	

Plant	Pest	Chemical Control	Non-Chemical Control
Willow (Continued)			
	Spider mites	Dicofol, Diazinon, or Disulfoton 15% G. Apply when mites appear, again in 10 days, and as needed.	Remove with hard stream of water.
	Oystershell scale	Dormant Oil, Acephate or Diazinon. Apply oil only as dormant treatment in spring, or other materials against crawlers in May-June.	
	Sawflies	Carbaryl or Diazinon. Apply when larvae appear but before heavy leaf feeding.	Remove by hand on small trees.
	Tent caterpillars	Diazinon, Rotenone, Chlorpyrifos or Acephate. Apply when worms or webs first appear, usually April.	*B. thuringiensis* spores.
Wisteria	Leafhoppers	Disulfoton 15% G, Chlorpyrifos or Carbaryl. Apply when leafhoppers appear and as needed.	
Yew (Taxus)	Black vine weevil	Lindane or Endosulfan. Spray foliage and soil around infested plants in June.	
	Fletcher scale	Dormant Oil, Carbaryl or Malathion. Apply oil only as dormant treatment in spring, other materials against crawlers in June-July.	
	Mealybug	Dormant Oil, Malathion, or Acephate. Apply oil only as dormant treatment in spring, or other materials against overwintering nymphs in May, and again in June and July.	In small numbers remove with hard stream of water.

Some Insect Pests of Trees, Shrubs, Woody, Annual and Perennial Ornamentals.

Aphid
(Union Carbide)

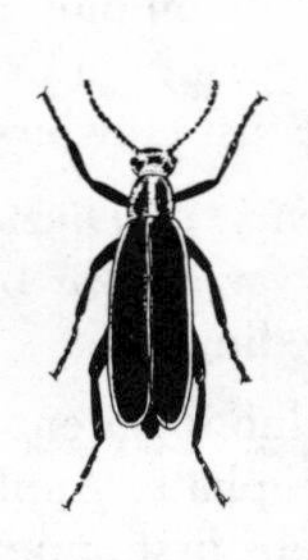

Margined Blister Beetle
(U.S. Public Health Service)

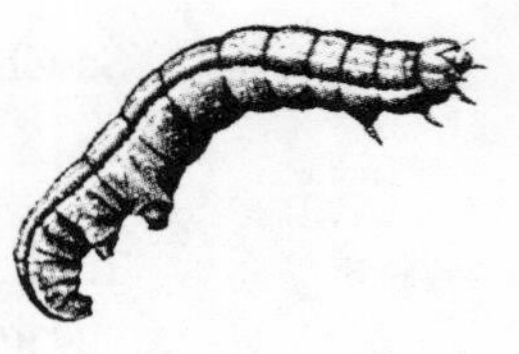

Cankerworm
(Union Carbide)

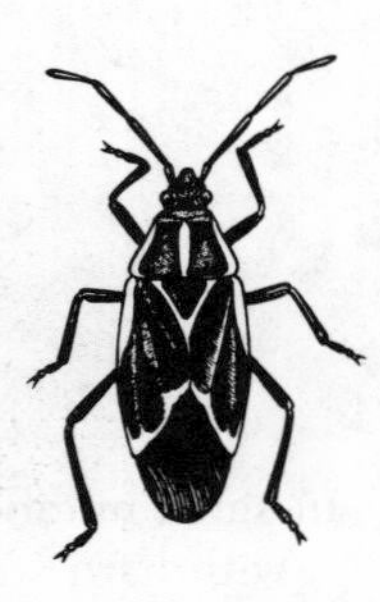

Boxelder Bug
(Univ. of Arizona)

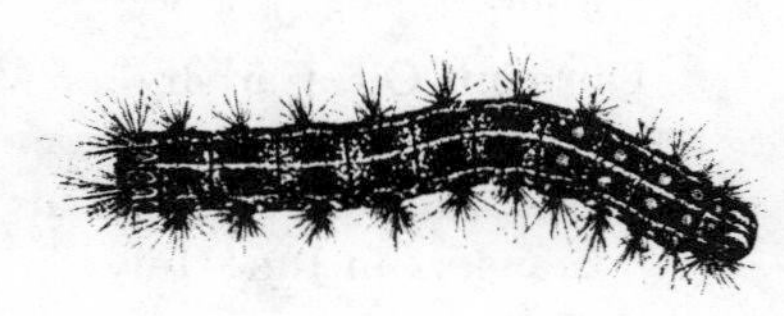

Gypsy Moth Larva
(Union Carbide)

Elm Leaf Beetle
(Union Carbide)

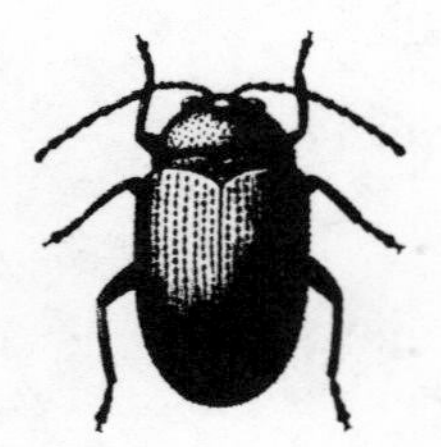

Flea Beetle
(Union Carbide)

Horntail
(USDA)

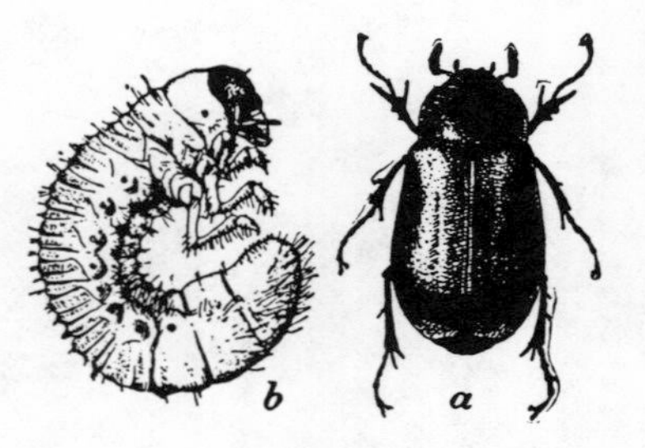

A May Beetle
A, adult; *B*, fully grown larva
(USDA)

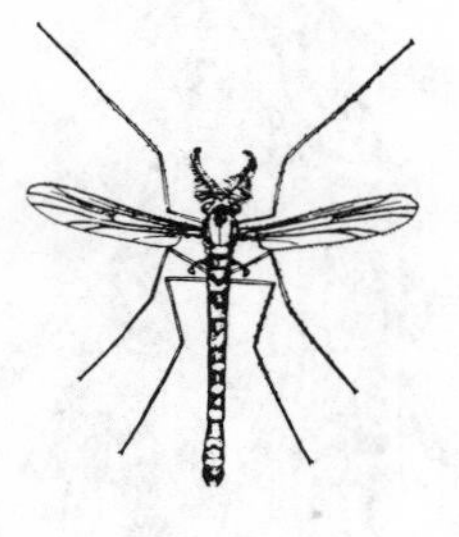

A Midge
(USDA)

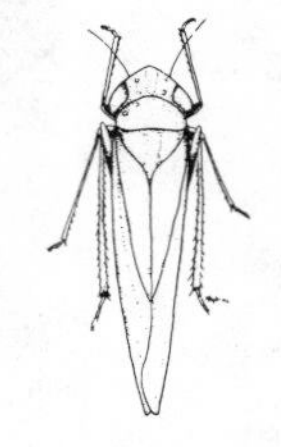

Leafhopper Adult
(Univ. of Arizona)

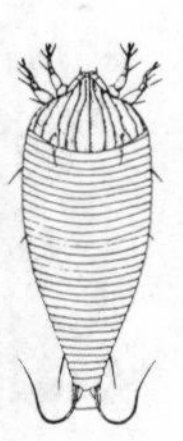

Eriophyid Gall Mite
(Enlarged 50 ×)
(USDA)

Birch Leaf Miner Adult
(Union Carbide)

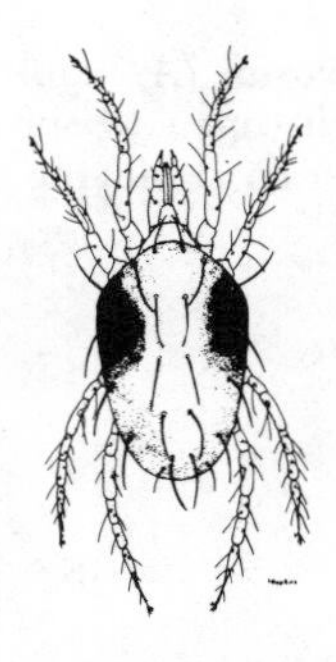

A Spider Mite
(Univ. of Arizona)

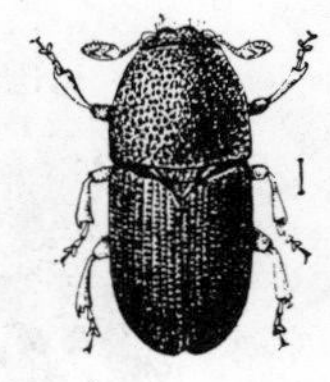

Shot-hole Borer Adult
(USDA)

Psyllid
Above: adult
Right: nymph
(Univ. of Arizona)

Peach Twig Borer
(Union Carbide)

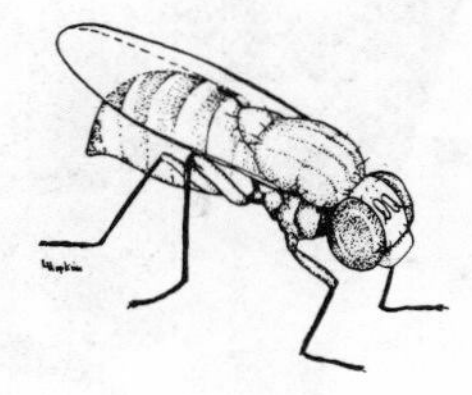

Leaf Miner Fly Adult
(Univ. of Arizona)

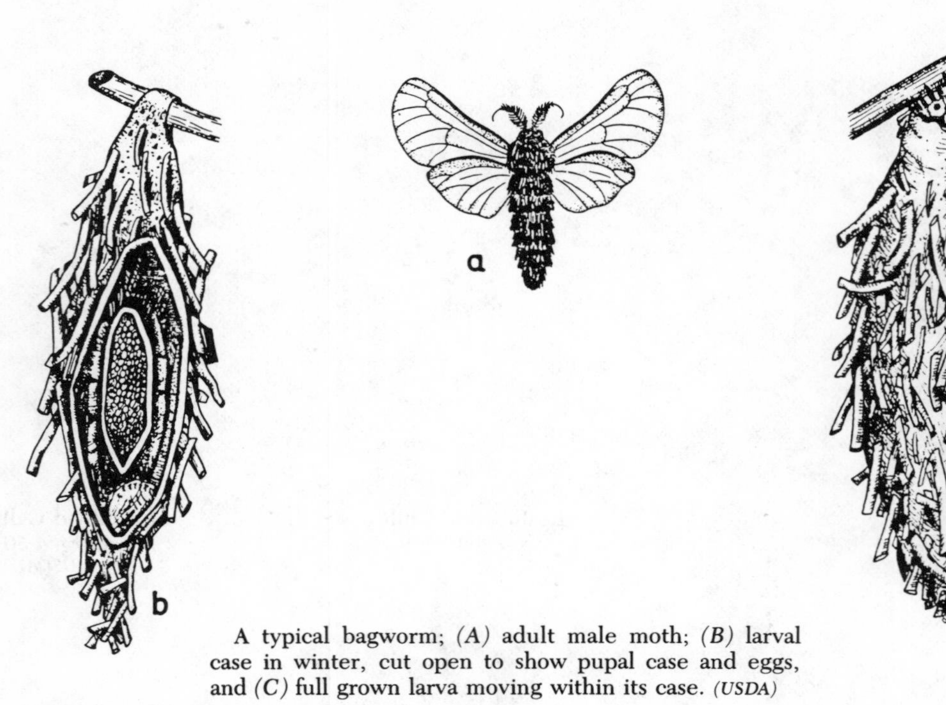

A typical bagworm; *(A)* adult male moth; *(B)* larval case in winter, cut open to show pupal case and eggs, and *(C)* full grown larva moving within its case. *(USDA)*

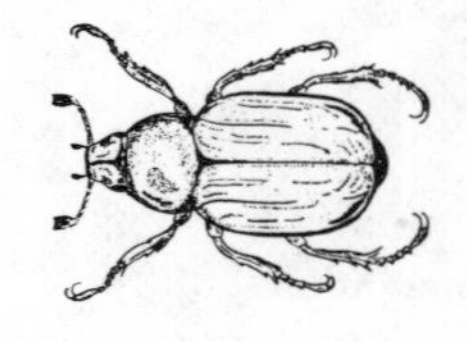

Garden Chafer
(Union Carbide)

Nests of the Eastern Tent Caterpillar; note the layers of silk.
(USDA)

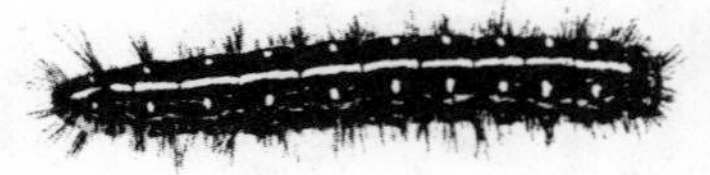

Eastern Tent Caterpillar
(Union Carbide)

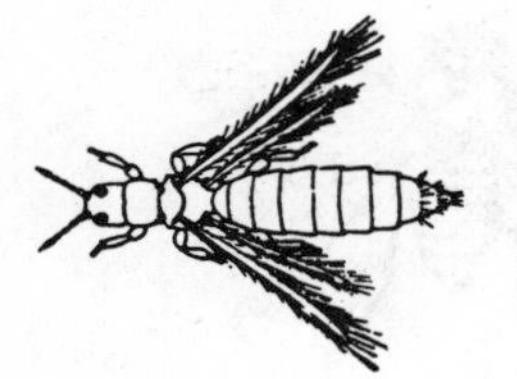

A Thrips
(Univ. of Arizona)

Azalea Lacebug
(USDA)

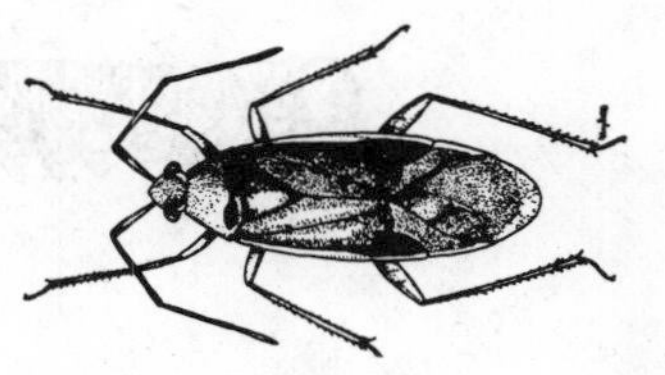

Plant Bug
(Univ. of Arizona)

White Pine Weevil
(USDA)

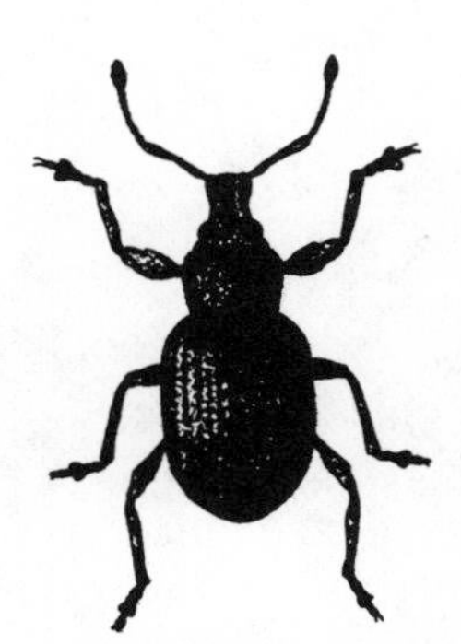

Black Vine Weevil
(USDA)

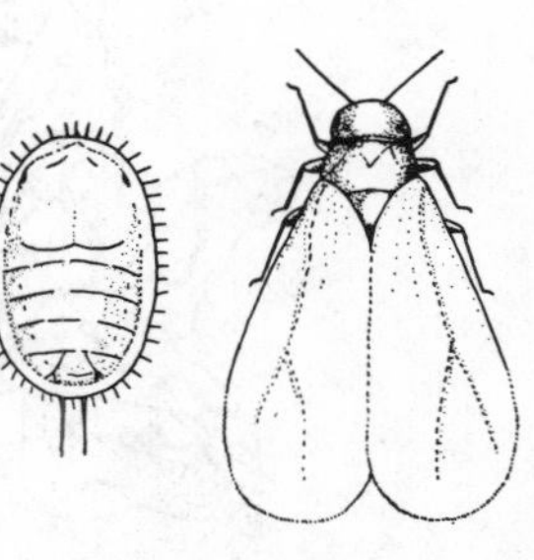

Whitefly. *A* — Nymph; *B* — Adult.
(Univ. of Arizona)

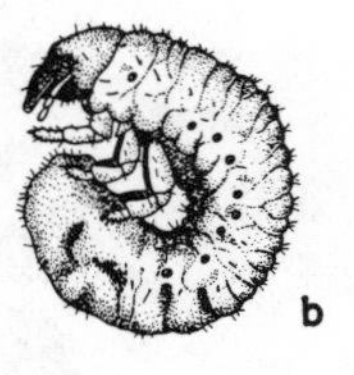

Japanese Beetle *(Popillia japonica)*
(a) Adult, *(b)* Larva (grub).
(USDA)

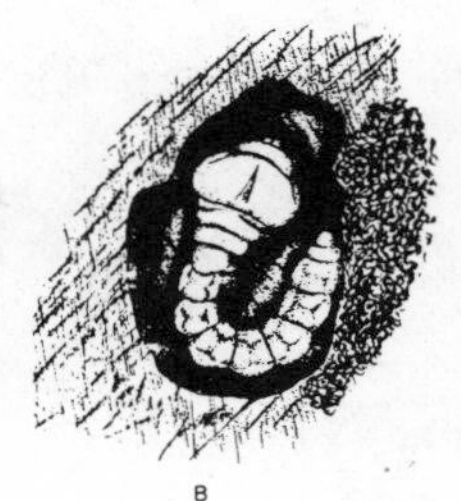

Flatheaded Appletree Borer
(Chrysobothris femorata)
A. Adult, *B*. Larva.
(USDA)

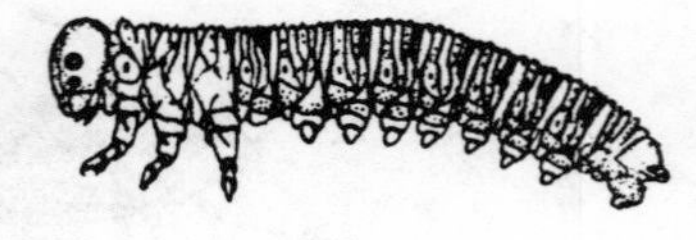

Sawfly Larva, *Allántus cintus*
(USDA)

Pear Sawfly, *Caliroa cerasi*
(USDA)

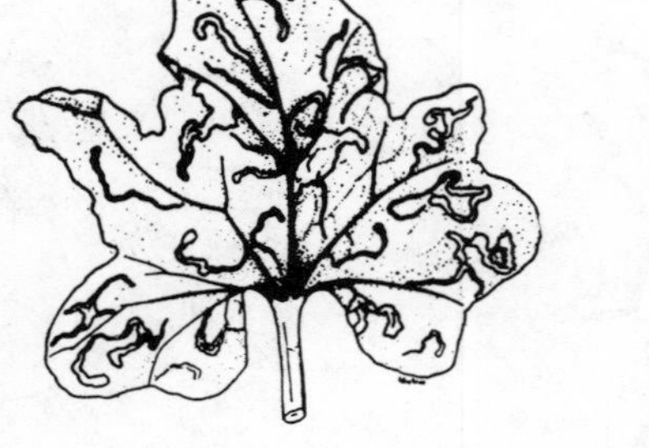

Leaf Miner Injury
(Univ. of Arizona)

California Orange Dog
(Union Carbide)

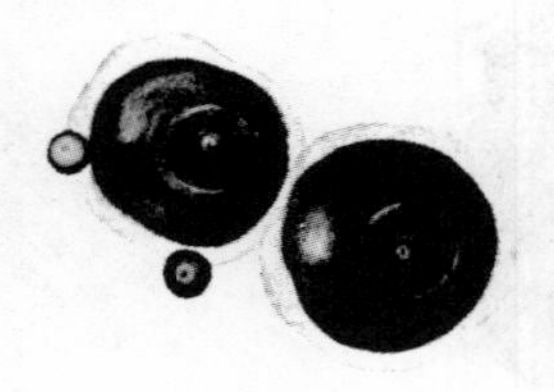

California Red Scale
(Union Carbide)

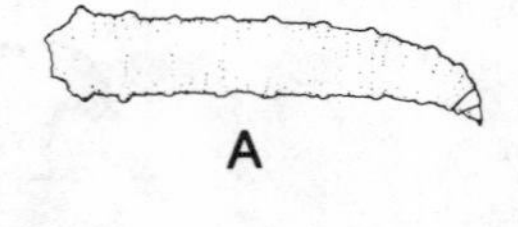

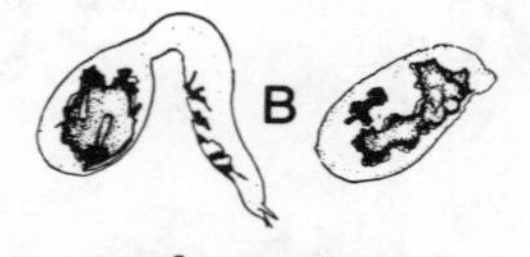

Seedcorn Maggot
A — Mature larva; B — Seed damage
(Univ. of Arizona)

TABLE 17. Insect and Mite Control Suggestions for Ornamental Annuals and Perennials With and Without Chemicals.[1]

Plant	Pest	Pesticide to Use	Non-Chemical Control
Ageratum	Whitefly	Malathion, Nicotine Sulfate, or Endosulfan. Spray 2 or more times at 5-day intervals.	
	Spider mites	Dicofol or Tetradifon. Spray as needed.	Remove with hard stream of water.
Chrysanthemum	Aphids	Malathion or Dimethoate. Spray as necessary. Disulfoton 1% G.	Remove with hard stream of water.
	Beetles	Carbaryl or Rotenone. Spray or dust as needed.	
	Caterpillars (including borers)	Carbaryl or Rotenone. Spray or dust as necessary.	Remove and burn large-stemmed weeds nearby.
	Plant bugs	Malathion, Rotenone, or Carbaryl. Spray or dust as needed. Disulfoton 1% G.	
Cockscomb	Spider mites	Dicofol or Tetradifon. Spray as needed.	Remove with hard stream of water.
Columbine	Aphids	Malathion or Dimethoate. Spray as necessary. Disulfoton 1% G.	Remove with hard stream of water.
	Leaf miner	Malathion, or Nicotine Sulfate. Spray 3 times at 5-7 day intervals, beginning when leaves are about half developed.	
Citrus	Scales	Malathion or Nicotine Sulfate. Apply as spray or with paint brush.	
	Whitefly	Malathion. Apply as spray.	
	Orange dog	Chemical control not normally required.	Remove by hand or apply *B. thuringiensis* spores.
	Spider mites	Dicofol, Tetradifon or Oxythioquinox. Apply when mites appear and as needed.	Remove with hard stream of water.
Dahlia	Aphids	Malathion or Dimethoate. Treat as needed. Disulfoton 1% G.	Remove with hard stream of water.
	Beetles and borers	Malathion and Methoxychlor. Mix and spray on 14-day schedule.	
	Leafhoppers	Malathion or Rotenone. Treat when first leaf damage is noticed and repeat as needed.	
	Spider mites	Dicofol or Tetradifon. Spray as needed.	Remove with hard stream of water.
	Thrips	Malathion. Spray when flowers begin to open.	Chemical control usually unnecessary.

[1] Refer to Table 35 for trade or proprietary names of recommended pesticides.
Many of the pests mentioned in this table are illustrated on pages 118-122.

Plant	Pest	Chemical Control	Non-Chemical Control
Day Lilies	Thrips	Malathion and Methoxychlor. Mix and spray 2 or 3 times at 10-day intervals, beginning just after flower buds form.	Destroy all plant residue in fall.
Delphinium (see Larkspur)			
Forget-Me-Not	Aphids	Malathion or Nicotine Sulfate. Spray or dust as needed.	Remove with hard stream of water.
	Flea beetles	Carbaryl or Rotenone. Spray or dust as needed.	
Hollyhock	Beetles	Carbaryl or Rotenone. Spray as needed.	Remove by hand.
	Leafhoppers	Malathion and Methoxychlor. Mix and spray as needed.	
	Spider mites	Dicofol or Tetradifon. Treat as needed.	Remove with hard stream of water.
Hydrangea	Aphids	Malathion or Nicotine Sulfate. Spray as needed.	Remove with hard stream of water.
	Leaf tier (caterpillar)	Carbaryl and malathion. Mix and spray as necessary.	Remove rolled leaves by hand.
	Spider mites	Dicofol or Tetradifon. Treat as needed.	See Aphids.
Impatiens	Aphids	Malathion or Nicotine Sulfate. Spray or dust as needed.	Remove with hard stream of water.
	Spider mites	Dicofol or Tetradifon. Treat as needed.	See Aphids.
Iris	Aphids	Malathion or Nicotine Sulfate. Spray as needed.	Knock off with water spray.
	Borers	Dimethoate. Spray twice at 2-week intervals, beginning when leaves are ¾ inch long.	Mulch heavily with bark chips.
Larkspur	Aphids	Malathion or Nicotine Sulfate. Spray as needed.	Knock off with water spray.
	Borers	Malathion and Methoxychlor. Mix and spray 3 times, beginning at early growth and at 2-week intervals thereafter.	Collect and burn large-stemmed weeds. Burn all larkspur and delphinium after fall frost.
	Spider mites	Dicofol or Tetradifon. Treat as needed.	See Aphids.
Lilies	Aphids	Malathion or Nicotine Sulfate. Spray or dust as needed.	Remove with hard stream of water.
Lupine	Aphids	Malathion or Nicotine Sulfate. As needed.	Remove with hard stream of water.
	Whitefly	Malathion or Nicotine Sulfate. As needed.	
Marigold	Spider mites	Dicofol or Tetradifon. Treat as needed.	Knock off with water spray.

Plant	Pest	Chemical Control	Non-Chemical Control
Nasturtium	Aphids	Malathion or Nicotine Sulfate. Spray or dust as needed.	Remove with hard stream of water.
	Spider mites	Dicofol or Tetradifon. Treat as needed.	See Aphids.
	Plant bugs	Malathion and Methoxychlor. Mix and spray as needed.	
Pansy	Aphids	Malathion or Nicotine Sulfate. Spray as needed.	
	Mealybugs	Malathion or Nicotine Sulfate. Spray in spring or as needed.	
	Spider mites	Dicofol or Tetradifon. Frequent treatment may be needed.	
Peony	Ants	Malathion, Rotenone, or Carbaryl. Spray as needed.	Ants are nuisance but not pest. Attracted to sweet nectars.
	Scales	Nicotine Sulfate. Spray when insects are first seen.	Destroy all plant residue in fall.
	Thrips	Malathion or Rotenone. Spray lightly with fine mist daily through bloom.	Destroy all plant residue in fall.
	Japanese beetle and Rose chafer	Carbaryl or Rotenone. Spray when insects are first seen and repeat weekly or as necessary.	Protect plants with gauze netting.
Phlox	Plant bugs	Malathion and Methoxychlor. Mix and spray as needed.	
	Spider mites	Dicofol or Tetradifon. Spray as needed.	Remove with hard stream of water.
Poppy	Aphids	Malathion or Nicotine Sulfate. Spray or dust as needed.	See Spider Mites.
	Leafhoppers	Malathion and Methoxychlor. Mix and spray at 5-day intervals or as needed.	
Shasta Daisy	Aphids	Malathion or Nicotine Sulfate. Treat as needed.	Remove with hard stream of water.
	Beetles	Carbaryl or Rotenone. Spray or dust as needed.	
Snapdragon	Aphids	Malathion. Spray as needed.	
	Beetles	Carbaryl. Spray or dust as needed.	
	Spider mites	Dicofol or Tetradifon. Treat as needed.	
Sweet Alyssum	Leafhoppers	Malathion. Spray as needed.	
Sweet Pea	Aphids	Malathion or Nicotine Sulfate. Spray as needed.	
	Leaf miner	Malathion or Nicotine Sulfate. Spray or dust weekly.	Remove and destroy infested leaves.
	Seedcorn maggot	Diazinon granules placed with seed at planting.	

Plant	Pest	Chemical Control	Non-Chemical Control
Verbena	Aphids	Malathion or Nicotine Sulfate. Spray as needed.	Remove with hard stream of water.
	Plant bugs	Malathion and Methoxychlor. Mix and spray as needed.	
	Leaf miner	Malathion or Nicotine Sulfate. Spray at first sign of mines.	
	Spider mites	Dicofol or Tetradifon. Treat as needed.	See Aphids.
Zinnia	Japanese beetle	Rotenone or Malathion and Carbaryl. Mix and treat as needed.	Protect with gauze netting.
	Spider mites	Dicofol or Tetradifon. Spray as needed.	Remove with hard stream of water.

HOUSEHOLD — THE UNWANTED, UNWELCOME HOUSE GUESTS

Controlling household pests is important throughout the year. There are several species of cockroaches which are obnoxious, odorous, and leave trails of fecal droppings on foods and house furnishings during their nocturnal frolicking. Termites (heaven forbid!) move into the structure from the ground and attack wood, books, and papers. Ants can contaminate food and cause homeowners great annoyance. Clothes moths and carpet beetles often damage or destroy woolen clothing, furs, rugs, and that record-holding trophy head-mount in the den. Fleas attack pets and then their masters. House dust mites, carpet beetle cast skins, cockroaches, and unknown others produce allergies. Flies and mosquitoes are tremendous nuisances outdoors and occasionally indoors.

If the householder can determine the identity of the pest and the extent of the infestation, has the proper pesticide or non-chemical control, and is confident of his ability to solve the problem, he won't need a pest control professional to do the job for him. When he follows the instructions properly, he will derive immense pride in a job well done and especially in the savings realized. So, let's have at it!

Ants

Ants of various sizes and species may be found in houses. They have constricted "waists," distinguishing them from termites, which have straighter bodies without such narrow constrictions. Those most commonly seen are wingless, non-reproducing adults of the worker caste.

They feed on a variety of substances including sweets, starches, fats, dead insects, and other animal matter, and are particularly fond of man's foods.

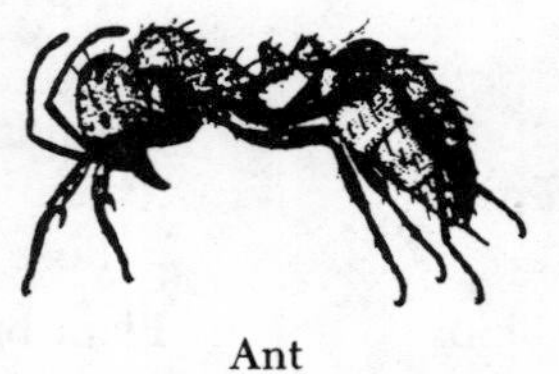

Ant
(U.S. Public Health Service)

Ants build outdoor nests in soil or other protected areas. Some species, particularly carpenter ants, nest in damp or unsound house timbers. Fire ants, which occur in the South, can inflict painful stings, build nests in gardens and under turf, with mounds of excavated soil extending above the grass level. Most ant species commonly travel in search of food along "trails" which are easily recognized when individuals are numerous.

Ants normally will not enter homes where food is not accessible. Control of ants in homes should begin with the removal of crumbs and other attractive material from floors and areas where food is prepared, stored, or consumed. Food should be stored in ant-proof containers.

Spot treatments of an approved residual insecticide may be applied to ant trails and other traveled areas for temporary relief. Such applications are not a substitute for locating and eliminating the nesting site. Approved ant baits, preferably in sealed containers placed in secluded locations beyond the reach of small children or pets, may be a helpful supplementary control measure.

Ants may be controlled by using an approved insecticidal spray applied to nesting sites, which may be near foundations or at some distance from invaded houses. Additional protection can be provided by an insecticidal spray applied to the lower foundation walls and to the soil area surrounding the house. Special attention should be given to door and window sills and other areas of ant entrance. Carpenter ants and other house-nesting ants may be controlled by applying an approved insecticide to the nesting site. Damaged areas may require structural repair.

Aphids

Aphids, or plant lice, feed on plant juices and are occasional pests of house plants or annoying temporary invaders on cut flowers. They may be winged or wingless, of differing sizes and colors. Feeding may weaken or disfigure house plants by causing leaves to curl or to become covered with honeydew. Potted plants should be inspected for aphids and treated outside, if necessary, before being brought indoors. Light infestations may be controlled by washing plants in lukewarm water or removing the aphids with tweezers or a cotton swab dipped in rubbing alcohol. On cut flowers, aphids are at most a temporary nuisance best avoided by not bringing infested blooms indoors or by removal by hand as described.

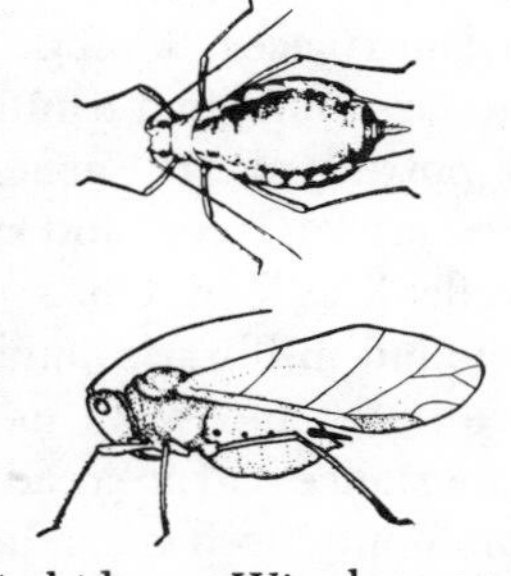

Aphids — Wingless and Winged
(USDA)

Assassin Bugs — see Kissing Bugs

Bark Beetles

There are several species of small (¼ inch) beetles, black or dark brown, and oval-shaped, commonly called bark beetles. Almost never are they generated within the home, but rather hitchhike into the home by means of firewood or are occasionally attracted to lights in the spring and early summer. They should be of no concern. Removing the firewood usually removes the source of the problem.

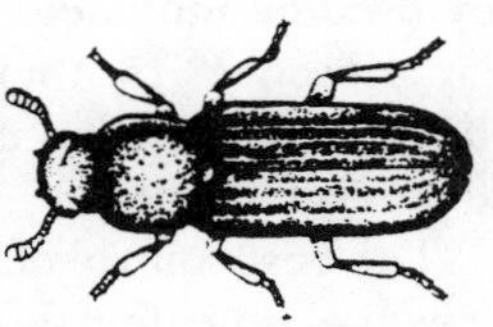

Bark Beetle
(U.S. Public Health Service)

Bean Weevils

Stored dry beans and peas in loose or uncovered containers may become infested with these short-snouted weevils, which can breed continuously in homes. Beans are destroyed by the internal feeding of the larvae with several individuals often found in a single bean. The adults, which emerge through holes bored in the seed coat are ⅛ inch long and mottled gray, and do not feed on beans.

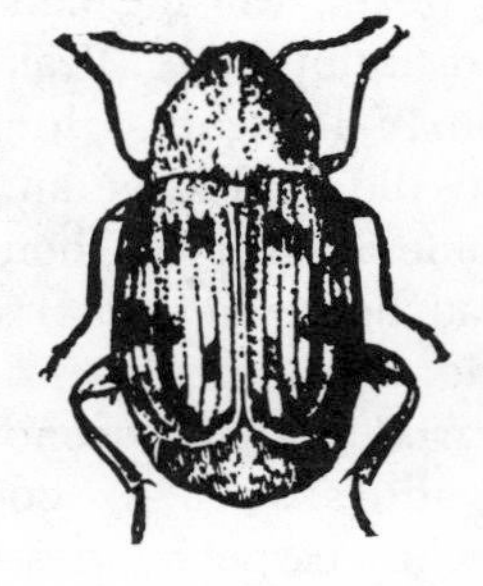

Bean Weevil
(Univ. of Arizona)

Storing beans in tight metal or glass containers and avoidance of spillage may prevent infestations. Small quantities of heavily infested beans should be promptly destroyed when salvage is not practical. Light weevil infestations may be controlled by heating the beans in an oven for 30 minutes at 130°F., or by cooling them for 4 days in a freezer at 0°F., or below.

Bed Bugs

Adult bed bugs are flat, wingless, reddish-brown insects, 3/16″ long and ⅛″ wide. The nymphs and adults feed on human blood, although pets and other animals may be attacked. Bed bugs usually hide during the day in floor cracks, behind baseboards, in cracks in wooden beds, inside mattresses and upholstered furniture, and in other similar situations. Infested homes have a sweetish, indescribable odor. Bed bugs are mainly night feeders, but during the day they may emerge from hiding to attack persons using infested furniture such as padded chairs or stools. Bed bugs may be widely transported through infested luggage, clothing, bedding, and furniture. They are commonly associated with, but not restricted to, homes with poor sanitation.

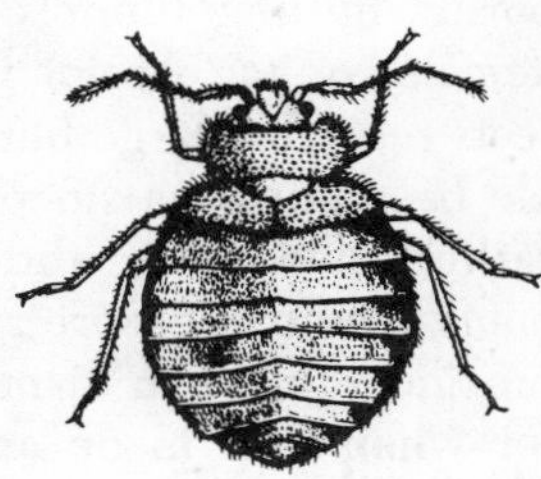

Bed Bug
(Univ. of Arizona)

When bedbugs have infested an occupied house they can be removed only by use of an insecticide. Prevention through good housekeeping and proper living conditions, is preferred.

Bees

Swarms of honey bees may enter cracks or other openings in walls of houses to establish new colonies. Under favorable conditions these colonies may exist for years, causing continued annoyance. Weak or poorly located colonies may die, and leave an agglomeration of wax, honey, dead bees, and other debris attractive to new swarms and to secondary invaders such as cockroaches, carpet beetles and wax moths. An objectionable odor will probably develop.

Honey Bee
(U.S. Public Health Service)

Control of honey bees in houses involves transferring or killing the bees, removal of the combs, honey and debris from the nest area, and sealing the wall to prevent further invasions of bees or other insects. With the cooperation of an interested beekeeper, who has the required skill and equipment, the bees may be transferred to a hive and taken away. A colony may be killed outright by blowing an insecticidal dust into the nest. This will also kill newly emerging young bees and new insect invaders from the outside.

Carpenter bees may weaken exposed, unpainted, or weathered boards of timbers of softer woods by boring nesting tunnels. These large bluish-black (female) or tan (male) insects resemble large bumble bees but have smooth rather than hairy abdomens. The tunnels are approximately ½″ in diameter when first made and may extend either vertically or horizontally. The wood is not eaten. The tunnels are used only as nesting sites and, if not disturbed, may be re-used, widened, and lengthened by further generations over a period of years. Carpenter bee tunnels are not usually found in wood with well-painted surfaces. Control includes use of an insecticide to destroy the developing bees within the tunnels, followed by repair, sealing, and painting the damaged wood.

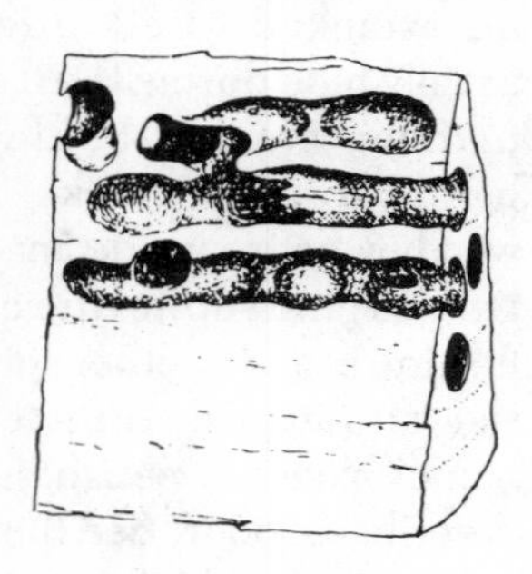

Carpenter Bee Nest
U.S. Public Health Service

Bee stings may be relieved by application of ice packs to the affected areas. Persons allergic to bee stings should remain away from areas where bees are active. Such persons should consult a physician regarding medication available on prescription for emergency treatment of stings.

Booklice

Booklice, sometimes called barklice, are psocids, tiny (1/16 inch) soft-bodied insects which closely resemble small, bleached-out aphids. They may occasionally become pests by feeding on paper, starch, grain and other substances in damp places. Sometimes they are found in the mulch of potted plants, around the bases of trees, and stone walls where lichens are available as food. One species of booklice is commonly found in libraries and museums, and in deserted bee hives and wasp nests. They are usually of no importance and are readily controlled with any of the commonly used household insecticides.

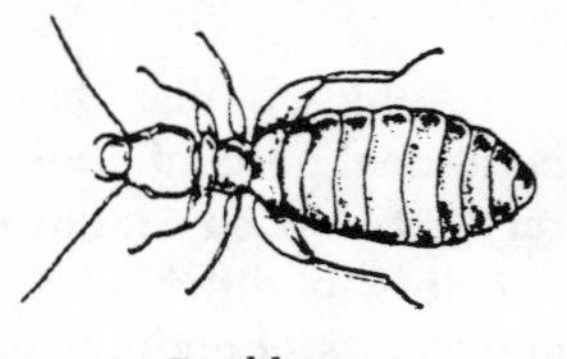

Booklouse
(U.S. Public Health Service)

Boxelder Bugs

These red and black plant-feeding insects may invade houses where the boxelder, or ash-leaf maple, is grown as a shade tree. Adults have three lengthwise red lines on the shield, or pronotum, behind the head and other conspicuous red lines on the front wings. These insects feed mainly on seed-bearing, or female, boxelder trees. In the fall the adults may enter homes in search of shelter. Although they do not feed they become nuisances by their presence, by their foul odor when crushed, and by their stains on curtains, furniture, clothing, and similar objects.

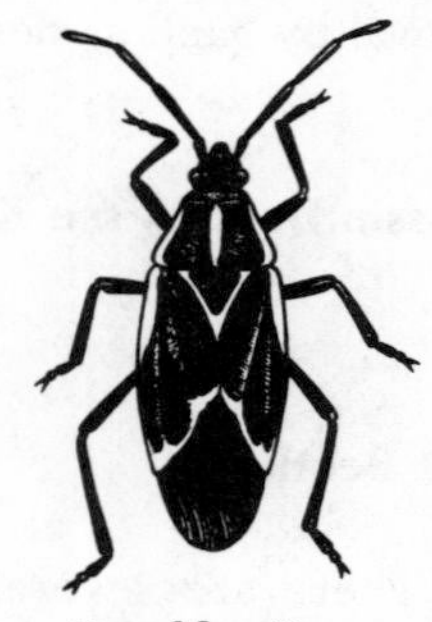

Boxelder Bug
(Univ. of Arizona)

The best control of boxelder bugs is elimination of seedbearing (female) boxelder trees. Staminate (male) trees are not attacked and should be chosen for planting. Insecticides may be applied to control outdoor infestations and to provide barriers surrounding homes. Control indoors can be achieved with a vacuum cleaner.

Carpet Beetles

Carpet beetle larvae feed on a variety of articles of animal origin and on dried foods high in protein. These include woolen rugs, blankets, clothing, upholstery, piano felts, feathers, furs, hides, bristles, powdered milk, cereals, spices, dog food, and bird seed. Irregular holes are eaten in rugs and stored woolens which may become localized patches of damage.

The contents of homes unoccupied for extended periods may be seriously damaged unless precautions are taken.

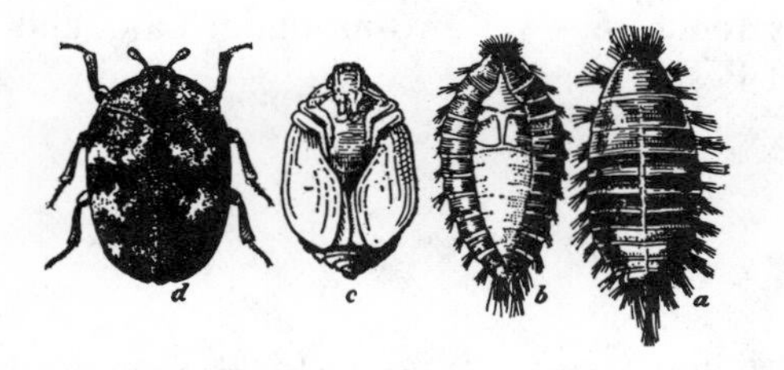

Carpet Beetle or Buffalo Moth. *(a)* larva; *(b)* pupa in larval skin; *(c)* pupa from below; *(d)* adult.
(USDA)

Larvae may be short and robust or wedge-shaped and more slender, depending on the species. They are about ¼″ long, with circular rows of stiff dark hairs around the body segments, and usually with tufts of longer hair at the rear of the body. They prefer to feed in dark, undisturbed situations rather than in actively occupied areas exposed to light.

The adults are about ⅛″ long, mottled gray, brown and white, and are attracted to windows and other lighted areas. Adults of most species feed outdoors on pollen and nectar.

Carpet beetles are best controlled by protective care of articles subject to attack and by eliminating surplus or discarded materials which might attract new infestations. Hides, furs, feathers, and woolens no longer of value should be destroyed. Lint and hair should be regularly removed by vacuuming cracks and crevices behind baseboards, moldings, and similar locations. Storage closets, indoors and out, should be regularly aired and inspected for susceptible material, including bodies of dead insects or mice. Rugs in continuous use, and subject to regular vacuum cleaning, are seldom damaged by carpet beetles provided heavy pieces of furniture are occasionally moved for examination of the vulnerable compacted areas beneath. Storage of cereals, spices, and dog foods in open containers for long periods should be avoided. Small amounts should be promptly used or discarded.

Woolens and blankets should be cleaned before summer storage and placed in tightly sealed containers. For storage over extended periods moth (PDB) crystals should be scattered on paper between layers of articles to be protected. In sealed containers one application per year is adequate. Plastic products, such as buttons, should be kept from direct contact with PDB crystals to prevent their disintegration by "melting." Cedar chests or cedar-lined closets give but limited protection beyond the tightness of their construction. Insects in infested materials may be killed by exposure to temperatures of 120°F, or higher, for at least 2 hours. In summer these temperatures are reached in uninsulated attics or during outdoor exposure to the open sun. When woolens are aired outdoors they should be thoroughly brushed, especially in seams, folds, and pockets to remove lint, eggs, and larvae.

Applications of residual insecticides are very effective. Some carpeting is mothproofed during manufacture. When new untreated carpeting is installed both sides of the underpad and the underside of the carpeting may be sprayed for long-term protection. Carpet beetles do not normally attack synthetic carpets but do feed on lint and proteinaceous foods imbedded in such carpets.

Clothing, blankets, and other susceptible materials may be protected by hanging them outside and spraying them with a stainless formulation of an approved household insecticide.

Centipedes

These "100 legged" relatives of insects have one pair of legs on each body segment. Most species live outdoors and only incidentally invade homes although the house centipede, 1½ inches long when full-grown, may regularly live indoors.

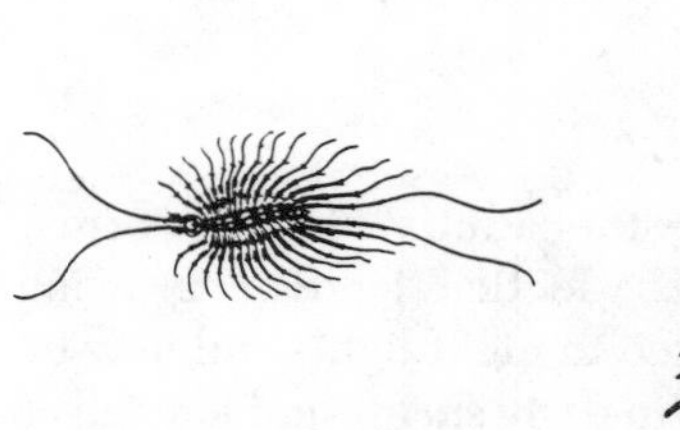

House Centipede
Scutigera cleoptrata
(U.S. Public Health Service)

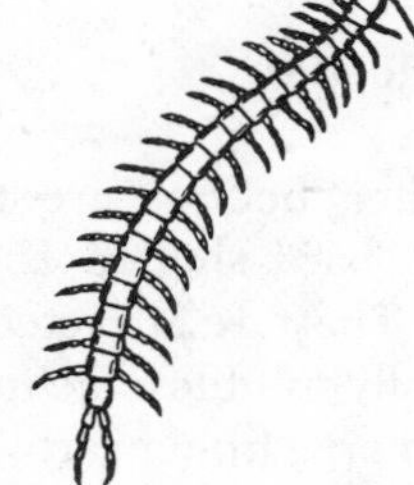

Common Centipede
(Univ. of Arizona)

They are usually considered nuisances because of their presence but may also be regarded as beneficial because they feed on insects, spiders and other small animals. They overpower their prey with prominent jaws and a relatively mild venom. Humans are rarely attacked, but if bitten, a temporary swelling, less painful than a bee sting, may result from the injected venom. The effects of such bites may be more emotional than physical. Deep puncture wounds from larger species should be examined by a physician.

Centipedes are active at night. In homes they are most commonly found in places of high humidity, such as basements, bathrooms and damp, dark storage areas. Outdoors they are usually found under stones, damp boards, and other debris. Invasions of homes may be reduced by eliminating these sources of shelter, especially from near foundations.

Chinch Bugs

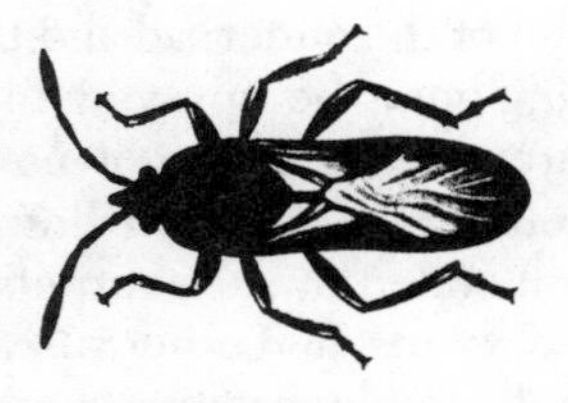

Chinch Bug
(Union Carbide)

Chinch bugs are pests of grain crops and grasses and are a serious pest of lawns in Florida. These small black bugs with clear wings may invade homes in the fall as cold weather begins, particularly near agricultural areas. The insects give off a vile odor when crushed and should be removed with a vacuum. Their fall intrusion is rather brief, and they are no more than a nuisance. (See Plant Bugs under Vegetable Garden Insect Pests)

Click Beetles

Click beetles are the adults of wireworms, brown, hard-shelled larvae that feed on growing plants. They are attracted to night lights and may be especially troublesome in early spring and late fall as they emerge from nearby cropland. They vary in size and color, being solid, striped, or spotted, and are identified by their clicking action when held between the thumb and forefinger. The click or snapping

Click Beetle
(USDA)

action is their rapid, spring-loaded movement of the abdomen hinging at the thorax. This is a defensive behavior and usually surprises the holder causing it to be released. They cause no harm within the home and can be avoided by using yellow bug-lights outside and by maintaining good, tight window and door screens. (See Wireworms under Lawn and Turf Insect Pests)

Clothes Moths

Clothes moths and carpet beetles are similar in that the damage of each is caused by the larvae, the products attacked are almost the same, and the same preventative and remedial control methods are used.

Two species of clothes moths are widely distributed in the United States. Larvae of the webbing clothes moths feed within silken tubes or webs spun over the infested material. Survival is much lower on clean wool than on wool soiled with food, beverages, or body stains. Infestations are thus centered in such areas. Larvae of the casebearing clothes moth have similar food habits but feed within a portable case rather than under a web.

Webbing Clothes Moth
Tineola bisselliella
(U.S. Public Health Service)

Infestations of clothes moths may be prevented or controlled by the procedures suggested for carpet beetles.

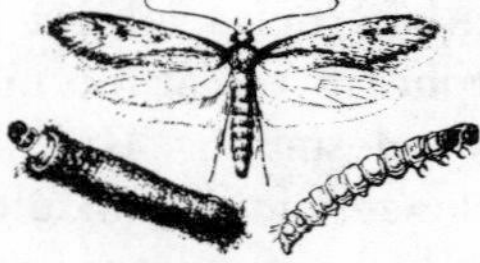

Case Making Clothes Moth
Tinea pellionella
(U.S. Public Health Service)

The following list of home remedies was proved useless by U.S. Dept. of Agriculture entomologists in the mid-1940's: allspice, borax, formaldehyde, lead-oxide, sodium bicarbonate, tobacco, black pepper, red cedar leaves, hellebore, lime, salt, lavender-flowers, cayenne pepper, eucalyptus leaves, lead-carbonate, quassia chips and sulfur.

Any success achieved from using any of the above would come from the fact that moths were absent or woolens so thoroughly cleaned and well stored that moths did not find them.

Clover Mites

These bright red mites invade homes in search of warmth or shelter, particularly in the early spring. They are objectionable by their presence, often in large numbers, and because of the reddish-brown stains made on floors, carpets, and other surfaces or fabrics when they are inadvertently crushed.

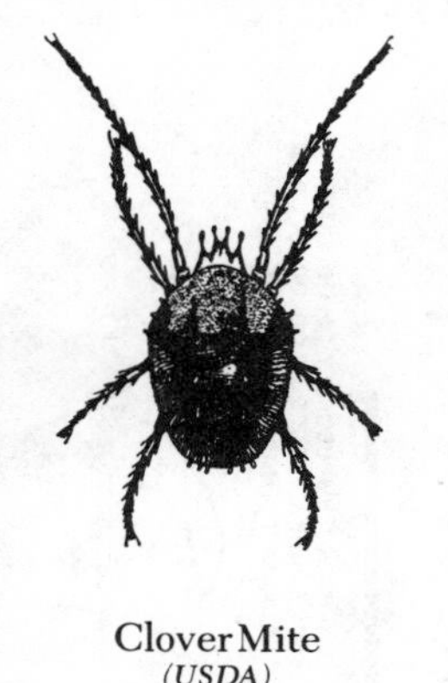

Clover Mite
(USDA)

Clover mites are about the size of a small pinhead. The front legs are longer than the others. They may become abundant in lush lawns and other vegetation near houses, especially when soils are rich in organic matter. After emerging from winter shelter they may become household pests when vegetation, including winter lawns, is maintained close to foundation walls. On warm, early spring days, they may crawl up the walls of houses, particularly white walls, and enter through doors or other openings.

They are controlled by maintaining a grass-free strip of loose soil 2 feet wide next to foundation walls. This area and adjacent doorways and window sills may be sprayed with a residual insecticide. Larger shrubs may be safely grown in this area although common weeds and annual flowers may attract clover mites.

Indoor infestations may be removed by a vacuum cleaner, using care to avoid surface staining from the bodies of crushed mites. Recommended spray treatments may also be needed.

Cockroaches or Waterbugs

Cockroaches have lived in or near human dwellings for thousands of years and are among the most familiar and objectionable of household pests. They are oval, flattened insects with a prominent shield extending forward over the head. Most species are active at night and hide in darkened areas, often in cracks, during the day.

Roaches and their egg cases may be brought into homes in boxes, cartons, grocery and produce bags and in household goods.

Light infestations are common in homes although heavy, continuing infestations are usually associated with sloppy housekeeping. These pests may carry on their bodies disease organisms which can be transferred from garbage, filth, and decaying organic matter to food, dishes, and kitchen utensils. Further contamination results from spots of regurgitated food and fecal matter on household articles. Articles with starch or glue sizing such as paper, fabrics, and book bindings, may be attacked. Cockroaches may seek shelter or food in closets, dressers, desks, electric clocks, and even TV sets. Contaminated articles, particularly foods, have a distinct roach odor.

Several species of cockroaches are major pests while others are only minor pests. These pest species may be distinguished by their size, appearance and habits (Figure 17). Young hatch from eggs deposited in groups within small satchel-shaped cases. These egg cases are often deposited in nearby debris although the females of the German cockroach and the field cockroach carry their cases attached to the rear of the body until hatching. Young roaches or nymphs are similar to adults in feeding habits.

The German cockroach is probably the most abundant and prolific pest species. The adults are pale yellowish-brown, about ⅝ inch long, with two blackish-brown lengthwise strips on the pronotum. Both sexes have wings but rarely fly. With ample food and moisture one female is capable of producing over a million descendants in a year. German roaches are usually found in or near kitchens and are general feeders, although fermented foods are particularly preferred.

The American cockroach is the largest and one of the commonest pest species in the southern states. It is reddish-brown, 1½ to 2 inches long, with darker blotches on a yellowish pronotum. It is commonly found in dark, moist areas near bathtubs, clothes hampers, basements and sewers. It may enter homes through plumbing traps. The American cockroach feeds on many substances but seems to prefer decaying organic matter.

The brown-banded cockroach is lightly smaller than the German cockroach and has a pale band across the base of the wings. It may have one or two similar bands towards the wing tips. It is a general feeder and tends to be more widely distributed over a

FIGURE 17.

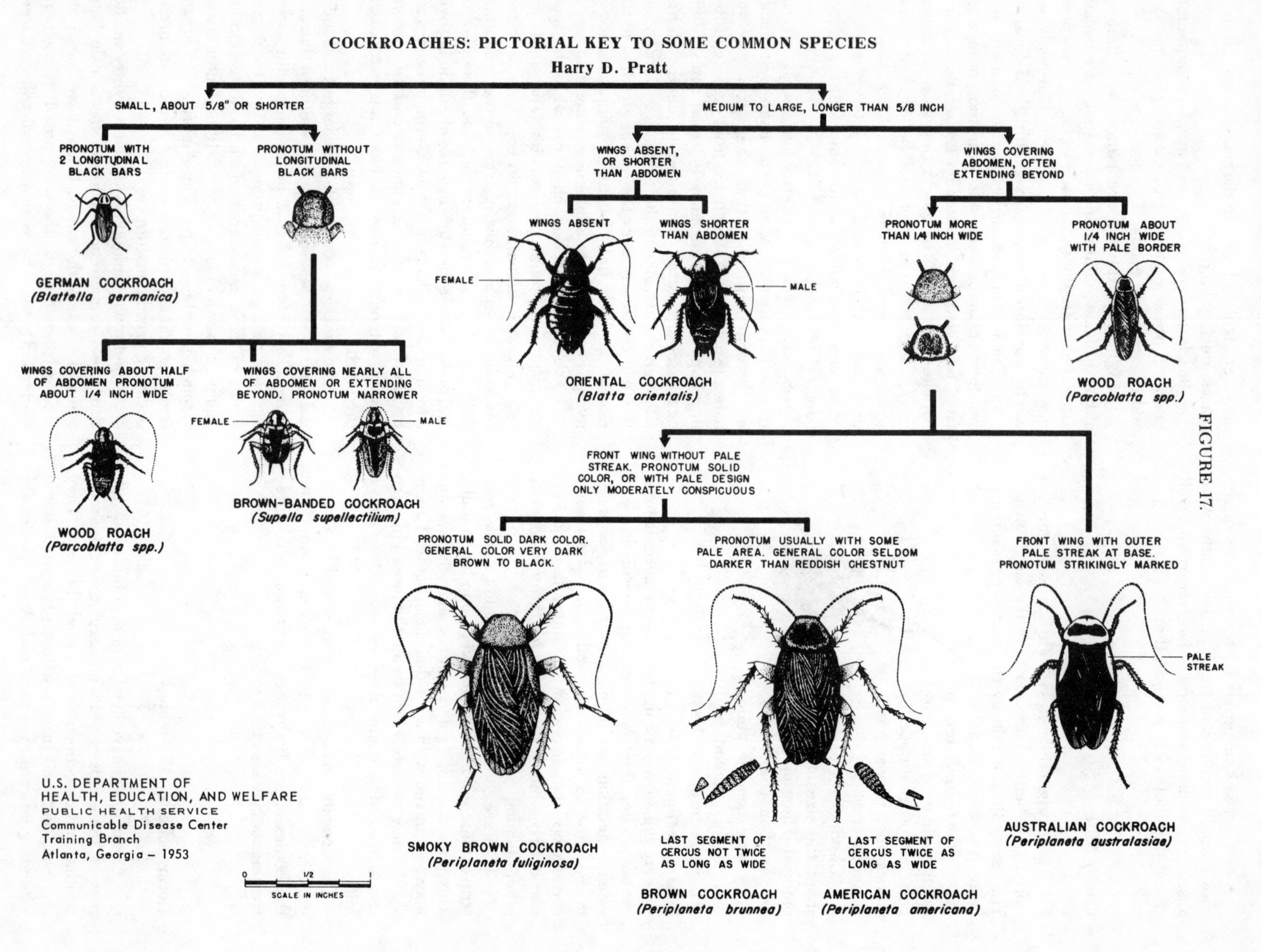
COCKROACHES: PICTORIAL KEY TO SOME COMMON SPECIES
Harry D. Pratt
SMALL, ABOUT 5/8" OR SHORTER
MEDIUM TO LARGE, LONGER THAN 5/8 INCH
PRONOTUM WITH 2 LONGITUDINAL BLACK BARS
PRONOTUM WITHOUT LONGITUDINAL BLACK BARS
GERMAN COCKROACH
(Blattella germanica)
WINGS COVERING ABOUT HALF OF ABDOMEN PRONOTUM ABOUT 1/4 INCH WIDE
WOOD ROACH
(Parcoblatta spp.)
WINGS COVERING NEARLY ALL OF ABDOMEN OR EXTENDING BEYOND. PRONOTUM NARROWER
FEMALE
MALE
BROWN-BANDED COCKROACH
(Supella supellectilium)
WINGS ABSENT, OR SHORTER THAN ABDOMEN
WINGS ABSENT
WINGS SHORTER THAN ABDOMEN
FEMALE
MALE
ORIENTAL COCKROACH
(Blatta orientalis)
WINGS COVERING ABDOMEN, OFTEN EXTENDING BEYOND
PRONOTUM MORE THAN 1/4 INCH WIDE
PRONOTUM ABOUT 1/4 INCH WIDE WITH PALE BORDER
WOOD ROACH
(Parcoblatta spp.)
FRONT WING WITHOUT PALE STREAK. PRONOTUM SOLID COLOR, OR WITH PALE DESIGN ONLY MODERATELY CONSPICUOUS
PRONOTUM SOLID DARK COLOR. GENERAL COLOR VERY DARK BROWN TO BLACK.
PRONOTUM USUALLY WITH SOME PALE AREA. GENERAL COLOR SELDOM DARKER THAN REDDISH CHESTNUT
FRONT WING WITH OUTER PALE STREAK AT BASE. PRONOTUM STRIKINGLY MARKED
PALE STREAK
SMOKY BROWN COCKROACH
(Periplaneta fuliginosa)
LAST SEGMENT OF CERCUS NOT TWICE AS LONG AS WIDE
BROWN COCKROACH
(Periplaneta brunnea)
LAST SEGMENT OF CERCUS TWICE AS LONG AS WIDE
AMERICAN COCKROACH
(Periplaneta americana)
AUSTRALIAN COCKROACH
(Periplaneta australasiae)
U.S. DEPARTMENT OF HEALTH, EDUCATION, AND WELFARE
PUBLIC HEALTH SERVICE
Communicable Disease Center
Training Branch
Atlanta, Georgia – 1953
0
1/2
1
SCALE IN INCHES

home, including warmer and drier areas, than the German cockroach.

The Oriental cockroach is shiny black or brown and about one inch long. The wings of the female are small, rudimentary, and functionless, and hardly more than small pads. The males' wings cover only about 75% of the abdomen. Neither sex flies, and they prefer dark, damp areas indoors and warm humid situations outdoors.

Roach control requires good housekeeping and thorough application of insecticides when required. There is no control unless it is complete. It is important that insecticide residues remain on treated surfaces for continued control action. Applications to warm areas, such as heater rooms, may need to be made more often than elsewhere because of more rapid disappearance of residues at warmer temperatures. In difficult situations, such as multiple dwellings, a community effort may be required. The slow-release paint-on formulation containing the insecticide chlorpyrifos is quite effective against roaches and other household pests. Boric acid as a powder or tablets is also very effective. (See Slow-Release section under Pesticide Formulations.)

Crickets

Our chirping friends are general feeders on most kinds of outdoor vegetation. When abundant they may enter homes for shelter or food, particularly when vegetation becomes dry or scarce, or at the approach of winter. They are attracted to lights outside of buildings where they may become nuisance pests when abundant.

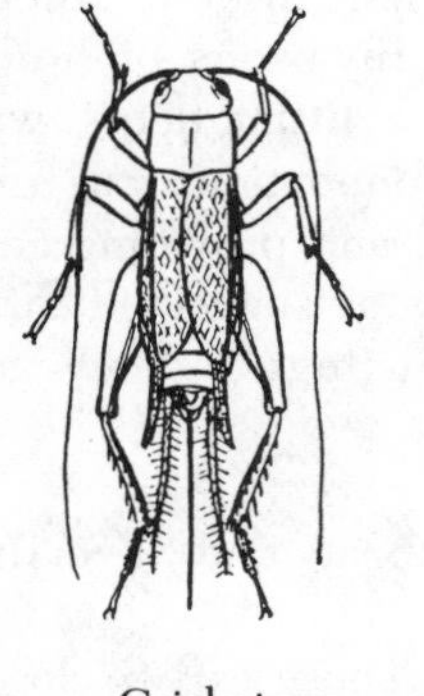

Cricket
(Univ. of Arizona)

Inside homes they may feed on clothing and fabrics spotted or stained by spilled food or perspiration. The resulting feeding holes have frayed edges.

Two species may become house pests, field and house crickets. Adults are ¾ to 1 inch long, brown to gray, and with thread-like antennae longer than the body. Females have a long, needle-like ovipositor at the rear of the body. The chirping call of crickets, which may be an annoyance if crickets are numerous, is produced only by the males.

Crickets may be excluded from homes by tightly-fitting doors, windows and screens. Foundation perimeter treatments with insecticide, particularly near doors and under lights, may be needed when crickets are abundant. At such times the use of yellow, non-attractive light bulbs is helpful. Residual sprays may be applied to baseboards and other surfaces in indoor areas where crickets are found. Roach sticky-traps work miracles.

Darkling Beetles

In the West and Southwest these outdoor insects may become nuisance pests by invading homes in search of darkness or shelter. Adult beetles are ¼″ long and dark brown to grayish black. They may be found in agricultural areas and in flower beds or leafy trash near foundations of houses. These beetles feed on living or decaying vegetable matter and may girdle stems of living or decaying vegetable matter and may girdle stems of living plants, especially seedlings, near the soil level. When infested crops or plantings become dry these beetles may invade houses, often to escape the summer heat. In summer, they may be found in large numbers under outdoor lights, especially near agricultural areas or extensive home plantings. The larvae, known as false wireworms, live below ground and are seldom seen.

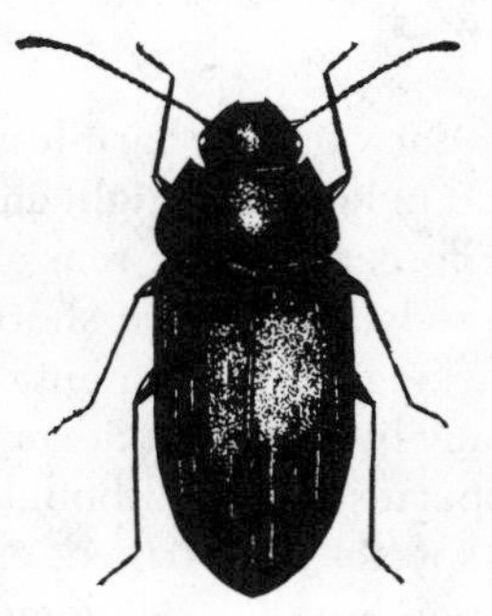

Darkling Beetle
(Univ. of Arizona)

Invasions of darkling beetles may be reduced by tight-fitting doors, windows, and screens. Dead leaves and rubbish should be removed from outside areas near foundation walls. An insecticide barrier surrounding the house, as used against other invading insects, may also be helpful.

Indoor infestations may be controlled with a fly swatter, vacuum cleaner, hand picking, or by surface applications of a residual-type insecticide to door sills, baseboards and dark corners where beetles may hide.

Drain Flies or Moth Flies

Drain or sewage flies, also known as moth flies are usually less than ⅛ inch long, with scaly wings held roof-like over the body. They may develop inside homes in the organic matter found in sink drains, plumbing traps, or garbage. They may also invade homes from nearby sewage plants or other areas with moist, decaying organic matter. These nuisance pests can be controlled by eliminating breeding places through good housekeeping, including regular treatment of sink drains and traps with a drain cleaner or flushing the overflow drains with scalding water.

Drain or Moth Flies
(USDA)

Earwigs

Earwigs are harmless nuisance invaders often found in homes at night and are easily recognized by the "forceps" at the rear of the body. Wings are short or lacking and movement is usually by crawling. Several species occur although two species, the striped, or riparian earwig (¾″ long) and the ring-legged earwig (⅝″ long) are most common in houses. They emit a substance with an offensive odor when disturbed.

Earwig
(Union Carbide)

Earwigs develop outdoors, often in flower beds and lawns near bases of houses. The striped or riparian earwig is predaceous on insects or earthworms. The ring-legged earwig feeds on decaying plant or animal matter.

Earwigs are best controlled outdoors by eliminating plant debris from areas close to foundations of houses and by insecticidal treatment of breeding areas.

Elm Leaf Beetle

This is a major foliage pest of elm trees. It occasionally becomes a household nuisance when adult beetles seek winter shelter in homes or when larvae seek protected areas for pupation in the spring and later in the season.

Young Larva

Pupa
(USDA)

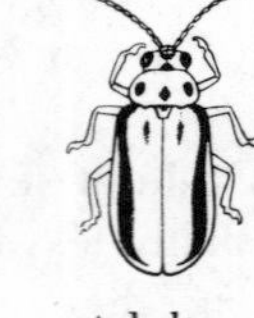

Adult

The buff-colored adult beetles are ¼″ long, with an olive green stripe along the outer edge of each wing cover. They commonly spend the winter in protected outdoor locations, but in colder areas, may also be found indoors behind curtains, under carpets, between books, and in similar locations. Mature larvae are ½″ long, dull yellow, and with two stripes along the back. They may seek protected areas for pupation in homes or porches.

Basically, their control involves treatment of outdoor infestations in elm trees. Invasions of homes may be reduced by tight-fitting doors, windows, and screens. Indoor infestations may be removed by vacuum cleaning or use of a recommended insecticide.

False Chinch Bugs

These are pests in the West and Southwest. Adults are less than ¼″ long, with grayish bodies and paler outer wings. These plant-feeding insects may invade homes in mid- to late spring, particularly near agricultural areas or fallow lands. These invasions are sporadic, of short duration, and usually of minor importance. They usually occur after winters with higher than average rainfall, when there is an abundant early spring growth of annual weeds and desert vegetation. Migration toward homes may occur after these plants mature and wither with the approach of warmer, drier weather.

Invasions of homes may be discouraged by tightly fitting doors, windows and screens, by keeping foundation areas free of weeds and annual vegetation, and providing an insecticidal barrier near the walls as suggested for other invading insects. (See Plant Bugs under Vegetable Garden Insect Pests)

Firebrats and Silverfish

These are two closely related, widely distributed species of household insects of similar appearance, habits, and importance. The firebrat prefers warm, dry areas while silverfish are moisture-loving.

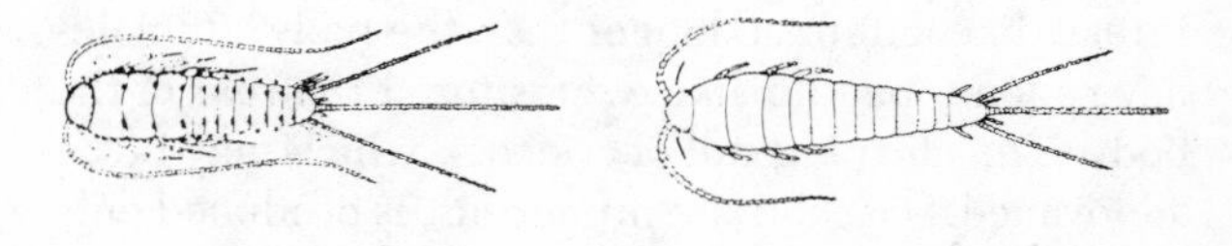

Firebrat　　Common Silverfish
(U.S. Public Health Service)

Adults are ½″ to ⅝″ long, mottled gray, flattened, boat-shaped, covered with scales, without wings, and with 3 long "tails" at the rear of the body. It hides during the day, runs swiftly and is most active at night. It commonly falls into bathtubs, from which it cannot escape. It eats and contaminates starchy foods, fabrics and paper, and may attack book bindings, leaving irregular roughened areas.

Infestations may be reduced by good housekeeping to remove starchy food scraps, elimination of hiding places, sealing cracks and other structural repairs, and by use of insecticides. Sprays should be applied to cracks and surface areas where they are observed and to floors or shelves under books and cartons in moist areas, especially garages and basements. The slow-release paint-on formulation containing chlorpyrifos is quite effective.

Fleas

Fleas found in homes are mainly pests of cats and dogs although humans may also be attacked. The wingless blood-feeding adults are brown to black, with laterally flattened bodies less than ⅛″ long and with hind legs adapted for jumping. The worm-like larvae are not blood feeders but feed on organic refuse in animal bedding, on floors, and in outdoor runways used by pets. Fleas may complete their development in the absence of animals. Persons entering infested homes unoccupied during vacation periods or between tenants may be attacked, particularly on the legs, by hungry, newly developed adult fleas that can survive for long periods without feeding and lose no time in attacking new arrivals. Bedding used by infested pets should be destroyed or thoroughly cleaned to kill the developing fleas. Floors, carpets, and upholstery in infested areas should be cleaned to remove all debris and animal refuse. Outdoor animal runs should also be kept free of debris and manure whigh might harbor developing fleas. After cleaning, the previously infested areas should be protected with an approved insecticide. Infested animals should be treated by one of the methods listed in the table on pet pest control, in addition to the use of a flea collar.

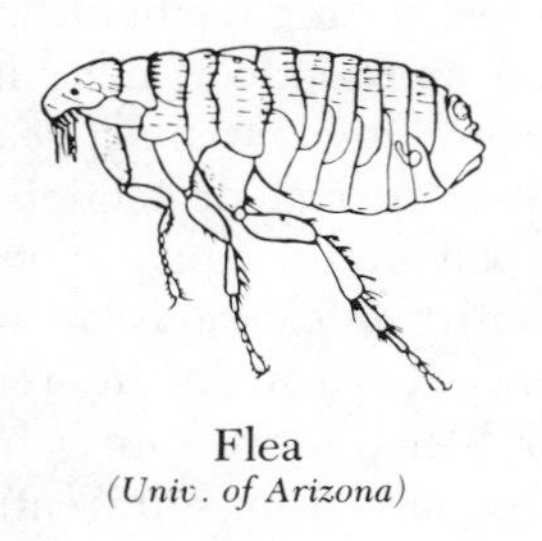

Flea
(Univ. of Arizona)

Flies

Several species of flies may be annoyances in and around homes, especially during the summer.

Female "filth" flies of various species may deposit eggs in manure piles, dead animals, meat, garbage and other decaying plant and animal matter. During warm weather flies quickly pass through egg, larval (maggot) and pupal stages (8-24 days, depending on species.) These dull, grey-black, or shiny, blue or green adult flies disperse freely and may enter homes in search of food or shelter. Their frequent movements between "filth" and our food and our persons may spread disease.

Of these flies, the common house fly needs little introduction. It is ¼ inch long, with gray to black markings. It is probably the most persistent nuisance pest and is an efficient disease carrier because of its ability to breed in a variety of substances. It has been implicated in the spread of typhoid, dysentery, tuberculosis, and other diseases.

Musca domestica
House Fly
(USDA)

The little house fly is a smaller, closely related species. It seldom alights on man or his food, and is less likely to contaminate food than the common house fly. This species is characterized by its habit of flying to and fro in the middle of a room or in shaded entryways.

Stable flies resemble very closely house flies, but bite with a sharp, blood-sucking proboscis that projects forward when the fly is at rest. Larval breeding sources include manure, especially when mixed with straw, and lawn clippings.

The false stable fly is slightly larger and darker gray than the house fly. It may enter homes and deposit eggs in well-decayed vegetable matter. It is not a persistent nuisance. Flesh flies are the largest of the gray-colored flies. They have a checkerboard pattern on the abdomen with the tip being red in many species. These flies breed in dead animals and excrement. Occasionally they alight on man or his food.

Fruit flies or vinegar flies *(Drosophila)*, are small, yellowish-brown and commonly associated with decaying fruit, vegetables or garbage cans. Eggs hatch in 24 hours and a life cycle takes from 8 to 11 days. These flies are most abundant around larval breeding sources but they may also enter the home as hitchhikers with fruit.

Blow flies are shiny, metallic blue, green, black or copper colored flies. They lay their eggs in exposed meat, garbage or pet droppings. The loud droning buzz created by flight is an annoyance in most confined situations. These flies may be particularly abundant during the spring and fall.

Phaenicia sericata
Green Blow Fly
(U.S. Public Health Service)

The first step in the control of all these flies is sanitation, the elimination of breeding places. Regular removal of garbage, livestock and pet manure, and all other decaying plant and animal matter is essential for successful fly abatement. Garbage cans should have tight-fitting lids. The use of plastic garbage or trash bags is very successful. Homes should have tight-fitting screens in good repair.

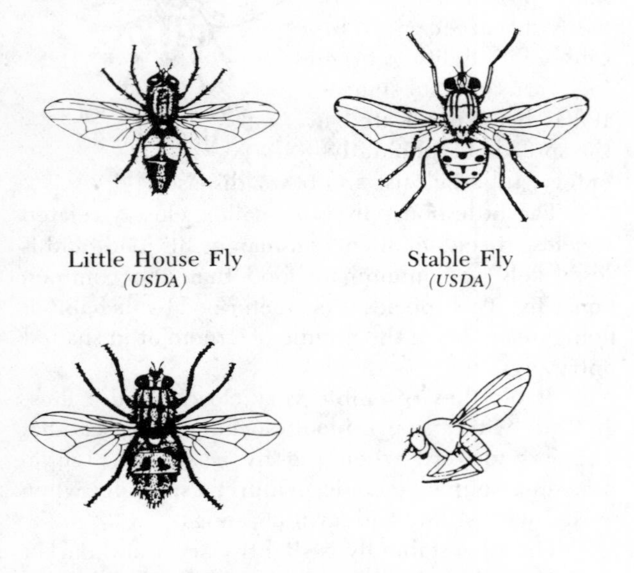

Little House Fly *(USDA)* — Stable Fly *(USDA)*

False Stably Fly *(USDA)* — Fruit or Vinegar Fly *(USDA)*

Fly swatters are still one of the best methods for control of the occasional flies that get into the tightest house. Keep one handy in the kitchen. Adhesive fly paper strips can help when hung discretely in selected areas.

In areas of high fly density, insecticides may be needed both inside and outdoors. Certain aerosol bombs may be used indoors as space sprays. Residual surface sprays are useful for killing flies outside the home. Insecticide-impregnated plastic pest strips are amazingly effective indoors and in well-protected areas. The use of "Big Stinky" and similar fly traps outdoors can also be of value.

Areas to be treated include porches, walls, garbage cans, carports and other surfaces where flies may rest. Bait formulations may be effective but should be kept out of reach of children and pets. Community action is usually necessary to achieve fly control in areas where breeding sources are extensive, such as trash dumps, animal feeding facilities, and pet areas.

Gnats

Gnats can ruin any outside activity. These annoying patio pests are tiny flies, less than ⅛ inch long, which are particularly attracted to the moist regions of the eyes, nose and mouth. They are most active during the warmer months, especially near fruit orchards and freshly cultivated land. Eye gnats belong to several closely-related species and may transmit the bacterium causing pink eye of domestic animals as well as man. The strong-flying adults are active during daylight hours. The larvae live in the soil and feed on decaying matter. There may be several generations a year.

Gnat
(U.S. Public Health Service)

Personal protection involves applications of a repellent to exposed body areas. It is difficult to control infestations on individual home properties since it is usually an area rather than an individual problem. More satisfactory control results may be produced from soil treatments to control the larvae on an area-wide basis. A community effort, much as for mosquitoes, may be required. The insecticide-impregnated plastic pest strips are very effective against gnats indoors.

Hornets — see Wasps

Kissing Bugs

Kissing bugs, which are blood-feeders, are also known as assassin bugs, conenose bugs, and Walapai tigers. They commonly live in outdoor nests of rodents, such as pack rats, and are more frequently reported from rural and desert locations than from urban areas. The adults are from ¾″ to 1¼″ long, and resemble large squash bugs, but with the head more slender than the rest of the body. Several species of varying color patterns, usually dark, are found in Southwestern desert areas.

Adult kissing bugs mature during the spring months, are most active at night, and are attracted to the lights of homes. Upon entry they may remove blood from sleeping persons without waking them, although painful swellings and occasional allergic reactions may later develop. During the day they may conceal themselves in bedding. Older homes and outbuildings may harbor pack rat nests in open crawl spaces beneath wooden floors and provide a source of kissing bugs which may attack persons living in the area.

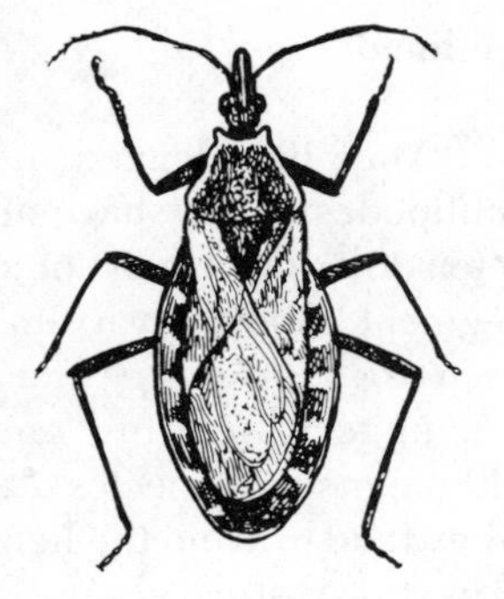

Kissing Bug
(Univ. of Arizona)

Preventive control includes elimination of rodent nests under buildings and in nearby desert areas. Pack rat nests have characteristic mounds of debris, often including remnants of cholla cactus. Invasions of homes may be reduced by using yellow bug lights outside and by tightly fitting doors, windows, and screens, sealing cracks in wooden floors and walls, and screening ventilation ducts.

Treatment of bites of kissing bugs may require the aid of a physician. Application of ice packs to the affected area may give partial relief.

Larder Beetle — see Pantry Pests

Lice

Three species of lice feed on man, and both males and females require blood meals. The head louse and crab louse are more commonly reported than the body louse. Lice are usually associated with crowded, unsanitary living conditions, including frequent contacts between persons. Recent trends toward greater cultural permissiveness and longer hair have led to a more uniform distribution of lice among all socio-economic groups.

The head louse or cootie is bluish-gray to whitish, wingless, up to ⅛″ long, usually found among the hairs of the scalp. The eggs, or nits, are attached to hairs close to the skin. Head lice may be spread through shared objects such as hats, hair brushes, combs, and towels. Infestations may be innocently obtained from upholstered furniture, bedding, and very commonly from contacts between children at school.

The crab louse, or crab, is a short, broad, thick-legged insect about ⅕″ long found in the crotch and other body areas with pubic hair. The eggs are also attached to hairs. Crab lice are spread mainly by personal contact but may also be received from bedding or from toilet seats (rarely) recently used by infested persons.

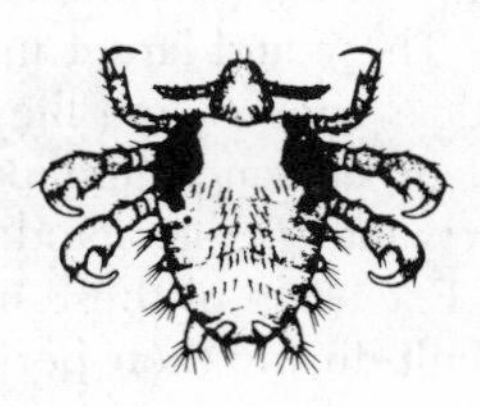

(Crab Louse)
(Univ. of Arizona)

The body louse resembles the head louse but is found mainly in seams of clothing, worn close to the body. The eggs are attached to clothing. In other countries, but not recently in the United States, it has been a vector of typhus and other diseases, particularly in times of war or disaster.

Louse control involves sanitation and personal hygiene based on a knowledge of lice and their methods of transfer. Infested persons should be promptly treated to destroy both the lice and their developing eggs. Several suitable louse ointments, lotions and shampoos are available and most require a physician's prescription.

Body Louse
(Univ. of Arizona)

When isolated from human contact nymphal lice usually survive not more than a day and adults not more than 5 to 6 days. Eggs may remain alive until hatching, from 8 to 10 days after being laid. Lice are eliminated when clothing is stored for 3 weeks.

Lice in clothing or bedding are destroyed by laundering or dry cleaning. Immersion in hot water, 125°F or higher kills adults in 5 minutes and eggs in 10 minutes. Thorough exposure of infested materials to the heat of the midday summer sun is also effective.

Light-Attracted Insects

Many insect species are naturally attracted to lights and may become nuisances both outside and inside homes, particularly in the warmer months.

Smaller flying insects, including tiny beetles, embiids, (web spinners), gnats, leafhoppers, and midges, may pass through screens and collect in lighting fixtures. Embiids are ½ inch long, dark and slender and, with some of the smaller beetles, are probably the largest insects able to pass through screens. Leafhoppers are tiny, wedge-shaped and usually green or yellowish. Gnats and midges are flies. These and larger insects such as moths, larger beetles, crickets and flies may become abundant at outdoor entrance and patio lights. Some of these insects may be attracted for considerable distances.

Exclusion of these insects from houses may be difficult during their periods of greatest abundance. Invasions may be reduced by tight-fitting doors, windows, and screens, use of light-proof window drapes and avoiding unnecessary use of lights inside or outside. Space sprays may be of temporary value. Dead insects in and beneath light fixtures should be removed to prevent secondary invasions of pests such as carpet beetles.

Insects are attracted to lights to varying degrees — some not at all. Despite exceptions, most insects are attracted in greater number to white incandescent, fluorescent, or blue mercury-vapor-type lamps then to yellow lamps. The use of yellow bulbs or "bug lights" for entrance or patio lighting is therefore helpful in suppressing light-attracted insects. White lights located at distances from human activity can be used to lure insects away from the party.

Long-Horned Beetles

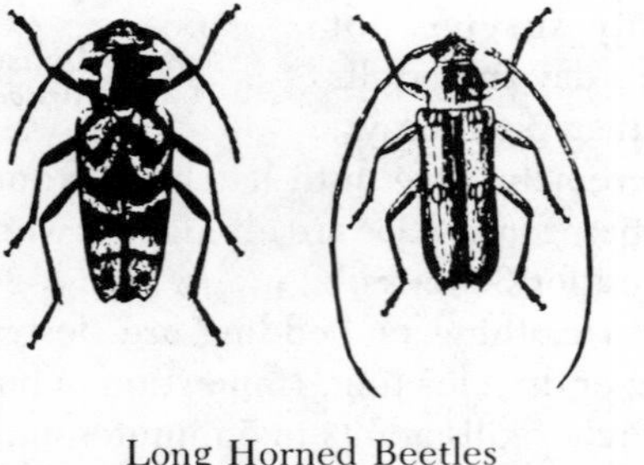

Long Horned Beetles
(Ohio State Univ.)

There are several long-horned beetles that may hitchhike their way into homes. They receive their name from their long antennae, longer than their bodies, held to the side. The larvae are wood borers and are found in dead wood. When firewood is brought into the home, adults sometimes emerge from the logs and are attracted to windows seeking an escape. They do not generate from house timbers unless the home is less than a year old. Long-horned beetles are normally not destructive, but may gnaw their way through curtains in their attempt to reach light, and should be removed when found. And, they can bite!

Millipedes

The "1000-legged" near-relatives to insects are millipedes. They have many body segments and often 30 or more pairs of legs, grouped 2 pairs to the segment. Species vary in length from less than an inch to 2 or more inches. Most millipedes are scavengers, feeding on decaying plant matter, although some species attack roots of living plants. They tend to avoid the light and may form a coil when at rest or disturbed.

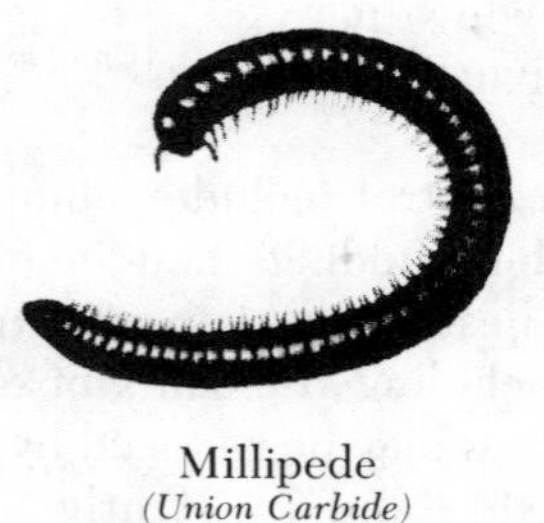

Millipede
(Union Carbide)

They occasionally find their way into homes as casual invaders although they do not feed or cause damage, except for possible stains when they are inadvertently crushed. These invasions may be discouraged by tight fitting doors and screens and by the removal of decaying vegetation, leaf litter, and similar materials from foundation areas of houses.

Mites

There are several kinds of mites that may attack man. These tiny relatives of insects have sucking mouthparts and may attack persons outdoors, as in the case of the mites known as chiggers, or may enter homes with birds and other pets, rats and mice, grains, cheese, dried fruits and infested clothing. Species of very tiny mites found in house dust may cause allergic reactions. Mites cause such human ailments as grocer's itch, mange, or "7-year itch." The clover

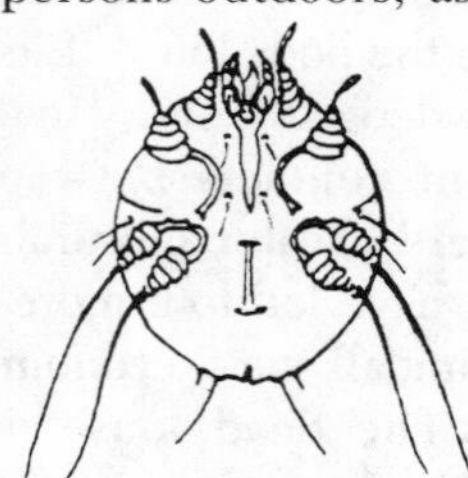

A Mange Mite
(Cornell Univ.)

mite is a plant-feeding outdoor pest that may invade homes but does not attack humans.

Tropical rat mites are a common problem when rats occupy attics. When a rat dies, large numbers of mites leave it and find their way into the living space of the home. Some people get a severe itch from the bites of these mites and suffer for years before finding the cause. (Refer to the section on rat control for means of controlling rats and keeping them out of buildings.) Bird mites can also be a problem, coming from bird nests, such as those built under the eaves of houses.

Treatment varies with the species of mite and in some cases may require the services of a physician. Control involves removal of mite food sources through good housekeeping or sanitation and for some species, the use of a suitable repellent on the person. Infestations on pets should be controlled as a part of the treatment process.

Also see "Chiggers" "Clover Mites" and "Spider Mites".

Mosquitoes, Midges and Black Flies

Several species of mosquitoes may become nuisances in and about homes. Adults are generally ¼ inch long, with two wings, long slender legs and a needle-like proboscis. Only females suck blood. Their bites may be painful at first, followed later by swelling and itching. Feeding typically occurs from dusk to dawn.

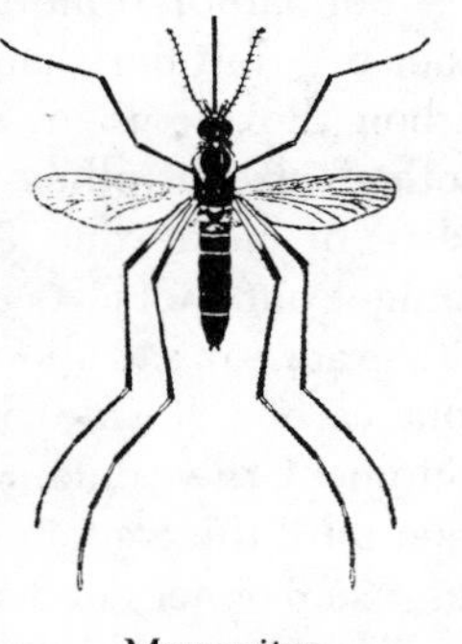
Mosquito
(Univ. of Arizona)

Mosquitoes are dependent on water for egg laying and development of their young. Eggs are deposited singly or in rafts on moist soil, and on the surface of standing water in pools, ditches, and discarded containers such as tin cans or tire casings. Eggs hatch into larvae or wigglers that feed on organic matter in the water. In a few days larvae stop feeding and transform into pupae or tumblers that later become adult mosquitoes. The complete cycle from egg to adult may take as little as 9 days during the summer.

The elimination of breeding sources is the first and best step toward controlling mosquitoes around the home. Low areas or ditches where standing water may collect should be eliminated. Remove or cover open containers that may hold water and permit breeding.

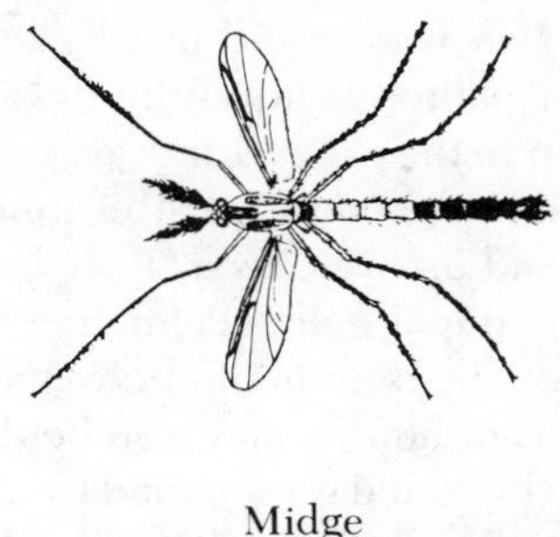
Midge
(U.S. Public Health Service)

Several species of small midges may be attracted in large number to lights. The males have typical plume-like antennae. Midges resemble mosquitoes, but their proboscis is short and not adapted for sucking blood. Midges are widely distributed and may be extremely abundant in areas near standing water. When attracted to lights they may create a major annoyance.

Black Fly
(U.S. Public Health Service)

Black flies are small, humpbacked, gray to green flies that bite most warm-blooded animals. They range from 1/25 to 1/5 inch long with clear wings and stiff horn-like antennae. They make very little sound as they swarm about the head and other parts of the body. Bites cause itching or swelling that may persist for several days.

Black flies are usually a problem only in mountainous areas near swift running water. Immature forms are found attached to objects such as gravel, rocks, or plants in flowing streams.

Protective clothing including boots, gloves, and head veils, along with repellents, provide the best protection from these biting flies. Area-wide control of adults and larvae is difficult and usually not practical. Residual or space sprays may give temporary relief in confined areas such as patios.

The repellents Deet or Rutgers 6-12, applied to exposed skin areas, give temporary protection from attacks of mosquitoes, black flies, stable flies and ticks. There are a host of home-remedy repellents, but the two suggested are far superior.

Palm Flower Caterpillar

This dark-cream-colored, cutworm-like caterpillar is about 1 inch long and usually feeds on the blossoms of fan palms. When fully developed, the caterpillars drop to the ground and search for protection while changing to moths. At this time they may invade homes and excavate oval, inch-long holes in carpets to form pupal cells. Palm trees in largely paved patios near houses should be watched for this insect. Invasions of houses may also be discouraged by tight-fitting doors and screens and by a foundation pesticide barrier similar to that used against other pests.

Pantry and Stored Product Pests

Pantry pests (Figure 18) are those that can live and develop in the food and other products of low moisture content stored in home cupboards. Materials attacked include whole grains and seeds, flour and cereal products, dried milk, dried dog and other pet food, dried fruit, dried or cured meats, baked goods, candy, nuts, chocolate and cocoa spices, chili powder and even drugs.

The rice weevil is dark brown, ⅛ inch long, with a prominent snout. It feeds on whole grains and hard cereal products such as macaroni.

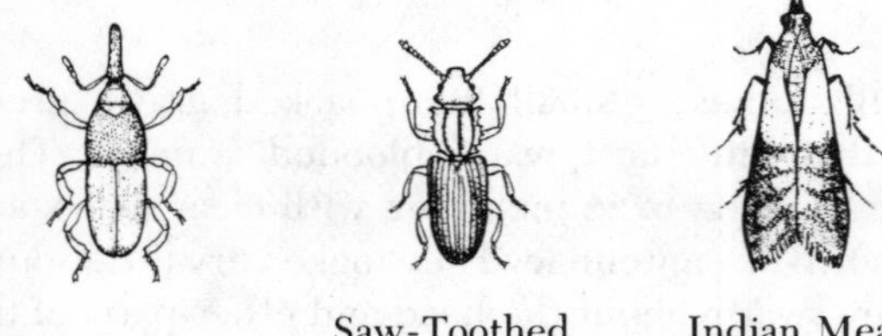

Rice Weevil — Saw-Toothed Grain Beetle *(Univ. of Arizona)* — Indian Meal Moth

More general feeders include flour beetles and saw-toothed grain beetles, each slender brown and ⅛ inch long, and the drug store beetle, slightly smaller, but with fairly long antennae. The saw-toothed grain beetle has prominent rows of saw-like projections on each side of the upper surface in the area behind the head. Dried fruits and nuts may be attacked by these general feeders and by the dried fruit beetle, recognized by its dark, oval body, ⅛ inch long, with 2 large amber-brown areas at the tips of the front wing covers. Larvae of Indian meal moths are general feeders on many stored foods and may be recognized by the conspicuous webbing spun over and through infested materials. Adult moths are ½ inch long with wings gray at the base and copper-brown at the tip. Foods stored in cupboards may also be attacked by such temporary invaders as ants, roaches, and firebrats. Also see "Bean Weevils."

Control of pantry pests requires elimination of existing infestations and prevention of new outbreaks by good housekeeping, frequent inspections, and applications of insecticides.

Shelves should be regularly and thoroughly cleaned to eliminate spilled food particles on exposed surfaces and from cracks and corners. Use of a vacuum cleaner with suitable accessories may be helpful. Small quantities of materials in open containers should be promptly used or destroyed, particularly if the house is to be vacated for the summer or for extended vacation periods. Usable but infested products may be sterilized by heating in an oven for 30 minutes at 130°F., or by exposure for a longer period outdoors in the heat of the noonday sun, or by placing the products in the freezer at 0°F. for 4 days. New food purchases should be inspected for pests before being stored. Uninfested or heat-sterilized dry foods should be stored in glass or metal containers with tight-fitting insect-proof lids. Coffee cans with tight-fitting plastic lids make good canisters. Clean surfaces of cupboards may be coated with an approved residual insecticide. The pesticide deposit should not be removed by washing although treated shelf surfaces may be covered with clean shelf paper.

For larger, emergency reserves of grain and other dry, but perishable products, fumigation with carbon dioxide is an exceptionally good and safe method of controlling stored grain insects. Heavy plastic or metal cans, 5-gallon capacity, make good storage containers. As the containers are being filled with grain, spread about 2 ounces of crushed dry ice (solid carbon dioxide) on 3 to 4 inches of grain in the bottom of the container, then add the remaining grain until the container is filled. For larger quantities, use 6 ounces of dry ice per 100 pounds of grain.

The vaporizing dry ice produces carbon dioxide fumes that are heavier than air, which displace the existing air in the container. **Allow from 30 minutes to an hour for complete vaporization before placing the lid on tightly. Premature sealing can cause bulging of the container. If plastic bags are used in the cans as liners, do not seal until vaporization is complete.** Carbon dioxide in closed containers destroys most

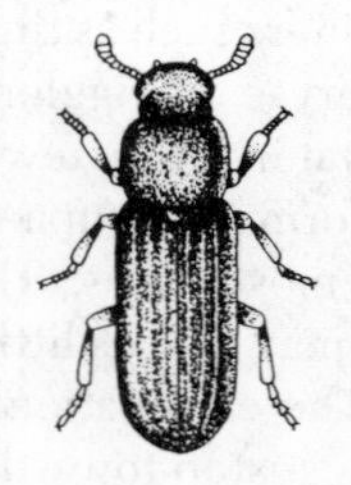

Confused Flour Beetle
(U.S. Public Health Service)

FIGURE 18.

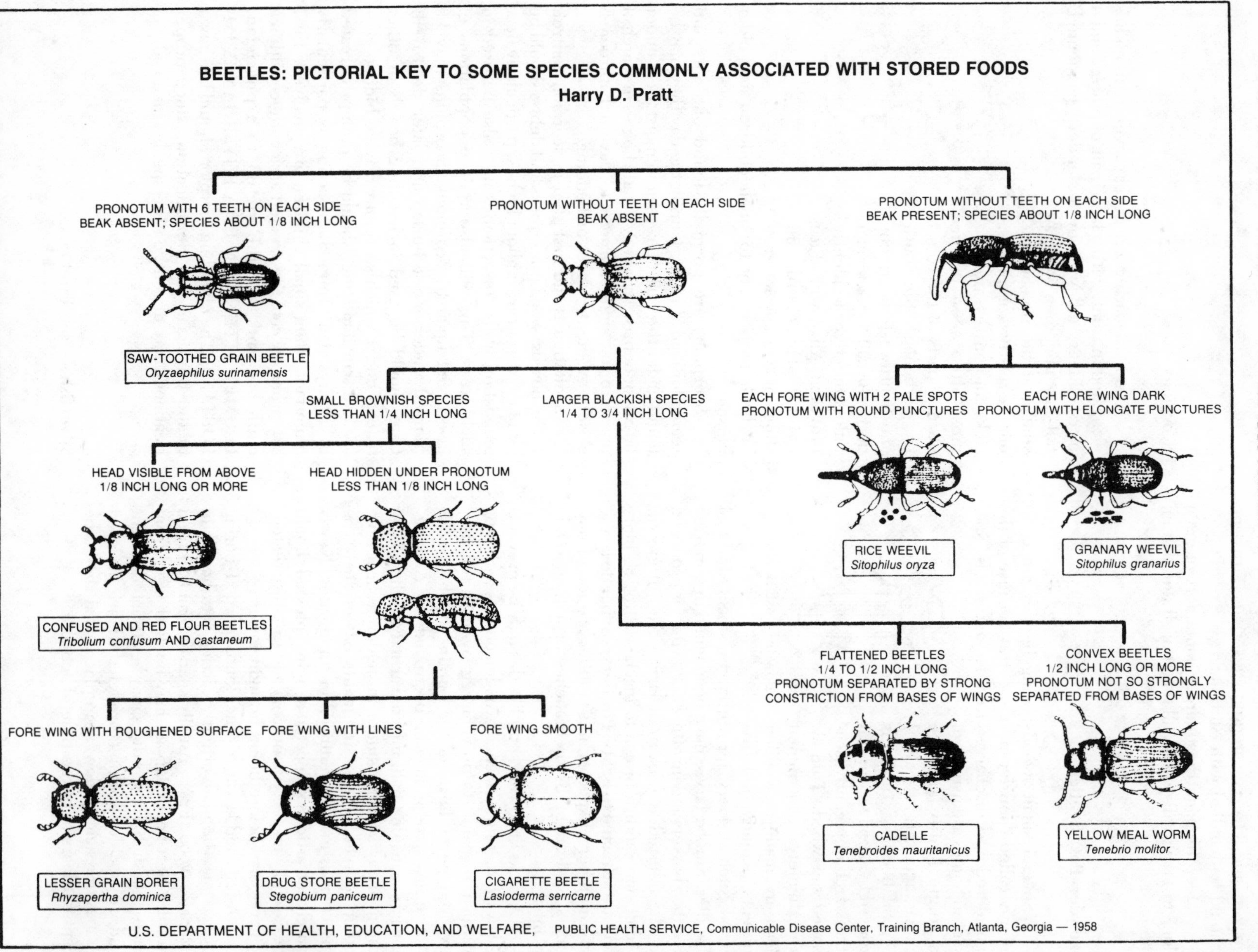
BEETLES: PICTORIAL KEY TO SOME SPECIES COMMONLY ASSOCIATED WITH STORED FOODS
Harry D. Pratt
PRONOTUM WITH 6 TEETH ON EACH SIDE
BEAK ABSENT; SPECIES ABOUT 1/8 INCH LONG
PRONOTUM WITHOUT TEETH ON EACH SIDE
BEAK ABSENT
PRONOTUM WITHOUT TEETH ON EACH SIDE
BEAK PRESENT; SPECIES ABOUT 1/8 INCH LONG
SAW-TOOTHED GRAIN BEETLE
Oryzaephilus surinamensis
SMALL BROWNISH SPECIES
LESS THAN 1/4 INCH LONG
LARGER BLACKISH SPECIES
1/4 TO 3/4 INCH LONG
EACH FORE WING WITH 2 PALE SPOTS
PRONOTUM WITH ROUND PUNCTURES
EACH FORE WING DARK
PRONOTUM WITH ELONGATE PUNCTURES
HEAD VISIBLE FROM ABOVE
1/8 INCH LONG OR MORE
HEAD HIDDEN UNDER PRONOTUM
LESS THAN 1/8 INCH LONG
RICE WEEVIL
Sitophilus oryza
GRANARY WEEVIL
Sitophilus granarius
CONFUSED AND RED FLOUR BEETLES
Tribolium confusum AND castaneum
FLATTENED BEETLES
1/4 TO 1/2 INCH LONG
PRONOTUM SEPARATED BY STRONG
CONSTRICTION FROM BASES OF WINGS
CONVEX BEETLES
1/2 INCH LONG OR MORE
PRONOTUM NOT SO STRONGLY
SEPARATED FROM BASES OF WINGS
FORE WING WITH ROUGHENED SURFACE
FORE WING WITH LINES
FORE WING SMOOTH
CADELLE
Tenebroides mauritanicus
YELLOW MEAL WORM
Tenebrio molitor
LESSER GRAIN BORER
Rhyzapertha dominica
DRUG STORE BEETLE
Stegobium paniceum
CIGARETTE BEETLE
Lasioderma serricarne
U.S. DEPARTMENT OF HEALTH, EDUCATION, AND WELFARE, PUBLIC HEALTH SERVICE, Communicable Disease Center, Training Branch, Atlanta, Georgia — 1958

adults and larvae, however, some eggs and pupae may escape. Dry ice can produce serious burns and should be handled with caution.

Deep freezing these larger containers of grain for 3 to 4 days will destroy all stages of all stored grain pests.

Powder-Post Beetles

Several small, wood-boring beetles are commonly called "powder-post" beetles because of the powdered frass discharged from larval tunnels. Species in homes are usually of a single family and are more accurately called "lyctid beetles." Injury is caused by larval feeding on the starch of recently-seasoned sap wood. Tunnels the diameter of a pencil lead are made in wooden objects, usually of hardwood, such as floors, furniture, gun stocks, and tool handles. Similar larvae have been found in imported bamboo baskets and furniture. "Powder" is produced in the infested articles until larvae are full grown. Adults are reddish brown to black, ⅛" to ⅕" long, and are attracted to lights. Powder-post beetle larvae of a number of species and families are found outdoors in dead limbs or branches of native woody plants.

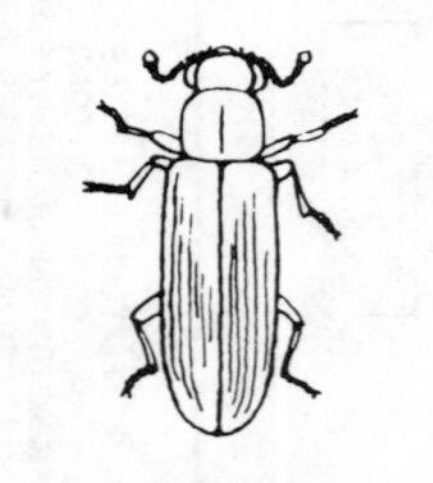

Powder Post Beetle
(U.S. Public Health Service)

Attacks of powder-post beetles may be prevented by protecting surfaces of wooden objects with wax, varnish, or paint, to prevent egg laying in the tiny, exposed pores forming the "grain" of the wood. Active investations may be destroyed by "cooking" the larvae inside small, infested articles in the heat of the noonday summer sun or in moderate heat from other sources. Use of fumigants or residual-type insecticides may be required in some cases. Newly purchased articles suspected of being infested should be carefully inspected outdoors and suitably treated, if necessary, before being brought inside.

Adults and larvae of other kinds of wood-boring insects may be inadvertently carried into homes in fireplace wood. Frass, usually coarser than that from powder-post beetle larvae, may be discharged from tunnels in such wood. Adults may be attracted to light and are capable of cutting through the fabric of closed drapes to reach windows. Fireplace wood should be kept outdoors until the time of actual use.

Roaches — see Cockroaches and Waterbugs.

Scorpions

These menacing creatures have stout bodies elongated in front, with a large pair of pincers and 4 pairs of legs, followed by a slender, segmented, tail-like abdomen with a stinger at the tip. Scorpions are relatives of spiders, ticks and mites and are most often seen in the warmer parts of the country, particularly the South and Southwest. Scorpions do not usually attack man unless directly or accidently provoked. All may produce painful stings and the stings of two species may be fatal, particularly to small children and older persons.

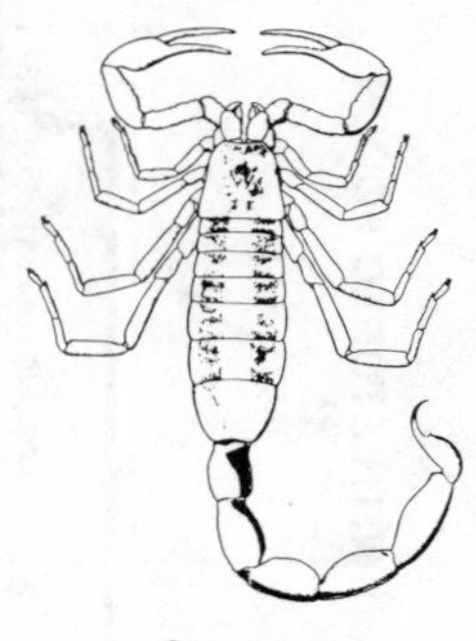

Scorpion
(Univ. of Arizona)

Scorpions are normally found outdoors and search for food at night. During the day they may be found under the bark of trees or in moist areas under boards or debris. They may invade homes in search of moisture and hide during the day in bathrooms, closets, garments, shoes or bedding.

Sanitation is the first step in scorpion control. Loose boards, wood piles, rocks and debris should be eliminated from areas about homes, particularly near foundation walls. Insecticides may also be used in such areas. This will also reduce populations of insects fed upon by scorpions. Scorpions may be trapped under moist burlap and later destroyed. Care should be used in handling boards or other objects under which scorpions may be hiding.

All scorpion stings should be promptly treated by a physician. Immediate action is particularly important when small children are involved. First aid: promptly cool the affected area with ice within a cloth bag and take the victim to a physician. PROMPT ATTENTION BY A PHYSICIAN IS IMPORTANT. The offending scorpion should be shown to the physician for identification, since proper treatment may depend on the species involved.

Silverfish — see Firebrats.

Solpugids

Solpugids, or sun spiders, are ferocious-looking but harmless relatives of spiders and are strictly confined to the desert Southwest. They are usually yellowish-brown, with bodies an inch or more in length, huge jaws, a pair of long, leg-like palps, and 4 pairs of legs. They have no poison glands and prey on insects and other small animals. Most species are active only at night. They are usually seen in patios and other lighted outdoor locations. They are not usually found indoors, except possibly in adobe buildings with dirt floors.

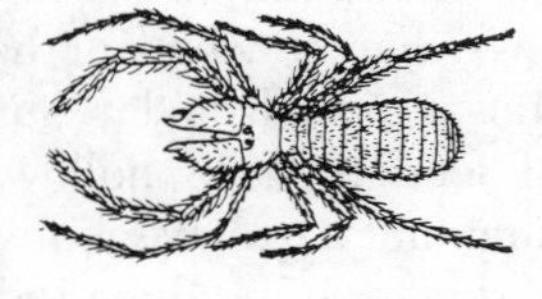
Solpugid or Sun Spider
(U.S. Public Health Service)

Sowbugs and Pillbugs

These are not insects but rather crustacea, and are about ½″ long, usually grayish, with 7 pairs of short legs and a hard, shell-like body covering. They breathe by gills and can live only in moist areas. Sowbugs have 2 short, tail-like rear appendages. Pillbugs can roll themselves into a tight ball for protection. They feed on decaying matter but may also attack living plants. They are commonly found outside homes in damp locations such as under stones, boards, flower pots, and plant debris. They hide during the day and are active at night. They may occasionally enter homes but are not pests except by their presence.

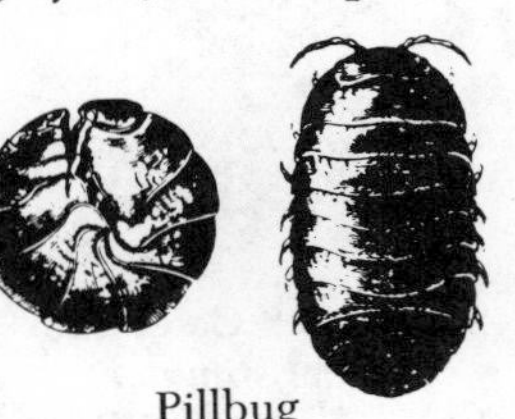
Pillbug
(U.S. Public Health Service)

Control measures are as for other house-invading pests, including tight-fitting doors, foundation cleanup, elimination of damp places of shelter, caulking cracks and seams, and insecticide barriers. (See Sowbugs under Vegetable Garden Insect Pests)

Spiders

Spiders are only distantly related to insects. Unlike insects, they have 4 pairs of legs but lack wings and antennae. A huge number of species of varying size and color are found throughout the U.S. and are usually regarded as pests merely because of their presence. Most spider species are really beneficial, since they feed on flies and other insects. The black widow is the only seriously poisonous spider commonly found about homes. The brown recluse spider, found in Eastern states and the Arizona brown spider are two closely related species of very similar appearance and are both very poisonous.

The adult female black widow has a rounded, glossy-black body about ½ inch long and an overall length of up to 1¼ inches. It is identified by a red to orange hour-glass-shaped marking on the under side. The male is much smaller, with white and pale brown markings and lacks venom. Insects and other spiders are aggressively attacked by the female black widow but humans are bitten only when she is accidentally touched or threatened. Black widows construct loose, irregular webs in dark protected corners of homes, carports, outdoor storage areas, and under outside chairs, benches, or privies. Bites are made by paired jaws with internal ducts from which a venom is discharged. A sharp pain, like that of a needle puncture, is first produced, followed by a variety of other discomforts, depending on location, as the venom penetrates the nervous system. Prompt treatment by a physician is recommended.

The brown recluse and Arizona brown spiders have a body ⅓ inch long and an overall span, including legs, of an inch or more. Their normal habitats are in isolated locations under pieces of wood, dead cacti, and similar situations but they may also move into dark places of nearby buildings. The Arizona spider does not appear to thrive in irrigated areas but may be brought into homes on wood or cactus skeletons from the desert. The Arizona brown spider and the related local species more frequently reported from California, where it is known as the "brown spider," are capable of inflicting painful bites on humans. The effects of these bites are reported to be less severe than those of the brown recluse spider found in more easterly states. Persons bitten by spiders should consult a physician for relief. Treatment will be aided if the offending spider is preserved for positive identification.

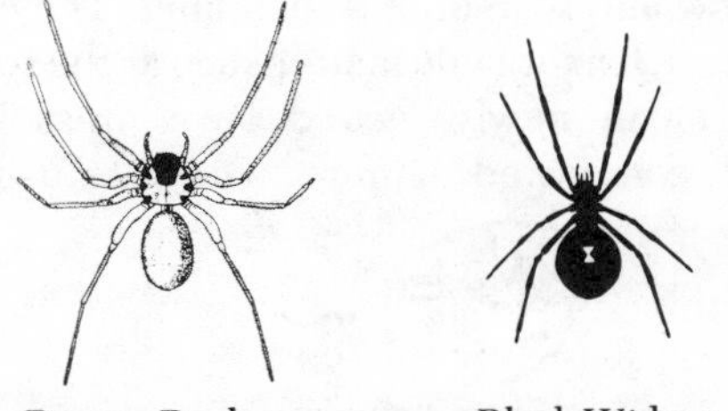
Brown Recluse Black Widow
(U.S. Public Health Service)

Among the non-injurious spiders commonly found in or near homes, and objectionable only by their presence, are patch spiders, giant crab spiders and daddy-long-legs. Patch spiders spin silken "patches" about ½ inch in diameter, particularly in slight depressions of plastered walls and in mortared joints of brick walls. Giant crab spiders measure as much as 2 inches across the legs and prey on insects but make no webs. Daddy-long-legs are recognized by their small, globular bodies ⅛ to ¼ inch long and their extremely long legs, extending for an inch or more.

Tarantulas are probably the most bizarre of the spiders, principally because of their size, and a few monster movies which have starred our furry friends. The tarantula is quite harmless, does not produce venom, and is not easily aroused. The males are usually seen in the late summer and fall, on the march, in search of females, which are somewhat more secluded and seldom seen. Tarantulas should never require control. Rather they should be permitted to move on in their nuptial pursuits.

Tarantula
(USDA)

Black widows and other spiders may be controlled by good housekeeping indoors and outdoors, including tight-fitting doors and screens, elimination of trash and debris, removal of webs by sweeping including the crushing of the spiders and egg sacs and by application of insecticides to infested areas. Night inspections with a flashlight of suspected outdoor hiding areas will reveal most black widows that hide during the day.

Springtails

Springtails (Collembola) are small, wingless insects with a jumping mechanism at the rear underside of the body. Species usually seen about homes are bluish gray and scarcely 1/16 inch long. They live in moist areas rich in organic matter such as the soil in flower pots or in newly-seeded lawn areas well covered with composted manure. They are usually but a temporary nuisance and are not found in hot, dry environments.

Springtails
(Ohio State Univ.)

Stinging Caterpillars

Hairy caterpillars, capable of causing skin irritations when touched or handled, may be found in patios, especially in late summer and early fall. These leaf feeders have hollow, brittle hairs or spines containing a toxic venom which, when injected into the skin, may produce reddened welts and blisters, accompanied by itching and burning sensations. Although nettling or stinging hairs are found in caterpillars of many species, those most often reported are the puss caterpillar and the larvae of the western grape leaf skeletonizer.

SOME OF THE MORE COMMON
STINGING CATERPILLARS

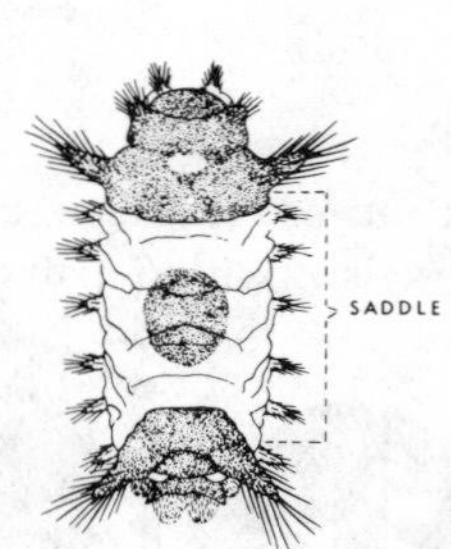

Saddleback Caterpillar
Sibine stimulae
(U.S. Public Health Service)

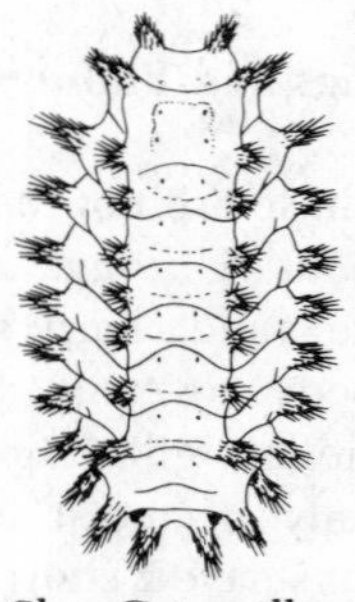

Slug Caterpillar
Euclea chloris
(U.S. Public Health Service)

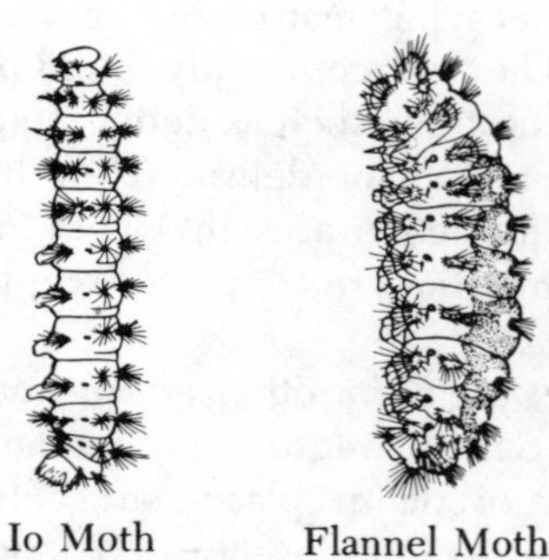

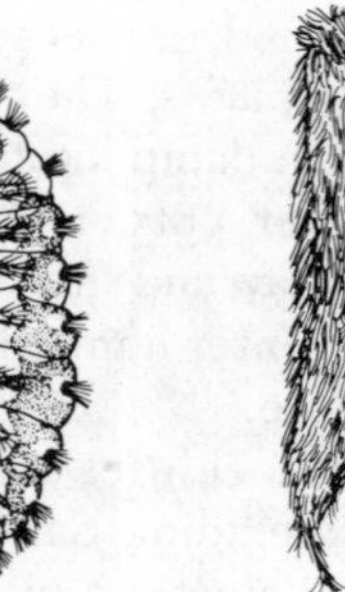

Io Moth Caterpillar
Automeris io
(U.S. Public Health Service)

Flannel Moth Caterpillar
Norape cretata
(U.S. Public Health Service)

Puss Caterpillar
Megalopyge opercularis
(U.S. Public Health Service)

The puss caterpillar is about 1 inch long, covered with a dense mat of soft, buff to gray hairs. It has a prominent hairy ridge or crest extending along the top of the body and ending in a slender tail. It is most often reported from communities where it feeds on oaks, mulberries, fruit trees, and shrubs. Curious school children are frequently stung by these caterpillars. Severe attacks have been accompanied by pain, swelling, and nausea. Discomfort may last several days.

Avoidance is the best procedure for dealing with stinging caterpillars. This requires recognition and education, especially with children. Affected areas should be promptly washed with soap and water. Cold packs may give partial relief. Severe cases, especially those involving children, should be seen promptly by a physician. Further relief may be given by use of one of several bland antiseptic ointments, containing a mild local anesthetic in a petrolatum base.

Termites

Termites are social insects that live in colonies and build their nests in wood or in the ground. They attack wood and paper because cellulose is their main source of food. Each colony is headed by a pair of functionally mature adults, called a king and queen, and thousands of pale, soft-bodied workers, plus a few soldiers that protect the colony against ants and other enemies. Winged termites are most active and visible during rainy seasons when new generations of adults emerge to mate and seek sites for new colonies.

HOW TO DISTINGUISH ANTS FROM TERMITES

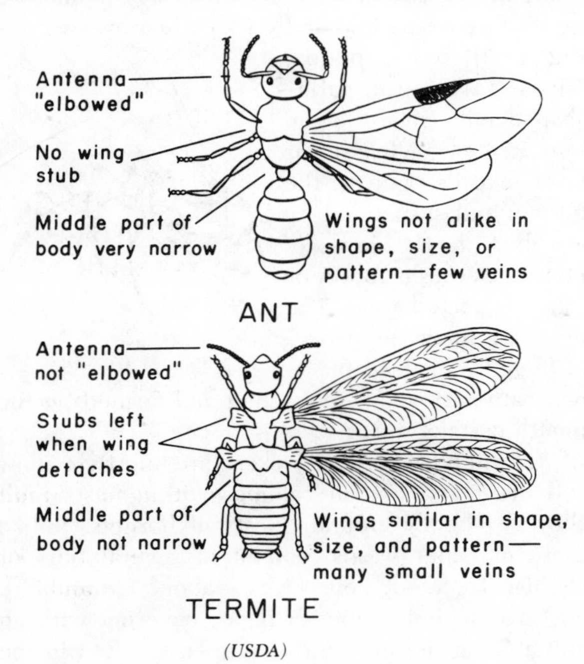

(USDA)

Subterranean termites form colonies deep in the ground, where needed moisture is present, and attack all types of dead plant material, paper, and wood in contact with soil. Most subterranean termites can also build mud tunnels, tubes, or shelters over brick or concrete to reach wood several feet above the ground. They frequently bore into baseboards, moldings, door frames and porch and carport structures.

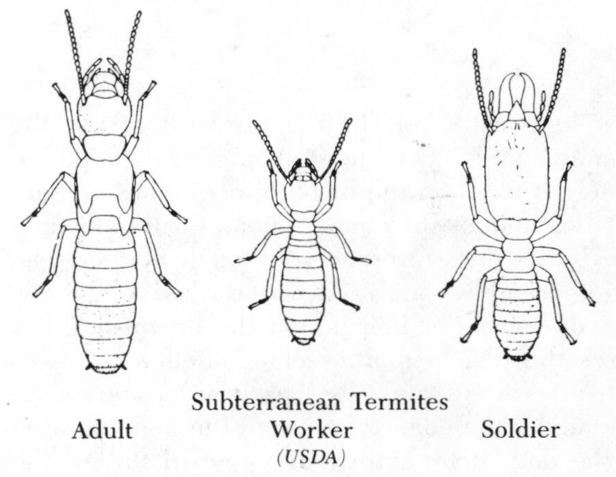

Subterranean Termites
Adult Worker Soldier
(USDA)

Dry-wood termites make clean tunnels or galleries in dry, sound wood and do not enter the soil for moisture. They usually cause less damage than subterranean termites but may attack floors, sills, door and window frames, rafters, and rarely, furniture. From their tunnels they sometimes eliminate piles of dry fecal pellets resembling fine sawdust or tiny seeds, the main distinguishing clue.

The ideal time to protect a house from termites is while it is being built. Steps at this time should include provision for adequate drainage away from the site, removal of stumps, roots, and wood scraps, insecticidal treatment of soil before footings or slabs are poured, professionally referred to as "pretreatment", and installation of screened vents in attics and crawl spaces to permit good ventilation.

The homeowner can increase his subterranean termite protection by observing a few simple precautions: Keep the under area of the home dry; don't let shrubbery block the breathing vents in the foundation; don't let sprinklers wet the stucco; avoid increasing the outside grade against the foundation; and avoid building window boxes against the house.

When it is suspected that a house is infested the first step is to positively determine that termites are actually present. When an infestation is found the most suitable control procedure should be used. Termites work slowly and there is time for a calm, considered decision. Termite control is more difficult in older houses but several techniques are available for treating such structures. The services of a licensed pest control operator are usually needed. Reputable firms will guarantee their work.

Ticks

These will likely be the brown dog tick, the "domesticated" tick most commonly found in homes. It feeds on the blood of dogs, causing both irritation and loss of vigor. In homes it is objectionable by its mere presence, although neither man nor other animals are normally attacked.

After each feeding period the brown dog tick leaves the dog to molt, seeking shelter in places such as crevices in walls, baseboards, and moldings, under the edges of rugs, and in bedding used by the dog. It tends to move upward toward the ceiling for shelter.

Newly emerged adults are reddish-brown, flattened, and about ⅛ inch long. As the female feeds she becomes about four times larger and bluish-brown in the swollen portion. After hatching from eggs commonly laid in indoor cracks and crevices the ticks pass through two immature feeding periods and one adult feeding period, each separated by a resting and molting interval off the dog. Length of the life cycle varies with temperature, from a minimum of about two months in summer to a much longer period in the cooler months. Dogs do not receive this tick directly from other dogs, but from coming near objects or surfaces from which newly hatched or recently molted ticks may "climb aboard". The brown dog tick can live as long as 1½ years without feeding. Dogs may thus become infested by entering homes or premises long after the departure of previous dog occupants. A tick-free family dog may be protected by confining it within an area not accessible to other dogs. Effective insecticidal control requires simultaneous treatment of the infested dog and the infested premises. For the latter, a residual insecticide is required to control the young ticks which may continue to hatch after the treatment. Infested animals should be treated according to one of the methods described in Table 20, Pest Control Suggestions for Pets, which should include a flea- or tick-collar for the pet(s). Ticks are virtually impossible to control without the aid of insecticides.

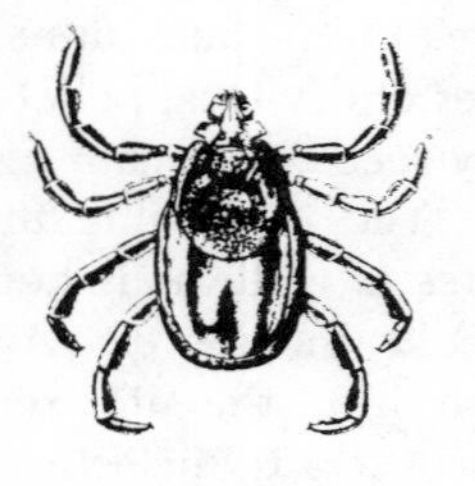

Brown Dog Tick
(USDA)

Wasps, Hornets, and Mud Daubers

Wasps include numerous species of stinging insects belonging to the same order as bees and ants. They are variable in size and color but all have constricted "wasp waists." Among the kinds found in the vicinity of homes and porches are those commonly known as paper wasps, hornets or yellow jackets, and mud daubers. They may be considered beneficial in that most species are predators and capture other insects or spiders to feed their young. Adult wasps are capable of inflicting stings, temporarily painful, although some persons may develop severe allergic reactions which can be fatal. This is even more common with honeybees since more people are exposed to and stung by bees. Wasps may be attracted to garden pools or to picnic and barbecue areas where they can be a nuisance.

Paper wasps live in colonies, which may be newly formed each year by individual overwintering fertilized females. Nests are umbrella-shaped and consist of an unprotected single circular layer of cells open at the bottom and suspended by a stalk from the eaves of houses, trees, shrubbery or other protected areas. The nests are made of "paper" of fibers from dead plants or weathered wood, moistened and formed by the mouth parts of wasps.

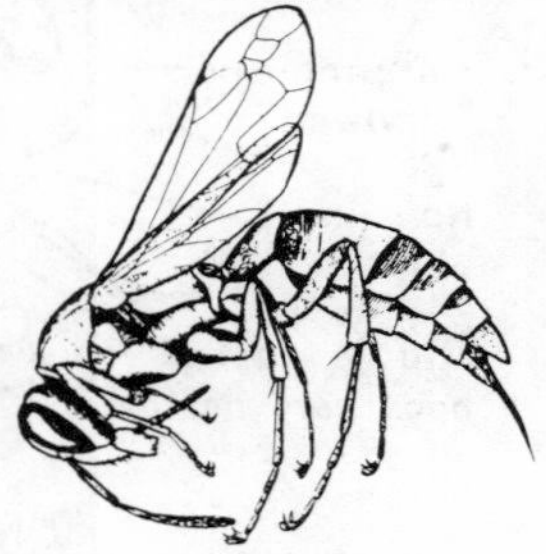

Wasp
(Univ. of Arizona)

Hornets or yellow jackets are mostly yellow and black and live in colonies within nests built above or below ground. Nests of hornets, unlike those of paper wasps, consist of several tiers or circular layers of cells. Nests above ground are enclosed within a grayish paper covering with an entrance on the underside, as seen so often in the comic strip, The Katzenjammer Kids.

Mud daubers construct individual cells of mud, usually in small groups side by side on walls of porches or carports. They stock these cells with paralyzed spiders which serve as food for the young.

Wasp control involves destruction of nests including the living contents of individual cells. Adult wasps may be killed by application of a fast-release, aerosol insecticide to the nest, preferably in the coolest portion of the night or at daybreak, to avoid stings. Foam-squirting aerosols are not satisfactory. Elimination of food scraps, including ripened or partially eaten fruit, from patio and barbecue areas will aid in reducing wasp numbers. A commercial, slow-acting bait is now available which foraging yellow jackets carry back to the nests where the developing young are poisoned.

The pain from wasp stings may often be eased by applications of ice packs to the affected areas. Persons who develop allergic reactions to stings should be rushed to a physician or hospital emergency room for treatment, and thereafter wear an emergency treatment information bracelet.

Waterbugs — See Cockroaches.

TABLE 18. Insect Control Suggestions for the Household With and Without Chemicals.[1]

Pest	Pesticide to Use	Non-Chemical Control
Ants	Chlorpyrifos,* Pyrethrins,* Diazinon,* Malathion, or Propoxur. Paint or spray door frames, cracks, baseboards and window sills. Dust in areas inaccessible by sprays.	Pour boiling water on nest. Eliminate food sources attracting ants. Follow good sanitation practices. Don't forget the leftover pet food.
Asiatic Oak Weevil	Diazinon or Lindane. Treat outside around foundation, basement windows, and other routes of entry.	Vacuum, sweep, or collect weevils by hand and destroy.
Assassin bugs	No chemical control.	Vacuum or sweep with broom and destroy. Do not handle.
Bark beetles	No chemical control.	Keep reserve firewood outside. Maintain tight door and window screens.
Bedbugs	Dichlorvos, Malathion or Pyrethrins. Apply to springs, slats and frames, and light treatment to mattress, tufts, and seams. Spray or paint baseboards, cracks and crevices.	Sun mattress, springs, bedding, slats and frame. Thorough cleaning will help but not eliminate. Old method was to rest bed legs in cans containing kerosene 1/4 inch deep.
Bees (honey)	Carbaryl, Pyrethrins, or Dichlorvos. Direct dusts or sprays into openings where bees have become established. Treat after dark when bees are calm and temperatures have dropped.	Check with County Extension Agent for name of beekeeper who will remove for a fee.
Booklice or Barklice	Propoxur or Pyrethrins. Treat damp areas where booklice have been seen.	Don't store cardboard boxes, books and papers in damp areas.

[1] Refer to Table 35 for trade or proprietary names of recommended pesticides.

Pest	Pesticide to Use	Non-Chemical Control
Boxelder bugs	Malathion or Diazinon. Spray directly on bugs. Best to spray bug clusters outdoors on tree trunks and other spots.	Vacuum or sweep with broom and destroy.
Carpenter bees	Carbaryl, Pyrethrins, or Dichlorvos. Direct spray or dust at and into wood galleries. Apply also to adjacent wooden areas and potential sites. Treat when dark and cool as for honey bees.	Dust diatomaceous earth generously into galleries as repellent. Fill holes and paint.
Carpet beetles	Propoxur, Diazinon or Malathion sprays; Napthalene or Paradichlorobenzene crystals. Spot treat rugs, carpets, and baseboards lightly in infested areas. Use 1 pound of either crystals per 100 cu. ft. of closet or storage area. PDB crystals preferred because odor is more readily removed by airing clothing.	Store only cleaned or well-sunned clothing. Dry cleaning kills carpet beetles (and moths). Avoid accumulations of dust and lint in corners, along moldings, in hot and cold air ducts. Vacuum such areas thoroughly, and frequently. Remove and destroy disposable vacuum bags.
Centipedes	Propoxur, Pyrethrins, Chlorpyrifos or Malathion. Apply directly to pests. Treat dark, moist areas in basement and garage.	Centipedes breed outside and may become pests indoors. Remove decaying grass and leaves from around house foundation. Foliage-free area around foundation makes ideal barrier.
Click beetles	No chemical control.	Maintain tight door and window screens. Beetles are attracted to white lights. Substitute yellow, non-attracting light bulbs.
Clothes moths	Dichlorvos or Malathion sprays; napthalene or PDB crystals. Follow same general procedures as described under carpet beetles. Store only cleaned clothing. Clothing may be protected by using a Dichlorvos aerosol. A No-Pest Strip hung in the closet also offers good protection.	Follow same procedures as given under carpet beetles. Cedar-lined closets and chests are effective in repelling moths. Remove clothing and expose to hot sunlight for 2 days.
Clover mites	Dicofol, Malathion, Propoxur or Pyrethrins. Apply Malathion or Pyrethrins onto mites in cracks and other areas where they hide. For outdoors, spray house siding up to windowsills and treat lawn out from house foundations 8-10 feet.	Establish a bare soil barrier about 2 ft. wide around foundation to discourage migration into dwelling.

Pest	Pesticide to Use	Non-Chemical Control
Cluster flies or Attic flies	Propoxur, Chlorpyrifos, Pyrethrins or Dichlorvos. Release Dichlorvos aerosol in tight enclosure, or hang No-Pest Strip in similar infested areas. Surfaces treated with Propoxur or Chlorpyrifos have fairly good residual activity.	Sticky fly ribbons and the old fly swatter are about your only non-insecticidal defense.
Cockroaches	Chlorpyrifos, Pyrethrins, Diazinon, Malathion, or Propoxur. Apply to baseboards, beneath sinks, cabinets, under and behind stove and refrigerator, openings where pipes enter walls and floors, around drains, dark-warm protected areas in basements, showers and drains, and roach runways under sinks and lavatories and floor cracks. (See Slow-Release Pesticides) Boric acid powder or tablets is also effective.	Sanitation, removal of food sources, is the key approach. Roach traps may be prepared by setting quart or pint jars containing beer or diluted (10:1) molasses and water in which roaches drown. Sticky traps are now available but only moderately effective.
Crickets	Propoxur, Chlorpyrifos, Diazinon or Malathion. Treat baseboards, in closets, under stairways, around fireplaces, in basements, and ground level floors, if needed.	Use non-attractive yellow light bulbs outside and roach sticky-traps inside.
Drain flies or Moth flies	Propoxur, Chlorpyrifos, Malathion, or Pyrethrins. Use Pyrethrins aerosol against adult flies. Clean drains thoroughly and apply liquid or crystalline drain cleaner to kill immature stages.	Sanitation, removal of breeding site, is key approach. Clean gelatinous film from drains. Pour large quantity of boiling water down drain and through overflow. Clean garbage containers regularly. Don't let wet lint accumulate under laundry machine, especially in basements.
Earwigs	Diazinon, Pyrethrins, Chlorpyrifos or Malathion. Apply spot treatment to cracks and crevices and directly onto earwigs.	Remove with broom and dustpan or vacuum. Eliminate dead vegetation, leaves, and excess mulch from around outside of foundation. Trap out of doors under boards or folded newspaper or dark plastic and destroy.
Elm leaf beetle	No chemical control.	Use vacuum, broom and dustpan, or remove by hand if they enter the home.
Firebrats	Propoxur, Pyrethrins, Chlorpyrifos, Dichlorvos or Malathion. Spray baseboards, window and door casings, cracks and openings in floors and walls for pipe and wire entrances.	Keep papers, cardboard boxes, and books in dry storage. Reduce humidity of problem area.

Pest	Pesticide to Use	Non-Chemical Control
Fleas	Diazinon, Rotenone, Propoxur, Dichlorvos or Malathion. Dust or spray animal's sleeping quarters and replace old bedding. Apply spray to floors, baseboards, and to walls to a 1-ft height in rooms where fleas are a problem. Treat dog or cat with Carbaryl or Rotenone dust as well as their outdoor resting areas. See Fleas under "Lawn and Turf", Table 15. Use flea collar on pet.	Keep pets outside at all times. Vacuum areas daily where fleas are a problem. You may find these methods rather ineffective.
Flies and Gnats	Diazinon, Propoxur, Pyrethrin, Malathion or Chlorpyrifos. Use Dichlorvos or Pyrethrin aerosol when flies are present. No-Pest Strip is an ideal choice for pesky flies. Outside, treat surfaces where flies rest, walls, around doors and windows.	Tight screens and good sanitation are the keys. Use ribbon flypapers. Empty garbage twice weekly, rinse and invert cans. Clean up pet dung daily from yard and place in plastic bag.
Flour, grain or cheese mite	Pyrethrins. Thoroughly clean, then treat infested shelves. Cover shelves with new shelf paper after spray dries.	If you are squeamish, discard infested foods and keep uncontaminated food in canisters, or coffee cans with plastic lids. Otherwise place dry materials in open pan in oven for 30 minutes at 130°F, or freeze for 3-4 days, and utilize in normal cooking procedures.
Ground beetles	No chemical control. Harmless incidental invaders.	Remove by hand, broom and dust pan or vacuum cleaner.
Hackberry psyllids	Pyrethrins. Adults can be killed with Pyrethrins aerosol. No-Pest Strips (Dichlorvos) work equally well.	Maintain tight door and window screens. Hackberry trees are the source, thus if persistent problem, their removal should eliminate pests.
Hornets	Pyrethrins, Carbaryl, Chlorpyrifos, Diazinon, Dichlorvos, or Propoxur. Apply spray directly to nests during coolest part of night. Repeat next night if survivors appear. Fast-release aerosols are best.	Wait until winter to remove hornet's nest. Otherwise there is no safe method of removal without the help of insecticides.
Horntail (pigeon tremex)	No chemical control.	Sweep up with broom and dust pan or vacuum, then destroy. They emerge from firewood, hardwood floors, studs, and subflooring, and do not reinfest wood in the home.

Pest	Pesticide to Use	Non-Chemical Control
Kissing bugs (see Assassin bugs)	Propoxur or Diazinon. Spray outside foundation perimeter and 3 ft. of adjacent soil. Brush or spray around window and door casings.	Maintain tight screens and weather stripping. Use yellow bug lights outside. Caulk cracks in walls, foundation, and around windows and doors. Remove associated rodent nests where they live.
Larder beetle	Pyrethrins. Treat shelving with oil spray or aerosol after removing dishes, utensils and food.	Discard infested products. Wrap and store cheese and meat products at normal refrigerator temperature. temperature.
Lice (head, body and crab)	Lindane (Kwell) 1% shampoo, lotion or cream. DDT (Topocide) 1% lotion, Pyrethrins + synergist (A-200 Pyrinate). Most lousicides are sold on a prescription basis by physicians. A-200 Pyrinate is sold over the counter, but requires several applications since it does not kill the eggs (nits).	For body lice only: clean underwear and outer clothing is important; laundering kills lice in 5 minutes, eggs in 10 at 125°F. water temperature. For head lice: Do not use someone else's hats, wigs, scarfs, combs or brushes.
Long-horned beetles	No chemical control.	Remove occasional beetles by hand, broom or vacuum. They usually emerge from firewood but will not infest wood in the home. Store firewood outdoors.
Millipedes	Carbaryl, Propoxur or Diazinon. Treat areas outdoors where millipedes may be present or enter home.	Millipedes are not poisonous. Remove by hand, broom or vacuum. Around outside of foundation remove excessive mulch.
Mosquitoes	(Indoors): Pyrethrins aerosol, Dichlorvos spray or No-Pest Strips. Spray according to label instructions. Strips can be hung in areas designated on label. (Outdoors): Spray yards and picnic areas, tall grass and shrubbery with Carbaryl, Pyrethrins, Dichlorvos or Methoxychlor. Spray entry sites (screens, doors and patios). Personal repellents: Deet or Rutgers 6-12.	Maintain tight screens and weather stripping. Use yellow non-attractive light bulbs at entrances. Remove or empty frequently any containers from premises that may hold rainwater (tires, cans). Clean out clogged roof gutters holding stagnant water. Add light-weight oil to surfaces of ponds, ditches, and even animal hoof holes in mud where mosquitoes may breed. Community effort may be required.

Pest	Pesticide to Use	Non-Chemical Control
Pantry pests (Includes Angoumois grain moth, bean weevil, cigarette beetle, confused flour beetle, drug store beetle, foreign grain beetle, granary weevil, Indian meal moth, mealworm, Mediterranean flour moth, rice weevil, sawtoothed grain beetle, and spider beetle).	Malathion, Propoxur, or Pyrethrins. Empty and clean kitchen shelving, then spray shelves lightly. Direct spray into cracks and crevices. Cover shelves with new shelf paper after spray dries.	If you are squeamish, discard infested products. If not, salvage them by heating in the oven for 30 minutes at 130°F, or placing them in the freeze-compartment at 0°F. for 3-4 days. Store such foods in canisters with tight lids. See Pantry and Stored Product Pests narrative for carbon dioxide fumigation of larger quantities of stored grain.
Powder post beetles	Dichlorvos. Spray, paint or dip with dilute solution to saturate the wood where appropriate.	Previously painted or otherwise finished wood surfaces usually remain free from infestation.
Silverfish	Propoxur, Pyrethrins, Diazinon, Chlorpyrifos or Malathion. Spray or dust baseboards, door and window casings, closets, cracks, and openings for pipes and wires to pass through walls and floors.	Store books and paper products in dry areas to avoid favorable habitat.
Sowbugs and Pillbugs	Propoxur or Chlorpyrifos. Apply around shrubbery, plants, foundation wall, window wells, and other routes of entry. Spray along doorways and basement windows.	Remove mulch, boards, and excess ground cover. Caulk cracks in foundation. Trap out of doors with boards, folded newspapers or dark plastic and destroy.
Spiders	Propoxur, Pyrethrins, Diazinon, Chlorpyrifos, or Malathion. Treat indoors with coarse droplet spray, hitting webs and probable hiding places. Outdoors, spray or dust around house foundation to reduce migration inside.	Most spiders can be kept out of the home by tight screens, weather stripping, and caulking. Most spiders are harmless and are generally beneficial. Exceptions are the black widow, brown recluse, and Arizona brown spider.
Springtails (Collembola)	Diazinon, Chlorpyrifos or Malathion. Control outdoors only by treating around foundation, especially moist areas.	Remove mulch and eliminate low moist areas around the foundation.

Pest	Pesticide to Use	Non-Chemical Control
Stonefly	No chemical control.	Remove individuals by hand, broom or vacuum. They accumulate in windows in early spring, not far from the water where they breed.
Termites (Subterranean)	Chlordane, or Dieldrin. Soil adjacent to the structural foundation must be soaked with 1% chlordane or 0.5% dieldrin preferably to the base of the footing. Dig a "V" shaped trench adjacent to the foundation 1 foot deep and no more than 6 inches wide. Apply one of the above insecticide solutions in the trench at the rate of 2 gallons per 5 linear feet of foundation. Treat the fill soil after refilling the trench. In some instances, the concrete floors, patios, and walks must be drilled at 1-foot intervals and injected with the insecticide solution under pressure, or by use of a long hollow rod inserted beneath slabs. Great care must be taken not to damage heat conduits, pipes and vapor barriers located in or beneath the slab.	Termites have not been successfully controlled with any methods other than chemical control. Metal plates installed during construction are not effective. Pre-treating the soil before slab construction is poured with one of the 2 chlorinated insecticides by a professional is the most effective and economical method of termite control — prevention. See Termites in preceeding narrative.
Termites (Drywood)	This calls for a professional fumigation job. There is nothing the homeowner can do other than call in a reliable pest control operator.	As with subterranean termites, there is no control without chemicals.
Ticks	Carbaryl, Chlorpyrifos, Diazinon, or Propoxur. Use only Carbaryl dust on dogs or cats. In the home use Diazinon or Propoxur spray. Treat pet's sleeping areas. Replace old bedding with clean, untreated bedding. "Outdoors use Diazinon 5%G or Carbaryl in pet areas." Personal repellents are Deet and Rutgers 612.	Examine animals' heads around ears and neck daily for engorged ticks. Apply fingernail polish remover or petroleum jelly to tick an hour before removing with tweezers.
Wasps	Pyrethrins, Carbaryl, Diazinon, Chlorpyrifos, Propoxur, or Dichlorvos. Apply directly to nests during coolest part of night. Repeat next night if survivors appear. Remove nests when no survivors remain. Or treat with new fast-release aerosol that gives immediate knockdown.	Wait until winter to remove nests. Otherwise, there is no safe method of removal without the help of insecticides.
Weevils	No chemical control.	Occasionally various weevils find their way into homes. These can be readily removed by hand or vacuum cleaner.

HOUSE PLANT PESTS

House plants may become infested with a number of insects and related pests, mostly feeders on plant juices. Aphids feed on stems, leaves and buds. Scale insects do not move about except when young, and are recognized by their shell-like coverings, usually measuring from 1/16 to 1/8 inch in length depending on species. Whiteflies are 1/6 inch long and wedge shaped, with powdery white wings. When disturbed they resemble clouds of small snowflakes. Mealybugs are soft-bodied, waxy-white, wingless insects that may secrete honeydew which, as with aphids, may support growths of sooty mold. Spider mites are barely visible to the eye and are most commonly seen on the under sides of leaves, where they may form silk-like webbing. Fungus gnats and springtails may develop in the rich soil used for house plants. Sowbugs and pillbugs may be found in moist areas beneath potted plants. Other outdoor plant pests may also be found on house plants.

Controls for Pests of House Plants

Several methods can be used to control insects or related pests of house plants (Table 19), some utilizing chemicals, some consisting of non-chemical alternatives. What the homeowner should use varies with the pest, the number of plants involved, the size of the infestation, and the personal inclination of the indoor gardener.

The simplest and easiest method, of course is to prevent the infestation from spreading from new plants to others. Examine cut flowers and new plants brought into the home to be certain they are free of all pests. New plants should be isolated for at least two weeks before placing them with other plants. During this interval you can observe the new plants and spot any infestations that develop.

Sterilized soil should be used for potting to prevent the development of infestations of soil pests such as springtails, psocids, and pillbugs or sowbugs.

Washing with soapy water and a soft brush or cloth may be all that is needed to remove aphids, mealybugs, and scale insects from broad-leaved plants. The wash should be made using 2 teaspoons of a gentle, biodegradable detergent to a gallon of water.

When concerned with only one or two plants, you can control aphids and mealybugs by removing them with tweezers or a toothpick. Caterpillars may be removed by hand and destroyed. Cutworms, snails, and slugs may be found in their hiding places during the daytime and destroyed, or picked from the plants at night as they feed.

Another easy way to control a light infestation of aphids or mealybugs on one or two plants is to wet and remove the insects with a swab that has been dipped in alcohol or kerosene. Use a swab made from a toothpick and a tuft of cotton or the commercial tufted swabs for that gentle touch.

Several aerosol insecticides have been registered by the EPA for use on plants within the home. They often consist of several active ingredients in combination, and are usually specially formulated to kill most plant pests.

When purchasing an aerosol for plant pest control, read the label on the can carefully to make certain that the spray is safe for plants. Some aerosols are designed as space sprays for flying and crawling insects in the home, and contain oils or other materials that will burn or kill foliage.

Aerosol sprays for plants contain small quantities of pyrethrins and other agents. They may be used to kill pests that can be hit directly with the spray, such as aphids and whitefly adults on plants, or whiteflies and fungus gnats swarming near the plants. Do not spray too close to the plants, for even the best designed aerosols will burn foliage hit with heavy spray concentrations.

The Pests

There are several pests of houseplants that occur and are seen in gardens, but usually do not capture the attention as they do indoors. Similarly they frequently appear as pests in the home greenhouse because of favorable environmental conditions such as temperature and humidity, and because of a lack of natural control agents that occur out-of-doors. The pests described here do not make up the complete list. Those not described can be found in the garden pest section.

Cyclamen Mites

Cyclamen mites, even when adults, are too small to be seen with the naked eye. They can be viewed under a magnifying glass as oval, amber or tan-colored, glistening, semitransparent mites. The young are even smaller and milky white, while the eggs are oval and pearly white.

These mites are found mostly in protected places on young tender leaves, young stem ends, buds, and flowers. They crawl from plant to plant where leaves touch; another means of spreading is the transfer of mites on hands or clothing while working and moving among them.

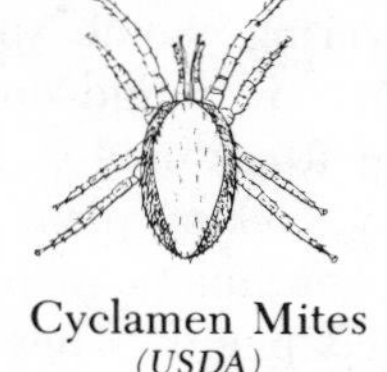

Cyclamen Mites
(USDA)

Damaged leaves of infested plants are twisted, curled, and brittle. Buds may be deformed and fail to open. Flowers are deformed and often streaked with darker color. Blackening of injured leaves, buds, and flowers is common.

Infested ivy will produce stems without leaves or with small deformed leaves. Infested African violets develop small, twisted, hairy leaves that may soon die.

To control, trim off badly injured plant parts where practicable. They can be "cooked" off infested plants by immersing infested plants, pot and all, for 15 minutes in water held at 110°F. The success of this method of treatment depends on careful control of the water temperature. In home greenhouses, control can be achieved by making 2 or 3 spray applications of an effective miticide, such as dicofol, at 10 day intervals.

False Spider Mites

Several species of false spider mites can infest plants in the home or greenhouse. They are flat, oval, dark-red mites too small to be easily seen with the naked eye. The young and eggs are bright red. All stages of these mites are found mostly on the undersides of leaves, generally along the veins of other irregularities on the leaves.

The damage caused by feeding of these mites results in finely stippled bronze or rusty-brown areas along veins or on entire leaves. The edges of infested leaves may die, or the leaves lose color and drop off. Infested plants become weakened and should be given additional attention to water and nutrition.

In the home or greenhouse mites can be controlled by making 2 or 3 applications or plant dippings with a recommended miticide, such as dicofol, at 10-day intervals.

Fungus Gnats

Fungus gnats are delicate, gray or dark gray, fly-like insects about ⅛th inch long. They are true flies, have only one pair of wings. They are attracted to light and when present in the home swarm around windows. The immature forms are white maggots that live in soil, and reach a length of about ¼ inch. These maggots are usually found in soils with surplus quantities of decaying vegetable or organic matter.

The gnats are only a nuisance, indicating the presence of maggots, likely in potted plants. The maggots cause injury to the root systems by burrowing in the soil. They may feed on the roots and crowns of plants. Severely injured plants grow poorly, appear off color, and may drop leaves.

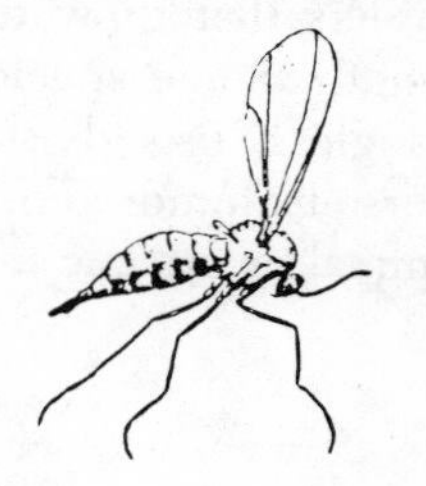

Fungus Gnats
(U.S. Public Health Service)

Maggot control can be achieved by avoiding overwatering of plants in the home and greenhouse. Soil drenches of an insecticide, such as malathion, will produce immediate control.

Mealybugs

There are several kinds or species of mealybugs that become pests on house and greenhouse plants. They are softbodied, and appear as though covered with dust or flour because of their waxy coats. They grow to be about 3/16ths inch long. Some mealybugs have waxy strands or filaments extending from the rear of their bodies. They are found at rest or crawling slowly on stems, where stems and leaves join, and on leaves, especially along veins on undersurfaces. Their eggs are laid in clusters enclosed in white waxy cottony or fuzzy material. Mealybugs are sometimes cared for by ants. The ground mealybug, a soil inhabitant, feeds on the roots of African violets and other plants.

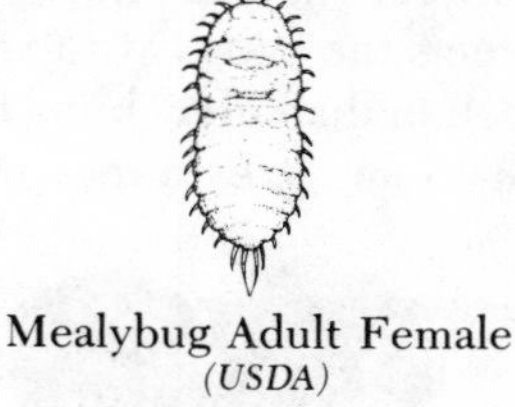

Mealybug Adult Female
(USDA)

Damage by mealybugs is caused by their sucking out plant juices, resulting in stunting or death of plants. Sooty mold grows on the honeydew excreted by some species of mealybugs. The ground mealybug damages the rootlets, causing the plants to grow slowly and wilt between waterings.

When only one or a few plants are infested, mealybugs can be controlled by washing, by hand-picking, or by using an alcohol-dipped swab to remove them. Treated plants should be isolated to avoid reinfestation.

Book Lice or Psocids

Psocids are softbodied, pale yellow to gray, oval insects that grow to about 1/32nd to 1/16th inch in length. Some species have wings while others are wingless. Psocids sometimes cluster in large numbers of a hundred or more. They feed on moist dead animal or organic matter, lichens, and fungi.

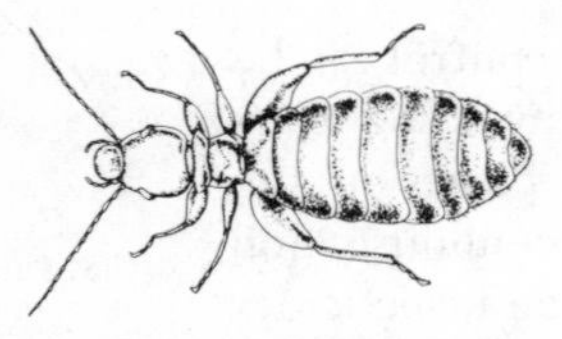

Booklouse or Psocid
(U.S. Public Health Service)

They may occur in large numbers in the soil or on pots and benches, especially in undisturbed locations in the home or greenhouse. Tiny, fast-moving psocids may sometimes be found on old books and papers stored in damp places.

Generally, psocids are only a nuisance, for though found on living plants, they are not known to feed on them. And, because they are only a nuisance, control measures are unnecessary.

Scales

There are many species of scales that commonly infest several species of plants in homes or greenhouses. Scale insects have a shell-like covering, or scale, that protects the body. Most species are about 1/16th to 1/8th inch in diameter, but a few species may span 3/8th inch. Some are round, others oval, and

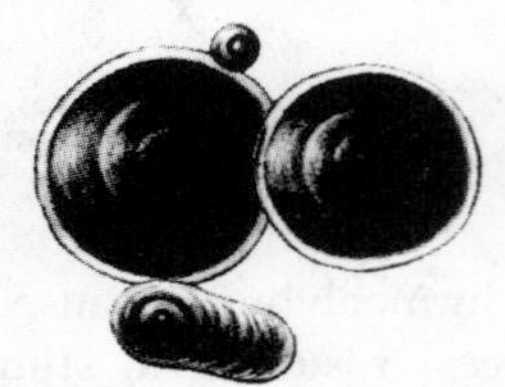

San Jose Scale
(Union Carbide)

some are shaped like oystershells. Color ranges from white to black, however the predominant species are browns and grays.

Some scales lay their eggs in white sacs secreted from beneath, and these can be mistaken for mealybugs if not examined closely for the presence of the tiny shell-like covering. Some species infest only plant leaves, others are found on both stems and leaves, while some attack only stems.

Scales feed by sucking plant juices with their piercing-sucking mouthparts, resulting in reduced growth and stunted plants. These insects excrete honeydew as do their cousins, the aphids, which is attractive to ants. The honeydew falls on lower leaves imparting a shiny or wet appearance to the foliage and provides a source of nutrition for the growth of sooty mold.

Control can be achieved by washing with soapy water if only a few plants are involved. However, control is difficult with insecticides, requiring repeated applications. It may be best to discard heavily infested plants and purchase non-infested replacements.

Whiteflies

There are several whitefly species that can be bothersome on house plants. The adults are about 1/16th inch long, have white, wedge-shaped wings and do resemble white flies. They are not flies, however, but are related to the aphids and scales. When infested plants are moved, the adults take flight, resembling small snowflakes or bits of paper ash swirling in the air, seemingly with no sense of direction.

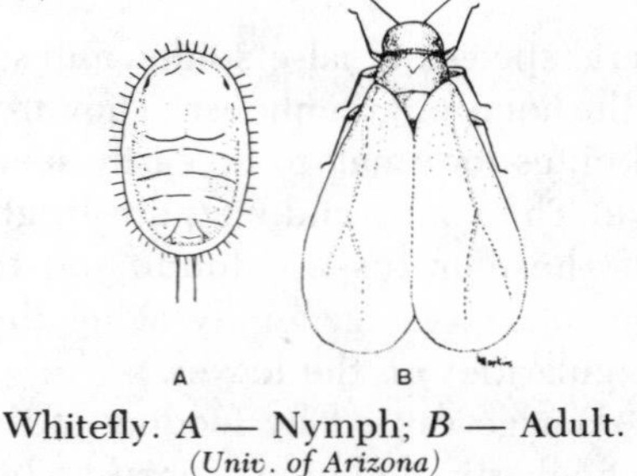

Whitefly. *A* — Nymph; *B* — Adult.
(Univ. of Arizona)

Immature whiteflies resemble scales and are mostly pale green to yellow or whitish, oval shaped, and flattened. Except for the newly-hatched crawlers, the immature stages are attached to the leaves, mostly on the undersides.

Plant damage is caused by both adults and young as they suck juices from the leaves. Infested leaves become pale, then yellow, and die or drop off. Surfaces of leaves may be covered with their sticky honeydew, which usually becomes blackened with sooty mold.

TABLE 19. Insect and Mite Control Suggestions for Houseplants, With and Without Chemicals.

Plant	Pest	Pesticide to Use[4]	Non-Chemical Control[3]
African violet	Mealybug	Malathion[1], Diazinon, or Nicotine Sulfate. Spray or dip[2] with WP formulation.	See Footnotes. Pick off by hand.
	Mites	Dicofol or Nicotine Sulfate. Spray or dip[2].	
Aralia	Scale insects	Malathion[1] or Nicotine Sulfate. Dip[2] or spray.	
	Mites	Dicofol or Nicotine Sulfate. Spray or Dip[2].	
Begonia	Aphids	Malathion[1] or Nicotine Sulfate. Spray or dip[2].	Remove by hand or with artist's brush.
	Mealybug	Malathion[1], Diazinon, or Nicotine Sulfate. Spray or dip[2].	Remove by hand or with artist's brush.
	Whitefly	Malathion[1], Nicotine Sulfate or Diazinon. Spray or Dip[2].	
	Mites	Dicofol or Tetradifon. Apply as needed.	
Citrus	Scales	Malathion[1] or Nicotine Sulfate. Dip[2] or use artist's brush to apply to scales.	Remove by hand.
	Whitefly	Malathion[1] or Dimethoate. Spray or dip[2].	
Coleus	Mealybug	Malathion[1], Diazinon, or Nicotine Sulfate. Spray or dip with WP formulation.	Remove by hand or with artist's brush.
	Whitefly	Malathion[1], Nicotine Sulfate or Diazinon. Spray or dip[2].	
Cyclamen	Mites	Dicofol. Spray or dip[2].	
Dracaena	Mealybug	Malathion[1], Diazinon, or Nicotine Sulfate. Spray or dip with WP formulation.	Remove by hand or with artist's brush.
Fuchsia	Mealybug	Malathion[1], Diazinon, or Nicotine Sulfate. Spray or dip[2] with WP formulation.	Remove by hand or with artist's brush.
	Whitefly	Malathion[1], Nicotine Sulfate or Diazinon. Spray or dip[2].	

[1] Use household formulation to avoid objectionable odor.
[2] Dip foliage and stem into solution equal to one-half the spray concentration.
[3] For a few plants with light infestations, most pests can be removed by spraying over the sink with a strong stream of water, or by dipping infested parts in soapy water, or by touching the insect with an artist's brush or cotton swab dipped in rubbing alcohol or kerosene. Avoid application of kerosene to plants.

CAUTION: **Nicotine and nicotine sulfate are highly toxic to man and his pets and should be used indoors with great care.**

[4] Refer to Table 35 for trade or proprietary names of recommended pesticides.

Plant	Pest	Pesticide to Use[4]	Non-Chemical Control[3]
Gardenia	Mealybug	Malathion[1], Diazinon, or Nicotine Sulfate. Spray or dip[2] with WP formulation.	Remove by hand or with artist's brush.
	Whitefly	Malathion[1], Nicotine Sulfate, or Diazinon. Spray or dip[2].	Remove by hand or with artist's brush.
Geranium	Whitefly	Malathion[1], Nicotine Sulfate or Diazinon. Spray or dip[2].	
	Mites	Dicofol. Spray or dip[2].	
Gloxinia	Aphids	Malathion[1] or Nicotine Sulfate. Spray or dip[2].	Remove by hand or with artist's brush.
	Mealybug	Malathion[1], Diazinon, or Nicotine Sulfate. Spray or dip[2] with WP formulation.	Remove by hand or with artist's brush.
	Mites	Dicofol or Pyrethrins. Spray or dip[2].	
Ivy (Boston and English)	Mites	Malathion[1], Pyrethrins or Dicofol. Spray or dip[2].	
Palms	Scale	Malathion[1] or Nicotine Sulfate. Dip[2] or use artist's brush to apply to scales.	Remove by hand.
	Mealybug	Malathion[1], Diazinon, or Nicotine Sulfate. Spray or dip[2] with WP formulation.	Remove by hand or with artist's brush.
	Mites	Dicofol or Pyrethrins. Spray or dip[2].	
Philodendron	Scale	Malathion[1] or Nicotine Sulfate. Dip[2] or spray. For small plants use artist's brush to apply to scales.	Remove by hand.
Pittosporum	Mites	Dicofol or Pyrethrins. Spray or dip[2].	
Podocarpus	Mites	Dicofol or Pyrethrins. Spray or dip[2].	
Roses (miniature)	Mites	Dicofol or Pyrethrins. Spray or dip[2].	
Rubber Plant	Mealybug	Malathion[1], Diazinon, or Nicotine Sulfate. Spray or dip[2] with WP formulation.	Remove by hand or with artist's brush.
	Scale insects	Malathion[1] or Nicotine Sulfate. Dip[2] or spray. For small plants use artist's brush to apply to scales.	Remove by hand.
Schefflera	Scale insects	Malathion[1] or Nicotine Sulfate. Dip[2] or spray. For small plants use artist's brush to apply to scales.	Remove by hand.
	Mites	Dicofol or pyrethrins. Spray or dip[2].	

Plant	Pest	Pesticide to Use[4]	Non-Chemical Control[3]
Miscellaneous Plant Pests	Fungus Gnats (midges)	Propoxur, Pyrethrins, Carbaryl, or Diazinon. Apply lightly to moist soil in pots. Hang No-Pest Strip in areas where gnats appear.	Maintain dryer flower pot soils.
	Pillbugs and Sowbugs	Propoxur, Carbaryl or Diazinon. Apply lightly to moist soil in pots and plants.	Reduce soil moisture in pots and planters. Remove by hand and destroy.
	Springtails	Malathion[1], Pyrethrins or Diazinon. Apply lightly to moist soil in pots and planters.	Take pots outside and expose several hours to warm, drying sun.

[1] Use household formulation to avoid objectionable odor.
[2] Dip foliage and stem into solution equal to one-half the spray concentration.
[3] For a few plants with light infestations, most pests can be removed by spraying over the sink with a strong stream of water, or by dipping infested parts in soapy water, or by touching the insect with an artist's brush or cotton swab dipped in rubbing alcohol or kerosene. Avoid application of kerosene to plants.
CAUTION: **Nicotine and nicotine sulfate are highly toxic to man and his pets and should be used indoors with great care.**
[4] Refer to Table 35 for trade or proprietary names of recommended pesticides.

PETS — Pest Control on Domestic Pets

How important is your dog or cat to you? What would you take for it? Pets are highly treasured objects of affection in the American home. Can you believe that more is spent annually for pet food than for baby food in the U.S.? Often dogs, cats, and birds as pets become so popular that they are practically considered members of the family. Because of this great recreational and companionship importance in our American way of life, it is essential that the health of these pets be protected.

Unfortunately, dogs, cats, birds, and other warm-blooded pets are subject to attack by fleas, ticks, lice, flies, mites and other annoying if not harmful pests. Their attack may persist throughout the year. Infested dogs and cats scratch to relieve irritation, often rubbing off patches of their coat around the neck, shoulders, and abdomen. This condition may worsen resulting in loss of appetite, weakness and increased susceptibility to disease. Birds become nervous, lose their appetite and appear old and ruffled.

With the proper selection, use and timing of insecticides, it is possible to greatly reduce if not totally eliminate these troublesome insect and mite attacks. The suggestions for control in Table 20 are up-to-date, effective, safe for the pet and his owner, and registered by the Environmental Protection Agency for use on pets at the time of publication. Nevertheless, it is possible that some of the chemicals recommended could lose their label status. **Consequently, it is the responsibility of those using the pesticides to determine the label status before use.**

Label precautions and directions are not guess-work. They are based on scientific data submitted by the manufacturer and reviewed by the Environmental Protection Agency as part of the product's registration procedures. **When controlling pests on your pets, treat only the pets listed on the pesticide label. Do not use any pesticides labeled for crops, ornamentals, or livestock on pets unless the label so states; use no more of the material than the recommended dosage and only at the time recommended; and, be especially cautious when treating young pets.**

Pests not described in the following section may be found elsewhere. Check the index.

Horse Bots

Horse bot flies have four stages: adult, egg, larva (bot) and pupa. The adults do not feed, since their sole purpose in life is to reproduce. They live at the most only a few weeks. There are 3 species, but their damage and life cycles are essentially the same. The eggs are attached by the adult female to hairs on the legs or near the mouth. By the licking of eggs, or by normal hatching near the face, the larvae reach the mouth where they spend a period. After slight development they move on into the horse's stomach where they attach with their mouthparts to remain for about 10 to 11 months. After reaching full development they release and pass to the soil through the feces, where they pupate. In the spring or early summer the adults emerge to mate and begin the 1-generation-per-year life cycle again.

Part of a horse's stomach heavily infested with bots. Note the lesions that these bots have caused. *(USDA)*

Damage by bot flies may be indirect, in that animals under attack may inflict damage on themselves or on anyone trying to handle them. Fright and irritation caused by egg-laying adults or newly hatched bots may result in animals acting out of control. Direct damage is produced by larvae feeding on the tissues of the mouth and stomach. Infested animals often suffer from colic or other gastric disturbances. The degree of damage done by the feeding of the bots is proportional to the number present. Several hundred larvae are commonly found in one animal.

Horse Bot Adult
(USDA)

Bot control in horses should consist of prevention of infection, as well as treatment of infections with effective insecticides. Infections of bots can be prevented during the fly season by removing bot fly eggs from hairs by clipping, or by applying warm water rinses (120°F.) to induce hatching of the eggs and subsequent death of the young bots. Chemical control should be under the advice of a veterinarian.

Horse Flies and Deer Flies

Horse flies and deer flies belong to the same group of robust, blood-sucking flies. They vary in size from ¼ inch to 1 inch in length. They are strong fliers, notorious pests of horses, cattle and many other warm-blooded animals. Only the females bite. The males feed on vegetable materials and cannot bite. They breed only in aquatic or semiaquatic habitats. The eggs are laid during the warm months in clusters on objects over water or marshy situations favorable for development of the larvae. Usually there is only one generation a year, overwintering as larvae in the wet soil. It completes its development in the early spring and emerges as the adult a few weeks later. The females begin seeking blood and the males feed on flowers and vegetable juices.

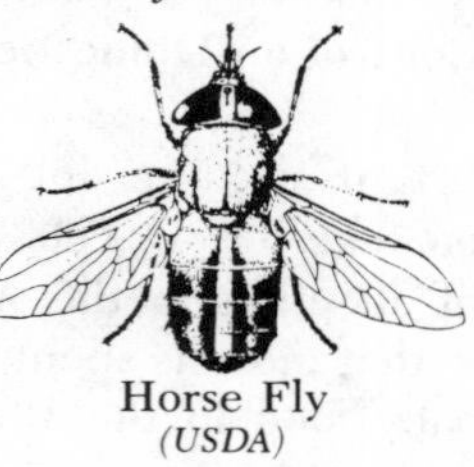

Horse Fly
(USDA)

Their bites cause serious annoyance and significant blood loss to domestic animals, particularly horses and cattle. Control of these biting flies can be very difficult because their developmental sites may include marshy or aquatic sites, as well as relatively dry soils. Thus, their control is achieved by regular application of an approved insecticide to the affected animals.

Lice

Chewing Lice

Biting or chewing lice feed on skin scales, skin exudations and other matter on the skin, and gnaw at the living epidermis. They are not blood feeders as are the sucking lice, but occasionally the irritation and scratching by the host results in bleeding.

Canaries, parakeets and other caged birds are frequently infested with a chewing louse, usually the red louse. The adults are about 1/16 inch long and reddish brown in color. Parts of the feathers, particularly the barbs and barbules, constitute a major part of the food of this pest. The irritation from the feeding of the louse causes the host to become quite restless, thus affecting its feeding habits and digestion. Young birds are particularly vulnerable. Bird lice tend to be more abundant where uncleanliness and overcrowded conditions exist. Control is best achieved with a dust, though aerosols are quite effective.

Chewing Lice

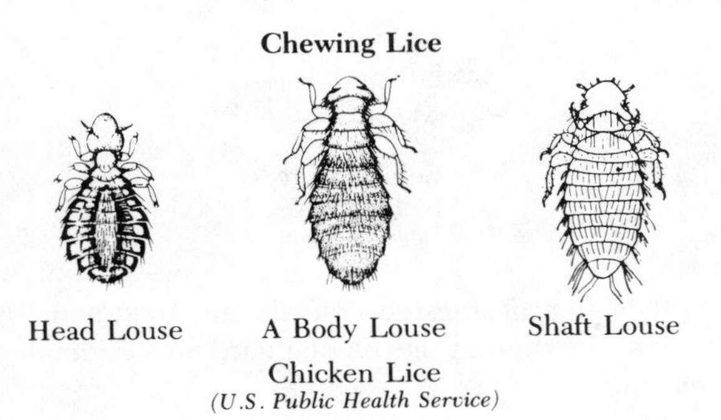

Head Louse — A Body Louse — Shaft Louse

Chicken Lice
(U.S. Public Health Service)

Dogs and cats have their very own special variety of biting or chewing louse. Dogs, particularly puppies, may suffer much irritation from the dog-biting louse, while cats, both kittens and adults, may become heavily infested with the cat louse.

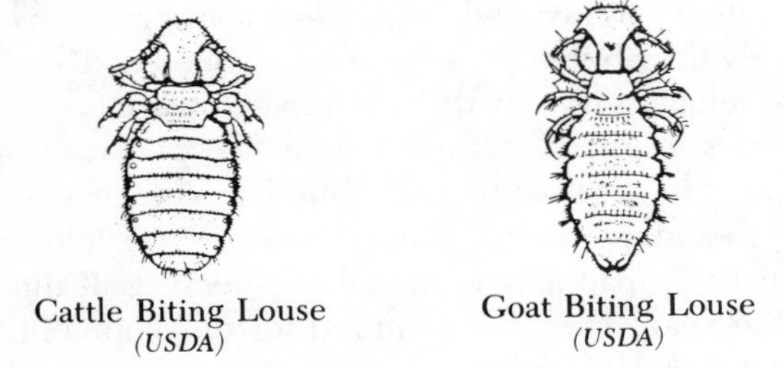

Cattle Biting Louse *(USDA)* — Goat Biting Louse *(USDA)*

Other domestic animals, cattle, horses, sheep, and goats, also are occasionally plagued with biting lice. Cattle are often heavily infested on the withers, root of tail, neck and shoulders by the cattle-biting louse. Horses, mules and donkeys, but horses more particularly, may suffer from the horse-biting louse when poorly or irregularly groomed. Sheep at times may show severe infestation of the sheep-biting louse, and goats are commonly infested with several species of chewing lice.

Spread from one animal to another is commonly by direct contact or by introduction of uninfested animals into quarters that have recently been vacated by infested ones. Control is dependent on sanitation practices, avoiding the introduction of infested animals into those that are uninfested, and the appropriate use of recommended insecticides.

Sucking Lice

The sucking lice are totally dependent on blood meals for survival, and nearly all domestic animals have their own variety of sucking louse, including humans (see index). All lice are wingless, have flattened bodies, and their legs are adapted for clinging to hairs and feathers. The young resemble the adults and have the same feeding habits and nutritional requirements. The entire life cycle is spent on the host. Eggs are attached to hairs and repeat generations occur throughout the year.

Sucking Lice

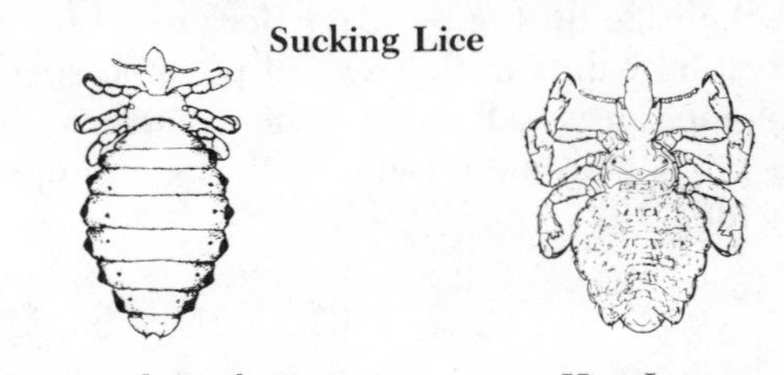

Short-nosed Cattle Louse *(U.S. Public Health Service)* — Hog Louse *(U.S. Public Health Service)*

Significant weight losses and anemia may result from heavy infestations by these lice. Swine have one species, the hog louse. Cattle are infested by five species: the long-nosed cattle louse, the little blue cattle louse, the shortnosed cattle louse, the cattle tail louse, and the buffalo louse. Horses, mules and donkeys are infested with the horse sucking louse. Sheep may be infested with the foot louse, or the sucking body louse. Goats may be infested by the goat sucking louse.

Nutrition is very important in keeping populations of sucking lice under control. The decreased ability of poorly nourished animals to groom themselves and interference with normal seasonal hair shedding favors survival of their lice. Spread from one animal to another is only by direct contact. Control of lice requires the use of an approved insecticide with 1 or 2 repeated applications at 2 to 3-week intervals.

Mites

Bird Mites

There are several mites that are known to infest caged birds, though none are specific for any

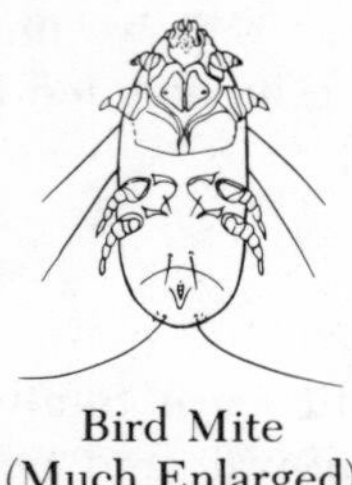

Bird Mite
(Much Enlarged)
(U.S. Public Health Service)

one bird. Their life cycles are similar to those described previously, and most are barely visible with the naked eye. Both the adults and nymphs are blood-feeders. Some species remain on the birds throughout their lives while others feed only at night and retreat to secluded resting places within the cage during the day. Because of the two basic resting habits of the several possible species, control measures call for a good treatment both of the birds and their cages with an appropriate insecticide.

Mange Mites

Mange Mite
(Much Enlarged)
(U.S. Public Health Service)

Mange mites or sarcoptic mites are also known as itch mites. Several species attack cattle, horses, hogs, sheep, goats, dogs, rabbits, and man. The mites burrow into the tender or soft areas of the skin where hair is usually sparse and often continue to spread until large portions of the body are affected. Nodules usually appear over and around the burrows. These burrows burst and ooze serum, which dries to form scabs. Intense itching causes the animals to rub and scratch, resulting in open sores which frequently are invaded by bacteria. This continual irritation causes the skin to become wrinkled and thickened as the infestation spreads. Transmission is usually by direct contact with mangy animals or with objects against which affected animals have rubbed. Overlapping generations of mites occur throughout the year at 2- to 3-week intervals.

Follicle Mites

Follicle mites also attack domestic animals and man, causing manifestations of disease similar to those of the mange mites, by burrowing into hair follicles and oil glands. The nodules formed vary in size from a pinhead to as large as a marble and are filled with pus resulting from secondary bacterial infection. Follicle mange is only a serious pest of dogs, causing what is known as red mange. It is transmitted by direct contact and does not itch severely.

Follicle Mite
(Much Enlarged)
(U.S. Public Health Service)

Isolation of infested animals and treatment by spraying or dipping is the standard recommended procedure.

Sheep Tick or Ked

The sheep tick or ked is really a wingless fly. In no characteristic does it resemble a fly, thus its name, sheep tick. It is reddish-brown, has a sac-like body, spiny and somewhat leathery. It is a widely distributed parasite of both sheep and goats. The female gives birth to a fully-grown larva that pupates in a few hours to a seed-like puparia. The pupae are most commonly found in the region of the shoulders, croup, thighs, and belly of infested animals. Pupae may be found on sheep at all times of the year. The time required for development is about 3 weeks in the summer but may be twice as long in the winter. The entire life of the ked is spent on its host; when off the sheep the keds die in about 4 days.

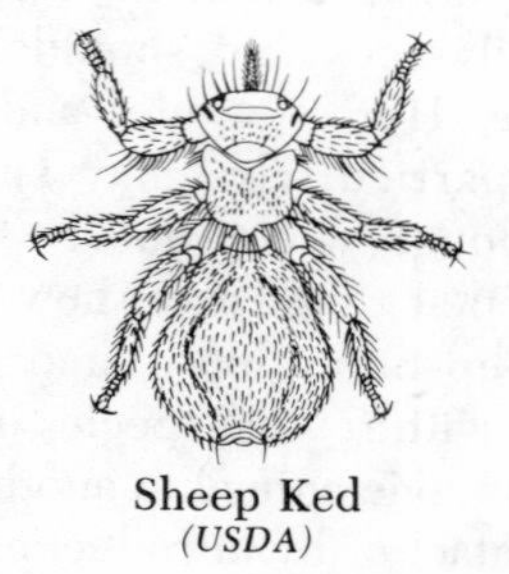

Sheep Ked
(USDA)

A few keds on the body of a sheep do not affect the animal measurably. In heavy infestations the animals rub themselves vigorously, bite the wool and scratch. Injury to lambs is especially marked by emaciation, anemia and general unthriftiness. Control is achieved only with the use of insecticides which may be applied as sprays or dusts.

Wool Maggots

Wool maggots or fleece worms are the maggots of blow flies. When the wool of sheep becomes soggy from warm rains, or soiled with urine, feces, or blood from wounds or from lambing, certain blow flies are attracted and deposit their eggs in the dirty wool, usually around the rump. The maggots feed in the wet wool causing the skin to fester and wool to loosen. The inflamed flesh with the maggots tunneling in it may become infected and the sheep develop blood poisoning. Cattle and other animals may become infested if they have putrid sores.

The adult blow flies are larger than the house fly and usually are metallic bluish-green color with bronze reflections. Flies overwinter as larvae or pupae in soil beneath carcasses or in manure. The flies appear very early in spring and from that time on breeding is continuous. The fly can complete a generation from egg to egg in about 3 weeks. In warm, rainy weather they may lay eggs in the wool, where the maggots develop and drop to the ground where they complete their pupal stage.

Infestation may be prevented by protecting animals from becoming drenched in summer rains and treating sores or dirty, urine-soaked wool. When once infested the use of insecticidal control is essential.

TABLE 20. Pest Control Suggestions for Pets[1].

Pet	Pest	Pesticide to Use	Directions
Birds	Lice, Mites (caged house birds)	Pyrethrins + synergist, aerosol. Rotenone dust (lice only)	Spray parakeets, canaries, and other caged birds as follows: For cage mite control, remove bird and spray entire cage thoroughly. For bird lice, leave birds in cage and spray directly on bird, holding can at least 24 inches from bird or any surface. Spray lightly, one shot 2 to 3 seconds. Treat no more than twice per week. Remove food from cage before spraying. Cages should be cleaned in hot water weekly.
Dogs	Fleas	Chlorpyrifos, 0.09% aerosol. Rotenone dust	Spray dog outdoors, holding aerosol 3-6 inches from dog, and cover thoroughly. A single 30-60 second treatment should give protection up to one month. Repeat as needed.
		Dichlorvos or Naled dog flea collar	Gives protection up to 3 months. Some animals may be sensitive to this product. Watch for signs of irritation. Collar should be buckled loosely.
	Mange mite (sarcoptic)	Refined wettable sulfur, turpentine, pine tar oil, and phenol mix.	Apply daily for several days. Work into the skin with firm finger massage. Treat all areas that respond slowly twice daily. The same for minor moist fungus in ears and on feet. For ear mange mites, lay dog

[1] Refer to Table 35 for trade or proprietary names of recommended pesticides.

Pet	Pest	Pesticide to Use	Directions
Dogs (Continued)			
	Manage Mite (Continued)		
			on side and distribute medications down into ear once daily for 1-3 days. On bald spots use daily until new hair growth is observed.
	Ticks and Fleas	Carbaryl 3-5% dust	Dust entire dog, including ears, legs and feet. Repeat treatment at weekly intervals. To control ticks and fleas in larger area, treat dog's sleeping quarters and surrounding area at rate of 1 lb./1000 sq. ft.
		Diazinon-Ronnel dip	Follow label directions for preparation of dip. Sponge on or dip animal until coat is wet. Allow animal to dry naturally.
		Pyrethrins + synergist aerosol	Hold can comfortable distance from dog and spray entire dog, avoiding the eyes. Follow label carefully.
	Lice and Fleas	Rotenone + Methoxychlor shampoo.	Wet dog then shampoo thoroughly, allowing lather to stand 5 minutes to kill any fleas or lice present.
	Fleas, Lice and Ticks	Pyrethrins + synergist aerosol	Give dog full coverage, weekly, until infestation subsides. Spray animal's bedding and living quarters regularly.
		Pyrethrins + synergist shampoo	Mix according to label instructions and give that dog the bath of its life. Allow lather to remain 10 minutes before rinsing.
Cats	Fleas	Dichlorvos cat collar	Follow label directions. Must be worn loosely.
		Lindane flea collar	Do not use on sick or convalescing animals. Lasts 6-8 weeks.
		Pyrethrins + synergist aerosol	Treat entire pet until hair is slightly damp. Do not spray directly into face.
		Rotenone foam shampoo	Apply to palm of hand then rub into pet's coat. Dry with towel, comb then brush.
	Fleas and Lice	Carbaryl 3% dust. Rotenone dust	Apply powder lightly over entire coat, while avoiding pet's eyes. Dust cat's bedding. Not to be used on kittens less than 6 weeks of age.
	Fleas, Lice and Ticks	Pyrethrins + synergist shampoo	Do not dilute with water, but sponge onto cat in original form. Rub dry with coarse towel. Not to be used on kittens less than 6 weeks of age.

Pet	Pest	Pesticide to Use	Directions
Horses and Ponies	Flies	Methoxychlor, 2% ready-to-use mist spray	Do not exceed 1 oz. per animal. Spray all areas of animal with attention to face, legs and under belly. Repeat as needed.
		Pyrethrins + synergist oil spray	Apply to entire animal as mist spray. Avoid coarse droplets.
		Methoxychlor + Pyrethrins Wipe-on	Apply by rubbing on hair coat, with attention to legs, shoulders, shanks, neck and face. Only light application required. Repeat daily as needed.
	Horse bots	Check with veterinarian.	
	Lice	Malathion spray (0.5%) or dust (4-5%). Rotenone dust	Spray coat until run-off. Or dust animal thoroughly. Repeat in 10 days and thereafter, if needed.
	Horn Fly	Malathion spray (0.5%)	Spray coat until run-off. Repeat in 10 days and thereafter, if needed.
Sheep, Lambs and Goats	Lice and Ticks (keds)	Diazinon (0.03%) or Malathion (0.5%) sprays. Rotenone dust	Spray animals thoroughly, wetting wool to the skin. Use about 1 gallon of finished spray per animal. Repeat at 2-week intervals, if needed. Do not treat animals less than 1 month of age.
	Wool maggots (fleece worms)	Ronnel (0.5%) spray	Spray animal thoroughly wetting wool to the skin. Repeat in 2 weeks, if needed.
Swine	Lice	Malathion or Methoxychlor, (0.5%) sprays.	After spraying, keep animals out of sun and wind for a few hours. Repeat in 2-3 weeks if needed.
		Malathion dust (4-5%)	Apply thoroughly to animals and pens. Repeat in 10 days and thereafter if needed.
	Mange (sarcoptic)	Malathion (0.5%) spray	Spray animals to point of run-off. Treat every 2-3 weeks. Check label regarding piglets under 1 month of age.
Beef Cattle	Horn Fly, Horse Fly, House Fly	Malathion or Methoxychlor (0.5%) sprays	Use about 2 quarts per animal and spray thoroughly.
		Pyrethrins + synergist oil spray	Use 1-2 oz. per animal. Apply as a mist to all parts of the body, especially to the back. Apply daily as needed.
	Face Fly Stable Fly	Dichlorvos (1%) spray	Apply 1-2 oz. per animal per day using hand sprayer, to all parts of the body, especially to forehead. Follow label precautions.
	Lice	Methoxychlor (0.5%) spray	Use about one-half gallon of spray per animal, spraying thoroughly. Repeat in 2-3 weeks, and again if needed.

Pet	Pest	Pesticide to Use	Directions
Beef Cattle (Continued)			
	Mange (sarcoptic)	Dichlorvos (0.5%) spray	Apply up to one-half gallon per animal, spraying thoroughly. Repeat within 10-14 days for lasting control.
Hamsters, Gerbils, White rats and White mice	Fleas	Carbaryl 3-5% dust or Rotenone dust	Apply to animals using a salt shaker with gentle rubbing. Repeat in 2-3 weeks. Frequent scratching is natural and does not necessarily indicate fleas.
	Mites	Malathion (0.5%) spray	Spray animals to point of run-off. Treat every 2-3 weeks. Avoid treating the very young or parents of the very young.
Snakes	Mites	Dusting sulfur	Treat the animal and its quarters with a light dusting using a salt shaker and gentle rubbing throughout the body length. Repeat as needed.

PLANT DISEASES

Some English seed they sew, as wheat and pease, but it came not to good, eather by ye badnes of ye seed, or lateness of ye season, or both, or some other defects.

Gov. Wm. Bradford,
History of Plimoth Plantation
April 1621

CONTROLLING PLANT DISEASES

Most plant diseases encountered around the home are caused by certain fungi or bacteria. Other organisms that cause plant disease are viruses, rickettsias, algae, nematodes, mycoplasma-like organisms, and parasitic seed plants.

There are hundreds of examples of plant diseases. These include storage rots, seedling diseases, root rots, gall diseases, vascular wilts, leaf blights, rusts, smuts, mildews, and viral diseases (Figure 19). These can, in many instances, be controlled by the early and continued application of selected fungicides that either kill the pathogens or inhibit their development.

Chemical control is not the only route to follow for disease management. In some instances it is of no value at all, as in the case of virus or mycoplasma-like diseases. These are either transmitted mechanically by insect vectors feeding on diseased plants then moving to non-infected plants or by propagation methods such as grafting, rooting of diseased cuttings, and the use of infected seed.

Other methods of disease control include the use of resistant varieties, planting times, cultural control, and simply good plant nutrition. Gardeners should use plant varieties or cultivars which are resistant or at least tolerant to certain diseases. These are usually developed for geographical areas of the country and selected for their resistance to local disease problems.

Cultural control may involve one of several methods of altering cultural practices in favor of the plant and to the disadvantage of the disease organism. For instance, Fusarium and Verticillium wilts are carried over in the soil and in decaying plant parts from year to year. A simple crop rotation system with switching of susceptible to non-susceptible garden species will avoid the problem almost completely. The early or late planting of vegetables to avoid the normal infective periods of certain diseases is another simple way to manage diseases. Watering or sprinkling only in the mornings may avoid producing the wet afternoon environments that promote downy mildew and certain fungal fruit rots. These techniques are too numerous to give in detail.

And, finally, balanced fertilization may often provide the additional plant vigor needed to outgrow a disease. This is certainly not always the case, but disease in general has a more profound effect on weak and undernourished plants than on hearty, vigorously growing specimens.

DISCOURAGING PLANT DISEASES THROUGH GOOD MANAGEMENT

Following several of the suggestions listed below will help reduce the incidence and severity of diseases in all gardens, vegetable or flower:

1. Choose a suitable location. Don't place shade-loving plants in exposed situations or sun-loving plants in the shade. Also avoid extremely wet or dry locations or use plants that are suited to these conditions. Most root diseases are favored by wet soils. Creating good drainage in garden soils can reduce the severity of these diseases.

2. Take additional time to plant a seedling with care — it pays dividends in reducing replacement and maintenance later in the growing season. Spend a little extra time working the soil into a good seed bed or planting condition.

3. Choose disease resistant varieties that are locally adapted for your area if available. Do not grow nonadapted varieties from other areas of the country unless you simply like to experiment.

4. Occasionally change the garden location, and always practice a rotation within the garden plot. Many diseases, especially soil-borne ones, are most serious when the same or related crops are grown in the same area year after year. Change the flower species in an area every year or two. Not only will this add variety to the landscape but it helps combat build-up of disease organisms in the soil.

5. Use disease-free seed and seedlings. Buy seed from a reputable seed company and transplants from a greenhouse operator who grows them from disease-free seed and in soil free of disease organisms. Many disease organisms can be carried to the garden on seeds, transplants, or transplant soil.

6. Apply fertilizer and lime only on the basis of soil testing results or established soil fertility management practices. Exceptionally weak or vigorous plants are more susceptible to some diseases than those grown on a balanced fertility and optimum pH (soil acidity) program.

7. Weeds, particularly perennial weeds, near gardens often are initial sources of viruses in the

DISEASE SYMPTOMS

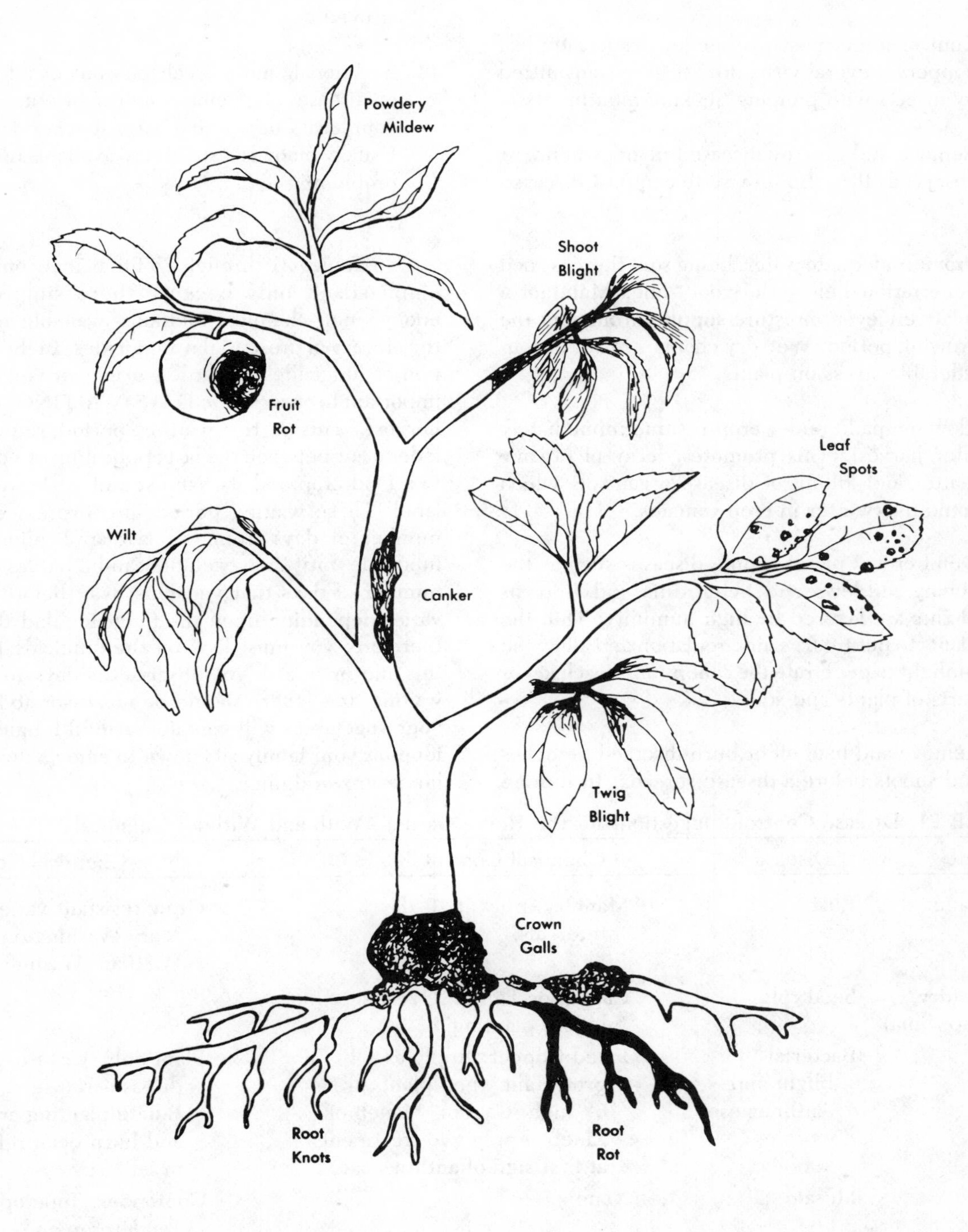

FIGURE 19. Disease Symptoms on Plants.

spring. Dense clusters of weeds within the garden create a microclimate that is ideal for development of fungus and bacterial diseases. And, weeds also compete for water and soil nutrients.

8. Control insect pests as needed, especially leaf hoppers. Several virus diseases are transmitted by insects with piercing-sucking mouthparts.

9. Remove and destroy diseased plants when first observed. It is also unwise to compost diseased plants.

10. Provide adequate water, being sure there is good penetration below the root zone. Maintain a relatively even moisture supply throughout the growing period. Wet-dry changes result in considerable stress on plants.

11. Plow or spade under crop residue immediately after harvest. This promotes decay of organic matter and killing of disease organisms which could overwinter in crop remains.

12. Don't crowd plants. Some diseases such as the downy mildews and Sclerotinia and Botrytis blights are favored by high humidity. Thin the plants to permit free air circulation and allow the sunlight to penetrate the canopy and reach lower parts of plants and soil.

13. Remove and haul off or burn diseased branches and shoots before a disease spreads. In routine pruning, always remove diseased or sickly growth first, and then prune to develop and shape the shrub or tree. Many disease organisms carry over the winter on fallen leaves; thus, diseased foliage should be collected and destroyed.

14. Additional, more localized control information on diseases of specific crops can be obtained from your local Cooperative Extension Service Agent. Usually many fact sheets are available on specific problems.

In Tables 21 through 25 fungicidal controls are emphasized, only because there simply aren't enough non-chemical methods available to match the efficacy of present-day fungicides. In the application of fungicides to garden plants, it is extremely important to observe the DAYS-WAITING-TIME to harvest. This is the waiting period required by federal law between the last application of a pesticide to a food crop and its harvest and is shown on the label. These waiting periods are expressed as the number of days from the last application of the fungicide until the vegetable can be harvested. The number of days that you must wait before harvest varies depending upon the fungicide and the crop; therefore, you must look on the fungicide label for this information. If you observe the days-to-harvest waiting time, there should be no reason to fear that your vegetables will contain harmful fungicide residues as your family sits down to enjoy a delectable, home-grown dinner.

TABLE 21. Disease Control Suggestions for the Home Garden, With and Without Chemicals[1].

Plant	Disease	Chemical Control	Non-Chemical Control
Asparagus	Rust	Maneb. Apply at 10-day intervals.	Grow resistant varieties as Mary Washington and Waltham Washington.
Beans (dry, snap, lima)	Seed rots	Captan or Thiram. Treat seed before planting.	
	Bacterial blight and Anthracnose	Fixed Copper. Apply as protectant when plants are 6″ high. Captan, Maneb or Zineb. Apply two treatments at first sign of anthracnose.	Plant only western-grown disease-free seed. Rotate planting areas and burn bean refuse in fall.
	Mosaic	None	Contender, Topcrop, Tendercrop and Tendergreen are resistant varieties.

[1] Refer to Table 36 for trade or proprietary names of recommended pesticides.

Plant	Disease	Chemical Control	Non-Chemical Control
Beans (Continued)			
	Sclerotinia white mold	Benomyl. Spray at green bud stage and again in 7 days.	Plant in well-drained area not planted to beans last year.
	Downy mildew	Maneb. Make spray application when mildew first appears.	Water or sprinkle only in mornings.
	Powdery mildew	Sulfur. Dust or spray with wettable sulfur, and again in 7-10 days.	
Beans (pole)	Rust and Halo blight	None	FM-1 is resistant.
Beets	Damping-off Seed rot	Thiram, Captan, or Zineb. Treat seed with or soak in Thiram solution. Use Captan or Zineb as seed treatment or apply Captan to soil before planting.	
	Curly top	None	Parma Globe is resistant.
	Cercospora leaf spot	Zineb, Captan, or Fixed Copper. Apply to foliage 2-3 times at weekly intervals starting at first symptoms.	Most varieties are resistant.
	Mildew		F.M. Detroit and dark red are resistant. Avoid overwatering.
Cabbage Cauliflower Collards Broccoli Brussels Sprouts	Damping-off	Thiram or Captan. Dust seed before planting.	Soil temperature must be at least 55°F for planting.
	Clubroot	Terraclor (PCNB). Add dilute solution to hole during transplanting.	Do not plant crucifers on same soil in consecutive years.
	Downy mildew	Maneb. Apply at mid-season or on first sign of mildew, and late-season treatment with Fixed Copper.	Plant hot-water-treated seed. Water or sprinkle only in mornings.
	Blackleg Black rot		Buy only hot-water-treated seed.
Carrots	Cercospera and Alternaria leaf spots	Maneb or Zineb. Apply when first signs of disease occur.	Use 2-3 year rotation to reduce fungus carryover.
	Carrot yellows (virus)	Control leafhoppers with recommended insecticide.	Aluminum foil mulch will repel leafhoppers.
Celery	Early or Late blight	Maneb. Spray every 7-10 days.	Soak seed at 118°F for 30 minutes.

Plant	Disease	Chemical Control	Non-Chemical Control
Cucumbers Muskmelons Watermelons	Damping-off	Thiram or Captan. Dust seed before planting.	
	Bacterial wilt		Control cucumber beetles which spread bacteria from plant to plant.
	Powdery mildew	Dinocap. Complete coverage spray at first signs of mildew and again in 7 days.	
	Angular leaf spot	Maneb. Spray at first symptoms.	Plant tolerant cucumber varieties: Pioneer, Premier, Carolina.
	Wilts (Fusarium and Verticillium)	No chemical control.	Avoid planting where problem has occurred before. Muskmelons resistant to fusarium are Iroquois, Gold Star, Harper Hybrid, Saticoy Hybrid. Summer Festival watermelon is resistant to fusarium and anthracnose.
	Curly top (virus)	No chemical control.	Transmitted by leafhoppers. Protect plants from leafhopper feeding.
	Scab	No chemical control.	Resistant cucumbers: SMR-58, Marketmore 70, Pacer, Slicemaster and Victory.
	Mosaic	No chemical control.	Resistant cucumbers: Marketmore 70, Pacer, Slicemaster, Tablegreen 65, Victory. Destroy perennial weeds.
Eggplant	Damping-off	Captan. Add to first water used on seed flats, or to hole during transplanting. Spray plants weekly, enough to run down stem.	
	Anthracnose and Phomopsis fruit rots	Maneb, Zineb or Ziram. Spray at 10-day intervals. Begin when first fruits are 2″ in diameter.	Use 4-year rotation. Plant hot-water-treated seed.

Plant	Disease	Chemical Control	Non-Chemical Control
Eggplant (Continued)			
	Wilt (Verticillium)	No chemical control.	Avoid planting in wilt problem area. Rotation free of peppers, strawberries and tomatoes. Plant resistant varieties.
Lettuce	Downy mildew	Maneb, Zineb or Fixed Copper. Apply every 7 days beginning at first signs of mildew.	Water or sprinkle only in mornings.
	Bottom rot (Rhizoctonia) and Drop (Sclerotinia)	No chemical control.	Rotate planting areas each year, with deep plowing.
Muskmelons (see Cucumbers)			
Mustard Greens	Alternaria leafspot and Downy mildew	Maneb or Zineb. Spray at 7-10 day intervals.	Water only in mornings. Do not sprinkle.
Onions, Garlic and Chives	Purple blotch	Maneb or Chlorothalonil. Spray at first signs and repeat at weekly intervals.	
	Downy mildew	Maneb, Captan or Zineb. Apply at first signs of mildew and at 7-day intervals as needed.	Water or sprinkle only in mornings.
	Smut	Captan or Thiram. Dust seed before planting.	Smut attacks only onions grown from seed. Plant disease-free onion sets.
Parsnips	Leaf diseases	Maneb or Zineb. Apply when spots first appear.	
Peas	Powdery mildew	Sulfur. Dust or spray at first signs of disease.	
	Wilt	No chemical control.	Plant resistant varieties. Plant early and rotate planting areas each year.
	Mosaic		Thomas Laxton 60 is resistant.
Peppers	Damping-off	Captan. Add to first transplant water.	
	Bacterial spot	Fixed Copper. Several applications are effective if started before disease appears.	Purchase disease-free seed or transplants.
	Wilt (Verticillium)	No chemical control.	Avoid planting where problem has occurred before.

Plant	Disease	Chemical Control	Non-Chemical Control
Peppers (Continued)			
	Mosaic		Yolo Wonder is resistant.
	Virus mosaic	Malathion. Spray to control the aphid vectors as they appear.	Remove and destroy perennial weeds and old crop residue.
	Anthracnose and Fruit rot	Maneb or Zineb. Spray at 7-day intervals through harvest. Begin when first fruits are 1″ in diameter.	
Potatoes	Scab	Sulfur. Add to soil to reach pH 5.2 before planting. Treat seed pieces with Captan dust before planting.	Resistant varieties: Russett Burbank, Russet Rural, Norland, Cherokee and Superior.
	Virus	Control vector aphids.	Plant certified seed potatoes.
	Early and Late blights	Maneb or Bravo. Spray every 7-10 days, beginning when plants are 6″ high and continue until vines are dead. Use fixed copper when severe.	Kennebec is tolerant to late blight.
	Rhizoctonia stem canker	Captan. Dust seed pieces before planting.	
	Seed piece rot	Captan. Dust seed pieces before planting.	
	Tuber diseases		Buy certified seed potatoes. Plant uncut tubers.
Pumpkins and Squash	Damping-off and Seed rots	Thiram or Captan. Dust seed before planting.	
	Bacterial wilt		Control cucumber beetles which spread bacteria from plant to plant.
	Powdery mildew	Dinocap or Benomyl. Apply spray or dust when mildew appears, and again in 10 days.	
	Mosaic (virus)	No chemical control. Control aphid vectors and weeds, instead.	Eliminate pokeweed and wild cucumber weed hosts. Kill other perennial weeds within 150 feet. Aluminum foil mulch will repel aphids.
	Curly top (virus)	No chemical control. Control leafhopper vectors and weeds instead.	Aluminum foil mulch to repel leafhoppers.

Plant	Disease	Chemical Control	Non-Chemical Control
Radishes (see Turnips)			
Rhubarb	Leaf spot	Maneb. Apply early as needed.	Remove and burn leaves in fall.
Spinach	Downy mildew	Maneb or Zineb. Spray at 7-10 day intervals.	Plant resistant varieties as Hybrid 612, Early Hybrid No. 7 and Chesapeake. Water or sprinkle only in mornings.
Squash (see Pumpkins)			
Sweet corn	Smut	No chemical control.	Pick and destroy galls. Late maturing varieties are more tolerant than early varieties.
	Leaf blight	Control usually not necessary.	
Sweet potatoes	Black rot, Scurf, Foot rot	No chemical control.	Buy certified plants. Use 3- or 4-year rotation.
	Wilt, Root-knot, Soil rot	No chemical control.	Plant resistant varieties.
Tomatoes	Damping-off and Seed rot	Captan. Use in first water when transplanting.	
	Early blight, Late blight and Anthracnose	Maneb or Bravo. Make first application when fruit appear, and continue at 7-10 day intervals or after rains.	Rotate tomatoes with beans or sweet corn.
	Blossom end rot	No chemical control.	Maintain adequate soil moisture by watering plants if needed.
	Wilts (Fusarium and Verticillium)	No chemical control.	Plant resistant varieties and avoid planting where problem has occurred before. Campbell 1327, Jet Star, Springset, Better Boy, Heinz 1350, Supersonic are resistant varieties.
	Bacterial spot	Maneb or Fixed Copper. Apply in early flowering stage, and as needed when disease appears.	
	Septoria	Maneb. Spray at 7-day intervals.	
Turnips and Radishes	Alternaria leafspot and Downy mildew	Maneb or Zineb. Apply to turnips at 7-day intervals. Fixed Copper. Apply to radishes at 7-day intervals.	Water only in mornings. Do not sprinkle.

TABLE 22. Disease Control Suggestions for Houseplants, With and Without Chemicals[1].

Plant	Disease	Chemical Control	Non-Chemical Control
African violet	Botrytis blight Crown rot Petiole rot Ring spot	Benomyl. Spray or dip[3].	Water by soaking from base rather than overhead watering. Plant sanitation[2].
Begonia	Bacterial leafspot	Zineb or Fixed Copper. Spray or dip[3] frequently.	
	Botrytis blight	Same as above.	
	Powdery mildew	Dinocap or Benomyl. Spray or dip[3].	
Ferns	Anthracnose	No chemical control.	Plant sanitation[2].
Gardenia	Bacterial leafspot	No chemical control.	Plant sanitation[2].
	Bud drop	No chemical control.	Avoid high night temperatures.
Geranium	Bacterial leafspot	No chemical control.	Plant sanitation[2]. Keep foliage dry.
	Botrytis blight	Zineb or Benomyl. Spray or dip[3].	
Gloxinia	Leaf and Stem rot	No chemical control.	Provide good soil drainage by omitting clay from potting mixture.
Ivy (Boston and English)	Leafspot disease	Basic Copper. Spray or dip[3].	Plant sanitation[2].
Palms	Leafspot disease	Basic Copper. Spray as needed.	
Pittosporum	Leafspot disease	No chemical control.	Plant sanitation[2].
Rubber plant	Anthracnose Leaf scorch Oedema Root rot	No chemical control.	Maintain good cultural methods.

[1] Refer to Table 36 for trade or proprietary names of recommended pesticides.
[2] Remove and destroy diseased plants and plant residue.
[3] Dip foliage and stem into solution equal to one-half the spray concentration.

TABLE 23. Disease Control Suggestions for Lawn and Turf With and Without Chemicals.[1]

Disease	Susceptible Turfgrasses	Pesticide to Use	Non-Chemical Control
Algae	All grasses	Zineb + Maneb. When algae appear.	Avoid overwatering and water only in mornings.
Brown Patch (Rhizoctonia)	Bentgrass Bluegrass Bermuda Fescue Ryegrass	Anilazine, Zineb + Maneb, or Methyl Thiophanate. Apply at 7-day intervals in July-August.	Avoid high-nitrogen fertilizer, and increase air circulation.
Copper Spot (*Gloeocercospora sorghi*)	Bentgrass	Anilazine or Methyl Thiophanate. Apply from late June to October, as needed.	
Dollar Spot (Sclerotinia)	Bentgrass Bluegrass Ryegrass Fescue Zoysia	Anilazine, Methyl Thiophanate, or Chlorothalonil. Apply from late June to October. Avoid continued use.	Mow at maximum recommended height. Increase nitrogen in soil. Maintain soil moisture. Water only in mornings. Plant resistant varieties.
Fairy rings	All grasses	Fungicides are ineffective. Requires eradication.	Provide adequate water and fertilizer. Remove infested sod and soil; replace with clean soil and reseed or sod.
Fusarium	Bentgrass Bluegrass (esp. Merion) Fescue Ryegrass	Methyl Thiophanate, Thiram, or Benomyl. Apply in July and August.	Water frequently but lightly during dry periods. Lime annually to soil pH above 6.2. Avoid excess nitrogen. Mow at maximum height. Remove clippings and avoid thatch.
Leafspot and Crown Rot (Helminthosporium)	Bentgrass Bluegrass Bermuda Fescue Ryegrass	Anilazine, Chlorothalonil or Thiram. Apply at 7-14 day intervals, April-June more frequently in cool moist spring.	Merion Bluegrass is a resistant variety. Remove clippings, raise cutting height of mower, and fertilize to maintain vigor.
Nematodes	All grasses	No chemical control available.	Maintain plant vigor through regular and proper fertilization.
Powdery mildew (Erysiphe)	Bermuda Bluegrass Fescue	Dinocap, or Methyl Thiophanate. Apply when mildew appears, July-September.	Reduce shading where possible. Water only in mornings. Resistant bluegrasses include Adelphi, Bonnieblue, Fylking, Glade, Park, Pennstar and Windsor.

[1] Refer to Table 36 for trade or proprietary names of recommended pesticides.

Disease	Susceptible Turfgrasses	Pesticide to Use	Non-Chemical Control
Pink Patch, (see Red Thread)			
Pythium blight (Pythium)	Bentgrass Bluegrass Ryegrass Bermuda Fescue	Zineb, Thiram or Terrazole. Hot, wet weather disease which should be treated immediately.	Maintain good growth with moderate fertilizer use, and soil pH at at 6.2-7.0. Avoid mowing when grass is moist.
Red Thread (Corticium)	Bentgrass Bluegrass Fescue Ryegrass	Methyl Thiophanate, Anilazine, Chlorothalonil, or Thiophanate. Apply in May-June and again in August-September.	Maintain adequate soil fertility. Apply lime to achieve soil pH of 6.5-7.0. Water only in mornings and avoid overwatering.
Rust	Bluegrass (esp. Merion)	Zineb, Anilazine, Chlorothalonil, Zineb + Maneb. Make 2-3 applications at 12-14 day intervals in July-August.	Good soil fertility may help decrease rust problem. Resistant varieties include Adelphi, Bonnieblue, Fylking, Glade, Park, Pennstar, and Windsor.
Slime molds (Myxomycete)	All grasses	Zineb or Zineb + Maneb. Treat in August-September.	Remove affected grass by mowing and raking.
Snow molds (Fusarium and Typhula)	Bentgrass Bermuda Bluegrass Fescue Ryegrass	Anilazine or Thiram. Apply before snow, midwinter and during spring thaw.	Avoid late fall fertilizing. Minimize thatch accumulation by raking matted grass in spring.
Stripe smut (Ustilago)	Bentgrass Bluegrass (esp. Merion) Fescue Ryegrass	Methyl Thiophanate or Thiram. Apply in late fall or early spring using extra water for penetration.	Water carefully to maintain vigor. Assure adequate phosphorus fertility. Resistant bluegrasses include A-34, Bonnieblue, Glade, Nugget, Sydsport, and Touchdown.
Toadstools and mushrooms.	All grasses	Dinocap or Thiram. Drench soils of trouble spots.	Maintain adequate fertilization. Remove fruiting bodies when observed and safely discard. They may be poisonous.
Yellowing	Bermudagrass and its hybrids St. Augustine (West and Southwest U.S.)	Usually not a disease, but lack of iron.	Treat with fertilizer containing iron or iron chelates.

TABLE 24. Disease Control Suggestions for Trees, Shrubs and Woody Ornamentals With and Without Chemicals[1].

Plant	Disease	Pesticide to Use	Non-Chemical Control
Andromeda (Pieris)	Leafspots	No chemical control.	Remove and destroy diseased leaves.
Azalea	Leafspots	Ferbam or Benomyl. Apply frequently beginning in spring.	Plant sanitation[2].
	Dieback	Bordeaux. Apply as new leaves appear.	Prune and burn infected twigs.
	Flower gall	No chemical control.	Remove galls and burn.
Barberry	Bacterial leafspot	Bordeaux. Apply 3 times at 10-day intervals, beginning when new leaves open.	Rake and burn old leaves in fall.
	Verticillium wilt	No chemical control.	Replace with another kind of plant.
Birch	Nectria canker	Bordeaux. Apply 4 times at weekly intervals beginning in early spring.	Prune infected branches and paint stubs with wound dressing.
Boxwood	Canker	Bordeaux. Apply 4 times through growing season.	Prune infected branches, and burn old leaves.
	Leafspots	No chemical control.	Plant sanitation[2] and fertilize to maintain vigor.
Cercis (see Red Bud)			
Clematis	Leafspots	Zineb. Apply 3 times at 10-day intervals, usually May.	Plant sanitation[2].
	Stem rot	Zineb. Apply at 2-week intervals during growing season; drench soil around crown.	
Cotoneaster	Fireblight	Bordeaux. Apply twice during flowering.	Prune dead twigs only when plant is dry.
Crabapple	Cedar-apple rust	Ferbam or Chlorothalonil. Apply 4-5 times at 7-10 day intervals when galls on nearby junipers produce jelly-like secretions.	Remove nearby junipers.
	Fireblight	Bordeaux. Apply twice during flowering.	Prune dead twigs only when plant is dry.
	Scab	Captan or Benomyl. Apply weekly beginning when buds turn pink.	Select resistant varieties.
	Powdery mildew	Benomyl or Dinocap.	

[1] Refer to Table 36 for trade or proprietary names of recommended pesticides.

[2] Remove and destroy, preferably by burning, diseased plants and plant residue. Rake and remove stems, twigs, and leaves in the fall and burn.

Plant	Disease	Pesticide to Use	Non-Chemical Control
Dogwood	Flower and Leaf Blight	Zineb or Benomyl. Apply after periods of wet weather.	
	Leafspots	Zineb or Chlorothalonil. Apply after periods of wet weather.	
	Crown Canker	No chemical control.	Look for borer damage and treat accordingly.
	Twig blight	No chemical control.	Plant sanitation[2].
Dutchman's Pipe	Leafspots	Zineb. Apply 3 times at 10-day intervals, usually May.	Plant sanitation[2].
Euonymus	Crown gall	No chemical control.	Remove and destroy diseased plants including roots.
	Downy mildew	Dinocap. Spray as needed.	Water only in mornings. Avoid sprinkling. Thin dense foliage.
Firethorn	Fireblight	Bordeaux. Apply twice during flowering.	Prune dead twigs. Fertilize lightly but frequently.
	Scab	Captan. Apply weekly beginning when buds begin to expand.	
Hawthorn	Cedar-apple rust	(see crabapple)	
	Fire blight	Bordeaux. Apply twice during flowering.	Prune dead twigs only when plant is dry.
	Leafspots	Benomyl. Apply when leaves are young, after periods of wet weather.	Resistant varieties are Cockspur and Washington.
Ivy (Boston and English)	Leafspot	Captan or Mancozeb. Make 3 sprays when flowers open and repeat weekly.	
	Powdery mildew	Benomyl. Spray 3 times beginning in July.	Thin by pruning for air circulation.
Juniper	Twig blight	Bordeaux or Benomyl. Apply 3 times at 14-day intervals, beginning with new growth.	Prune twigs 2 inches into live wood and burn. Select resistant varieties.
	Cedar-apple rust	None.	Remove galls in early spring. Plant resistant varieties.
Laurel	Leafspot	Benomyl. Apply at 10-day intervals from beginning of new growth until leaves mature.	Remove and destroy diseased leaves.

[2] Remove and destroy, preferably by burning, diseased plants and plant residue. Rake and remove stems, twigs, and leaves in the fall and burn.

Plant	Disease	Pesticide to Use	Non-Chemical Control
Leucothoe	Leafspot	Zineb, Benomyl or Ferbam. Apply 3 times at 14-day intervals when new growth begins.	Plant sanitation[2].
Lilac	Bacterial wilt	Bordeaux. Apply 3 times at 10-day intervals when new growth appears.	Remove and destroy wilted twigs.
	Powdery mildew	Dinocap or Benomyl. Apply weekly when mildew appears.	
London Plane	Anthracnose	Zineb. Apply 3 times when buds break and at weekly intervals.	Rake and burn all leaves in fall.
	Powdery mildew	(see lilac)	
Mahonia	Leafspot	No chemical control.	Plant sanitation[2].
	Scorch	No chemical control.	Water during dry periods and protect from wind.
Maple	Anthracnose	Zineb. Apply 3 times when buds break and at weekly intervals.	Rake and burn leaves after frost.
	Scorch	No chemical control.	Water during dry periods and protect from wind.
	Tar spot	Ferbam. Apply 3-4 times beginning when buds break and at 2-week intervals.	
	Verticillium wilt	No chemical control.	Water during dry periods and fertilize properly.
Mountain Ash	Cytospora canker	None.	Prune diseased branches during dry weather.
	Leafspot	Zineb or Ferbam. Apply at 10-day intervals from new growth to mature leaves.	Rake and burn dead leaves.
	Fireblight	Bordeaux. Apply twice during flowering.	Prune dead twigs, and do not fertilize.
Pachysandra	Blight	Captan or Mancozeb. Apply 3 times at weekly intervals beginning with new growth.	Clean up and thin bed.
Periwinkle (Vinca)	Blight	No chemical control.	Thin for better ventilation.
	Canker and Dieback	No chemical control.	Prune and burn infected stems.
Peach and Cherry	Leafspot	Captan. Spray 3 times at 2-week intervals, beginning with petal fall.	Rake and burn dead leaves.
	Peach leaf curl	Dormant spray of Bordeaux in April.	

[2] Remove and destroy, preferably by burning, diseased plants and plant residue. Rake and remove stems, twigs, and leaves in the fall and burn.

Plant	Disease	Pesticide to Use	Non-Chemical Control
Peach and Cherry (Continued)			
	Twig blight	Captan. Spray twice at 10-day intervals, beginning just before blossoms open.	Prune infected limbs and paint stubs with wound dressing.
	Black knot	No chemical control.	Prune galls in winter.
Pieris (see Andromeda)			
Pine	Blister rust (white pine)	No chemical control.	Prune cankers on twigs and remove wild currant bushes.
	Needle blight	Bordeaux. Apply twice when new growth starts and at 2-week intervals.	
	Twig blight	Bordeaux or Benomyl. Apply 3 times, when new growth starts and at weekly intervals.	Plant sanitation[2]. Fertilize and water.
Privet	Anthracnose and Twig Blight	Zineb or Ferbam. Apply at weekly intervals until controlled.	Remove and destroy infected branches.
Pyracantha (see Firethorn)			
Quince	Crown gall	No chemical control.	Dig up and destroy infected plants and roots.
	Fireblight	Bordeaux. Apply twice during flowering.	Prune dead twigs.
	Leafspots	Zineb. Apply when leaves are half grown.	
	Cedar-apple rust	Ferbam or Chlorothalonil. Apply 4-5 times at 7-10 day intervals when galls on nearby junipers produce jelly-like secretions.	Remove nearby junipers.
Red Bud (Cercis)	Canker	No chemical control.	Prune cankers and paint wounds with orange shellac and wound dressing.
Rhododendron	Leafspot	Benomyl or Ferbam. Apply at 10-day intervals from new growth to mature leaf.	Plant sanitation[2].
	Dieback	Bordeaux. Apply 2 times at 14-day intervals. Hard to control.	Prune and burn infected plant parts.
Roses	Blackspot	Folpet or Benomyl. Apply at weekly intervals.	
	Botrytis blight	Captan or maneb. Spray 3 times at 7-14 day intervals.	

[2] Remove and destroy, preferably by burning, diseased plants and plant residue. Rake and remove stems, twigs, and leaves in the fall and burn.

Plant	Disease	Pesticide to Use	Non-Chemical Control
Roses (Continued)			
	Crown gall	No chemical control.	Remove bushes and burn Relocate rose bed.
	Powdery mildew	Dinocap or Benomyl. Apply when mildew appears and at weekly intervals throughout season.	Select resistant varieties.
	Rust	Mancozeb. Spray 3 times at 10-day intervals, beginning May 15-June 1.	
Spruce	Cytospora canker	No chemical control.	Prune diseased limb where it branches from trunk and apply wound dressing.
Sycamore	Anthracnose	Zineb. Apply 3 times when buds break and at weekly intervals.	Rake and burn all leaves in fall.
Vinca (see Periwinkle)			
Willow	Black canker and Twig blight	Mancozeb, Maneb, or Zineb. Apply 3 times at 14-day intervals. Begin when leaves are 1/4 inch long.	Prune and burn diseased twigs.

TABLE 25. Disease Control Suggestions for Ornamental Annuals and Perennials With and Without Chemicals[1].

Plant	Disease	Pesticide to Use	Non-Chemical Control
Ageratum	Root rot	No chemical control.	Use only sterilized potting soil.
Chrysanthemum	Leafspot	Folpet. Spray weekly if leafspot was severe previously.	Plant sanitation[2].
	Powdery mildew	Dinocap. Apply after periods of wet weather.	
	Rust	Zineb. Spray at first symptoms and as needed.	Remove and bury infected leaves.
	Virus	Malathion. Treat weekly to control aphid vectors.	Avoid propagating from unknown sources. Plant sanitation[2].
Cockscomb	Leafspot	Zineb. Apply after periods of wet weather.	Remove and destroy badly spotted leaves.
Columbine	Crown and Root rot	No chemical control.	Sterilize soil.
	Leafspots	Benomyl or Captan. Apply as needed.	Plant sanitation[2].
	Rust	Zineb. Apply when pustules first appear and as needed.	
Dahlia	Gray mold (*Botrytis*)	Zineb or Benomyl. Apply after periods of wet weather.	Plant in areas with good air movement.
	Powdery mildew	Benomyl. Apply when mildew appears, usually August.	Burn or bury all dead plant material in fall.
	Stem rot and wilt	No chemical control.	Sterilize soil.
	Virus	Malathion. Treat weekly to control aphid vectors.	Destroy roots.
Day-Lilies	Leafspot	No chemical control.	Plant sanitation[2].
Delphinium (see Larkspur)			
Forget-Me-Not	Downy mildew	Dinocap or Bordeaux. Spray as needed.	Water only in mornings. Do not sprinkle.
	Wilt	No chemical control.	Sterilize soil.
Hollyhock	Anthracnose	Benomyl. Spray after periods of wet weather.	
	Leafspots	No chemical control.	Plant sanitation[2].
	Rust	Zineb. Apply twice weekly beginning when infection appears.	Plant sanitation[2].
Hydrangea	Bacterial wilt	No chemical control.	Plant sanitation[2].

[1] Refer to Table 36 for trade or proprietary names of recommended pesticides.

[2] Remove and destroy, preferably by burning, diseased plants and plant residue. Rake and remove stems, twigs, and leaves in the fall and burn.

Plant	Disease	Pesticide to Use	Non-Chemical Control
Hydrangea (Continued)			
	Leafspots	Zineb. Apply at 2-week intervals and after rains.	
	Powdery mildew	Benomyl. Apply when mildew appears and at 2-week intervals as needed.	
Impatiens	Bacterial wilt and Damping off	No chemical control.	Sterilize soil.
	Leafspots	Zineb. Apply at 2-week intervals and after rains.	
Iris	Bacterial leafspot	No chemical control.	Plant sanitation[2].
	Bacterial soft rot	No chemical control.	Cut rotted areas from rhizomes. Dry in direct sunlight for 1 day.
Larkspur	Bacterial blight	Streptomycin. Follow label instructions carefully.	
	Crown rot	No chemical control.	Sterilize soil.
	Leafspot	Zineb. Apply at 2-week intervals and after rains.	
	Powdery mildew	Dinocap. Apply when mildew appears, usually August.	
Lilies	Bacterial soft rot	No chemical control.	Plant sanitation[2].
	Botrytis blight	Benomyl. Apply at 10-day intervals where disease was severe previously.	Plant sanitation[2].
	Stem canker	Terraclor. Dip infected bulbs.	
	Wilt	No chemical control.	Sterilize soil.
Lupine	Crown rot	No chemical control.	Sterilize soil.
	Powdery mildew	Benomyl. Apply when mildew appears, usually August.	Burn all dead plant material in fall.
Marigold	Blight	No chemical control.	Plant sanitation[2].
	Leafspots	Zineb. Apply at 2-week intervals and after rains.	
	Root rot Stem rot Wilt	No chemical control.	Sterilize soil.
Nasturtium	Bacterial leafspots	No chemical control.	Plant sanitation[2].
	Bacterial wilt	No chemical control.	Sterilize soil.

[2] Remove and destroy, preferably by burning, diseased plants and plant residue. Rake and remove stems, twigs, and leaves in the fall and burn.

Plant	Disease	Pesticide to Use	Non-Chemical Control
Nasturtium (Continued)			
	Leafspots	Zineb. Apply at 2-week intervals and after rains.	
Pansy	Anthracnose	Benomyl. Apply twice at 5-day intervals when disease appears.	Remove and burn infected leaves.
	Leafspots	Zineb. Apply at 2-week intervals and after rains.	
	Scab	Zineb. Apply after periods of wet weather.	Use sterilized soil.
	Downy mildew	Bordeaux. Apply to densely growing beds at 2-week intervals.	Thin plants for ventilation. Water or sprinkle only in mornings.
Peony	Anthracnose	Benomyl. Apply 3 times at 5-day intervals in early season.	Plant sanitation[2].
	Gray mold (*Botrytis*)	Benomyl. Apply 3 times at 2-week intervals. Begin when tips break through soil, soaking soil.	Plant sanitation[2].
	Root rot	No chemical control.	Plant sanitation[2].
	Leafspots	Zineb. Apply several times during growing season as needed.	Plant sanitation[2].
Petunia	Virus diseases	No chemical control.	Purchase virus-free seed, and destroy infected plants.
Phlox	Leafspots	Zineb or Bordeaux. Apply after periods of wet weather.	Plant sanitation[2].
	Powdery mildew	Benomyl. Apply as needed.	Plant resistant varieties and use plant sanitation[2].
Poppy	Bacterial blight	No chemical control.	Plant treated, disease-free seed. Sterilize soil.
Shasta Daisy	Leafspots	Zineb. Apply at 2-week intervals and after rains.	Remove and burn infected parts.
Snapdragon	Anthracnose	Zineb. Spray as needed, usually July.	Plant disease-free seed.
	Rust	Zineb. Apply at 2-week intervals until plants reach 15 inches.	Water in morning.
	Downy mildew	Zineb. Apply as needed.	Water only in mornings. Do not sprinkle.

[2] Remove and destroy, preferably by burning, diseased plants and plant residue. Rake and remove stems, twigs, and leaves in the fall and burn.

Plant	Disease	Pesticide to Use	Non-Chemical Control
Snapdragon (Continued)			
	Powdery mildew	Benomyl. Apply as needed.	
Stock	Leafspots	Zineb. Apply at 2-week intervals and after rains.	
	Wilt	No chemical control.	Sterilize soil.
Sweet Alyssum	Downy mildew	Bordeaux or Zineb. Apply as needed. Buy treated seed.	Plant disease-free seed. Water only in mornings.
	Wilt	Terraclor or Captan. Soak soil before planting or after rainy periods.	Sterilize soil.
Sweet Pea	Anthracnose	Benomyl. Apply at 2-week intervals during growing season.	Plant disease-free seed.
	Root rots	Benomyl. Soak soil.	Sterilize soil.
	Leafspots	Zineb. Apply at 2-week intervals and after rains.	Remove and destroy diseased leaves.
Verbena	Bacterial wilt	No chemical control.	Sterilize soil.
Zinnia	Powdery mildew	Benomyl. Apply as needed, beginning in July.	
	Root and Stem rots	No chemical control.	Sterilize soil.
	Virus	No chemical control.	Plant virus-free seed, and destroy infected plants.
	Leaf spots	Maneb. Spray as needed.	

ALGAE

Controlling Algae

The algae are a group of simple, fresh-water and marine plants ranging from single-celled organisms to green pond scums and very long seaweeds. For our purposes they become problems in swimming pools, aquaria, pet drinking troughs, fish ponds, greenhouses and recirculation water systems.

Swimming pools are a good place to begin. Only chemical control will handle the day-to-day problem of algae. Like all other plants, algae need water, food, light, and a certain temperature range. In swimming pools, the nutrients soon arrive: human urine, body oils, dust and trash blown in and settled on the bottom, fertilizer inadvertently sprinkled beyond the lawn into the water, an over-supply of chlorine stabilizers, and finally nutritional residues of algae themselves. Summer brings the other prerequisites of light, and temperature range.

Algal control is accomplished by (1) keeping the acid/alkaline range (pH) near the ideal, which is 7.4 to 7.6, and (2) maintaining a chlorine level of 0.5 to 1.5 ppm using a chlorine-based algicide that inhibits algal growth and development. This range of pH and algicide level should be monitored twice weekly to avoid algal problems.

Other algicides may be used which are effective against disease-causing bacteria. After the algae are controlled, it is also necessary to prevent the development of harmful bacteria or pollution using a disinfectant. This is achieved through the maintenance of a chlorine level. Since chlorine is also a good algicide, maintaining the proper disinfectant level also serves to prevent algal growth.

That strong smell of chlorine issuing from the YMCA or other public pool and athletic foot baths is probably released from the ever-popular calcium hypochlorite or chloride of lime, not only a good algicide but an excellent disinfectant as well. It contains 70% available chlorine and is the source of most bottled laundry bleaches. In addition to this, there are three other chlorine-based inorganic salts available for pool chlorination: sodium hypochlorite (NaC1O), lithium hypochlorite (LiC1O) and sodium chlorite ($NaC1O_2$). For a source of inorganic chlorine, this author recommends only calcium hypochlorite (70% available chlorine) in its most economical form, usually granular, purchased by the drum.

Soluble forms of copper make excellent algicides. (Remember, however, that they do not control disease-causing bacteria.) Any of the copper-containing algicides are equally effective, and longer lasting, than the chlorine materials. The copper content of swimming pool water may eventually become phytotoxic to grass, plants, shrubs and trees surrounding the pool that may be splashed or drenched occasionally. This same problem applies as well to all other algicides of greater potency.

Algae Control In Aquaria and Fish Ponds Without Chemicals (Almost)

Chemicals are not the only answer for algae control in aquaria and fish ponds, as most fish enthusiasts know. Bottom-feeding scavenger fish help to reduce algae production by feeding directly on algae, and on fecal material of other fish, which serves as a nutritional source for algae. Two or three species of aquatic snails serve equally well, and a heavy population can keep a large aquarium immaculate. A word of caution. Snails multiply rapidly, and can overpopulate the tank in 2-3 months, thus requiring an occasional removal of excess young.

If the proper balance of scavengers and fish are maintained, tank-cleaning can be a seldom-if-ever affair for the proud hobbyist. However, if everything seems to suddenly go wrong and an algae "explosion" occurs, empty the aquarium or pond after transferring the fish to a temporary holding tank, and give it a good scrubbing. Use a stiff bristle brush that will reach the corners and cracks, and a mild detergent. After rinsing, add a 1:1 dilution of bottled laundry bleach and water to the tank and scrub thoroughly. Rinse three times, emptying the contents after each rinse, and invert to drain and dry, if possible. Then refill following any intricate instructions required for certain exotic species of tropical specimens. Scrub the shells of snails with an old toothbrush and tap water to remove excess algae before returning them to the tank.

Mildew, Slime, and Disease-Causing Organisms

Mildew and slimes in shower stalls, on shower curtains, in the drains and overflow openings of bathtubs, sinks and lavatories, and in greenhouses and other moist situations, are various species of fungi and can be easily controlled if the right steps are taken. Fungi are microorganisms whose requirements are water, food and the proper temperature range. Unlike algae, they do not require light and, consequently, appear to thrive where light, especially sunlight, is not available.

Basically we can control these mildew and slime pests by making their environment unfit for survival. It is unlikely that their water or food supply can be removed for more than a brief period, but if this is feasible, it would be the logical process. Scrubbing followed by drying out and keeping dry will prove the demise of the problem.

If this cannot be achieved, the next step is to rely on the use of disinfectants, some of which have already been discussed as algicides. These include the inorganic chlorine compounds, quaternary ammonium halides, and the sodium potassium dichloroisocyanurates.

NEMATODES

Controlling Nematodes

A famous nematologist friend of mine said recently, "The average home gardener won't suspect that he has nematodes unless he has rootknot on some of his plants." Nematodes are microscopic roundworms that are very common and widespread in soil, water and other habitats. Most are free-living and feed on microorganisms such as bacteria, fungi, and algae. About 200 species of the more than 15,000 described species are known to be plant parasites or plant feeders. Most of these feed in or on roots and other below ground plant parts. Some of the more common nematodes and their host plants are shown in Table 26.

The symptoms of nematode injury to plants vary. Since nematodes damage root systems, the visible or above-ground symptoms are the same as those caused by a failing root system, such as lack of vigor, wilting, early leaf fall, yellowing foliage, stunting, and die-back. Infected plants may have galls, decay, and rough, stubby, or black and discolored roots. These symptoms are usually blamed on poor soil or on lack of water and fertilizer. Some of the more common nematodes are the root-knot nematodes, which cause the characteristic galls or knots on roots, easily seen with the naked eye.

If you suspect that nematodes are damaging your crops, send diseased plants to the nearest Extension Plant Pathologist (See Appendix). Most effective nematicides must be applied by a certified applicator using specialized equipment.

The non-chemical methods are really the only control methods I can recommend. They include crop rotation, fallow and dry tillage, and sanitation. In a rotation system, a new planting site should be chosen if an area is available, and not returned to for 3-4 years. Fallow and dry tillage allows the soil to remain uncropped, while removing all vegetation including weeds during the summer and cultivating frequently. In areas of moderate to high rainfall this method is basically impractical.

Sanitation calls for pulling up and disposing of all roots of annual plantings as soon as they are harvested, or when flower blooming is finished. The roots should not be used for compost since they will merely redistribute these wormy pests. Regardless of the care taken in carrying out all other precautions, your efforts are for the most part wasted if nematodes are brought back into the garden on infested plants or tools. Thus, it is necessary to buy clean transplants.

Despite my unwillingness to recommend chemical control, this is the best method. Several types of nematicides and methods of application are available. Some nematicides may be injected into the soil, drenched, sprayed or sprinkled onto the soil surface, applied into furrows along the row to be planted, or applied during watering or irrigation. Some nematicides may be applied prior to planting (preplant), or as with one popular material (DBCP)[1], applied adjacent to living plants (postplant).

Since DBCP (Fumazone and Nemagon) was the only nematicide that could be applied next to the roots of living plants it will be mentioned, despite its unavailability. Certain plants may be damaged or killed if exposed to DBCP and are referred to as susceptible plants. These are beets, carnations, chrysanthemums, dwarf palms, garlic, potatoes, lantana, onions, pepper and sweet potatoes.

Other nematicides which are available to the home gardener but which are not recommended for use by the author are sodium methyldithiocarbamate (Vapam, VPM) and mixtures of dichloropropene-dichloropropane (DD, Vidden D, and Telone). DBCP was also used to treat potted plants monthly in very dilute form during the normal watering.

In all cases it is essential to follow the manufacturer's label instructions carefully.

[1] At the time of this writing, all registered uses for DBCP (dibromochloropropane) have been cancelled by the Environmental Protection Agency.

TABLE 26. Some of the More Common Nematodes and Host Plants Which They are Known to Attack.

Plant	Nematode[2]
Beans and peas	Root knot nematodes Pea cyst nematode Sting nematodes
Citrus	Burrowing nematode Citrus nematode Lesion nematodes
Corn	Lesion nematodes Sting nematodes Stubby root nematodes
Grapes	Dagger nematodes Lesion nematodes Root knot nematodes
Ornamentals	Depending on the plant, almost any nematode
Potatoes	Potato cyst or golden nematode Potato rot nematode Root knot nematodes
Stone fruits (Peaches, Plums, etc.)	Lesion nematodes Root knot nematodes Stubby root nematodes
Tomato	Lesion nematodes Root knot nematodes
Turf	Lesion nematodes Root knot nematodes Turf sting nematode

[2] See Table 27 for common and scientific names.

TABLE 27. The Common and Scientific Names of Some of the More Important Nematodes.

Common Name	Scientific Name
Bulb and Stem Nematodes	*Ditylenchus spp.*
Bulb or Stem Nematode	*D. dipsaci*
Potato Rot Nematode	*D. destructor*
Burrowing Nematode	*Radopholus similis*
Citrus Nematode	*Tylenchulus semipenetrans*
Cyst Nematodes	*Heterodera spp.*
Cabbage Cyst, Cabbage Root or Brassica Root Nematode	*H. cruciferae*
Pea Cyst or Pea Root Nematode	*H. goettingiana*
Potato Cyst, Potato Root or Golden Nematode or Potato Eelworm	*H. rostochiensis*
Soybean Cyst Nematode	*H. glycines*
Sugar Beet Cyst or Sugar Beet Nematode or Beet Eelworm	*H. schachtii*
Dagger Nematodes	*Xiphinema spp.*
False Root Knot Nematodes	*Naccobus spp.*
Foliar or Leaf Nematodes	*Aphelenchoides spp.*
Chrysanthemum Foliar Nematode	*A. ritzemabosi*
Lance Nematodes	*Hoplolaimus spp.*
Lesion or Meadow Nematodes	*Pratylenchus spp.*
Needle Nematodes	*Longidorus spp.*
Pin Nematodes	*Paratylenchus spp.*
Ring Nematodes	*Criconemoides spp. and Criconema spp.*
Root Knot Nematodes	*Meloidogyne spp.*
Northern Root Knot Nematode	*M. hapla*
Peanut Root Knot Nematode	*M. arenaria*
Southern Root Knot or Cotton Root Knot Nematode	*M. incognita*
Sheath Nematodes	*Hemicycliophora spp. and Hemicriconemoides spp.*
Spiral Nematodes	*Helicotylenchus spp. and Rotylenchus spp.*
Sting Nematodes	*Belonolaimus spp.*
Peanut Sting Nematode	*B. gracilis*
Turf Sting Nematode	*B. longicaudatus*
Stubby Root Nematodes	*Trichodorus spp.*
Stunt or Stylet Nematodes	*Tylenchorhynchus spp.*

WEEDS

. . . lest in gathering the weeds you root up the wheat along with them. Let both grow together until the harvest; and . . . I will tell the reapers, Gather the weeds first, and bind them in bundles to be burned, but gather the wheat into my barn.

A Parable,
Matthew 13:29-30

THE WEEDS

When mankind ceased being the nomadic hunter, and settled down periodically to become the farmer, he encountered the third plague, weeds. And until recently he controlled weeds the same old ways, year-in and year-out, for thousands of years: hand pulling, hoeing, and cultivation. And, dear gardener, those are the choices you have, with the addition of mulching and herbicides.

It is not intended that this brief introduction to weed control also include a brief course in weed identification. That is much too great an assignment. Besides, there are numerous weed identification manuals available through every state Land Grant University, usually from the Cooperative Extension Service. (See Appendix for addresses).

On the following pages are illustrated 30 of the most common and widely distributed broadleaf and grass, perennial and annual weeds in North America. In many regions of the nation most of these can be found within a short radius. These illustrations were taken from "Weeds of the North Central States," Circular 718 of the University of Illinois Agricultural Experiment Station. (Slife, Bucholtz, and Kommedahl, 1960).

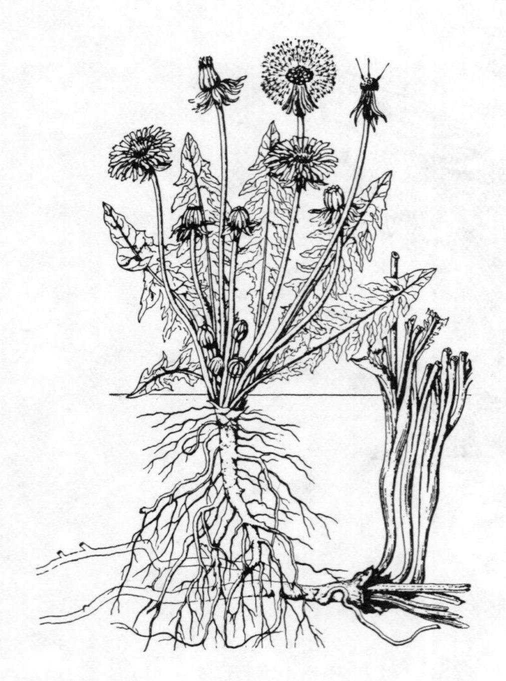

Dandelion *(Taraxacum officinale)*.

Canada Thistle, Creeping thistle *(Cirsium arvense)*.

Cocklebur, Clotbur *(Xanthium pennsylvanicum)*.

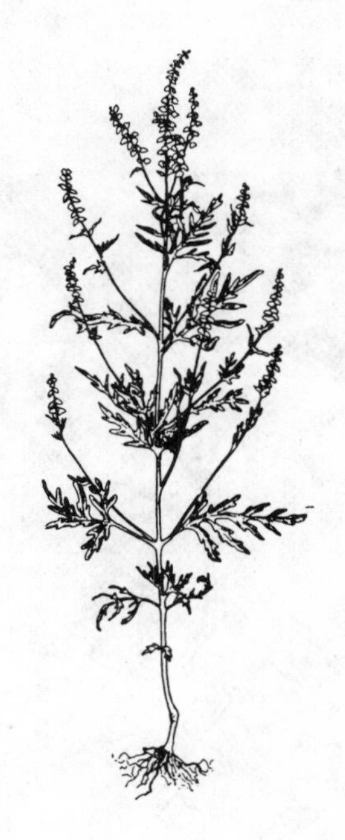

Common Ragweed *(Ambrosia artemisiifolia)*.

Wild Onion *(Allium canadense)*.

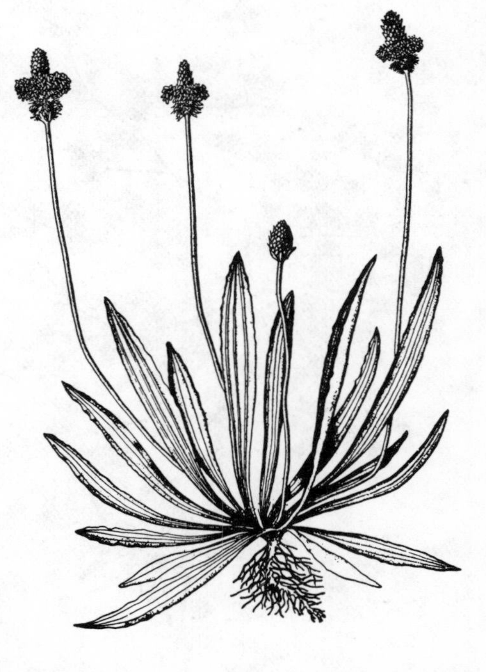

Buckhorn Plantain, Ribgrass *(Plantago lanceolata)*.

Buffalo Bur *(Solanum rostratum)*.

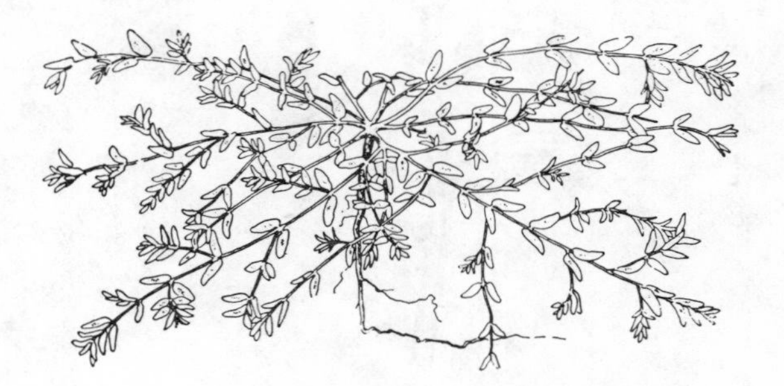

Prostrate Spurge, Milk purslane *(Euphorbia supina)*.

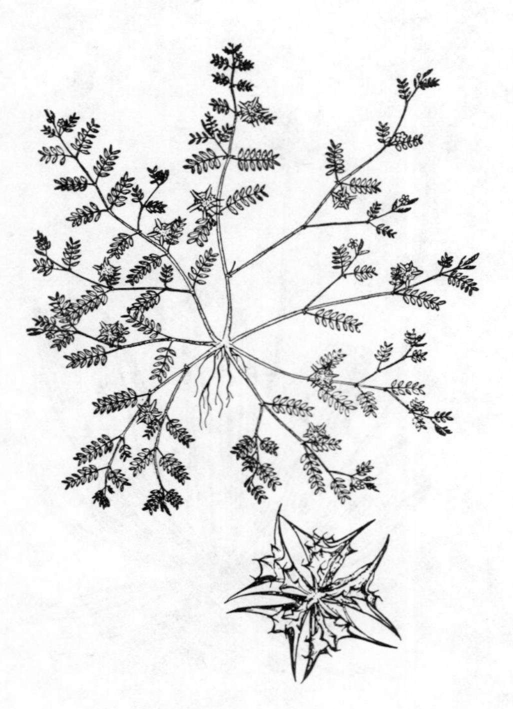

Puncture Vine, Caltrop *(Tribulus terrestris)*.

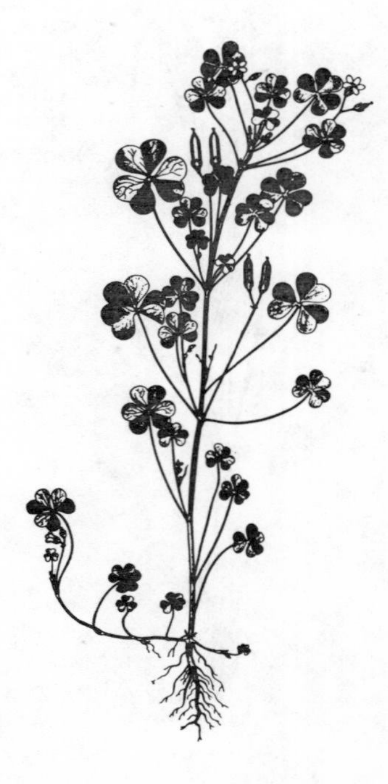

Yellow Wood Sorrel *(Oxalis europaea)*.

Shepherd's Purse *(Capsella bursa-pastoris)*.

Rough Pigweed, Redroot *(Amaranthus retroflexus)*.

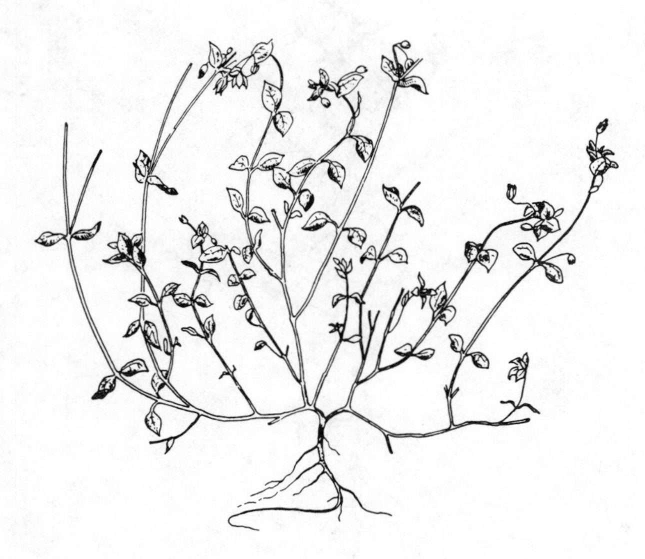

Common Chickweed *(Stellaria media)*.

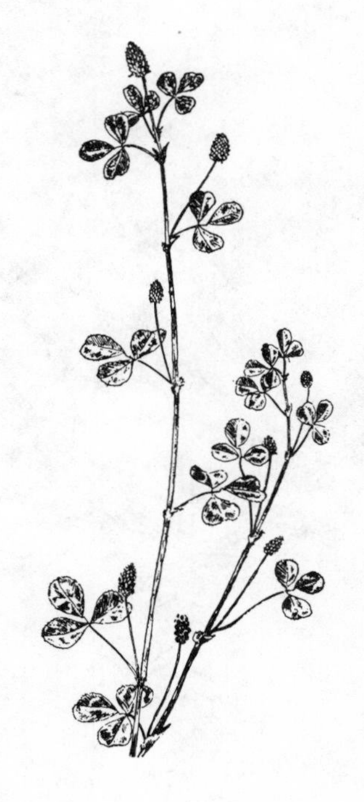

Black Medic, Yellow trefoil *(Medicago lupulina)*.

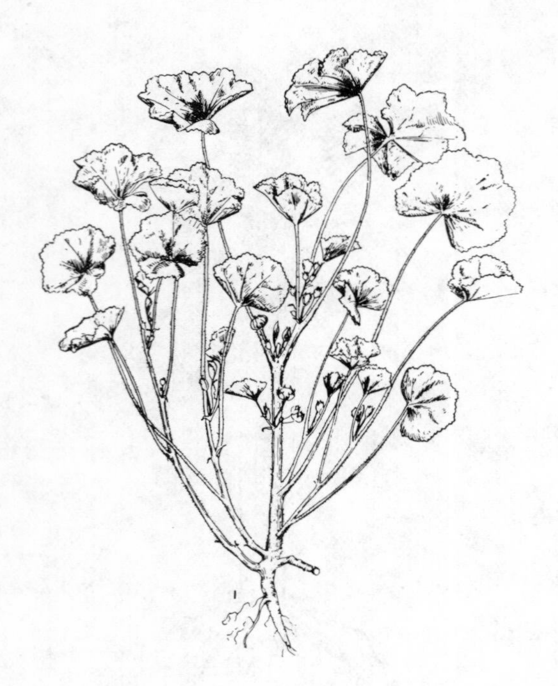

Roundleaved Mallow, Cheeses *(Malva neglecta)*.

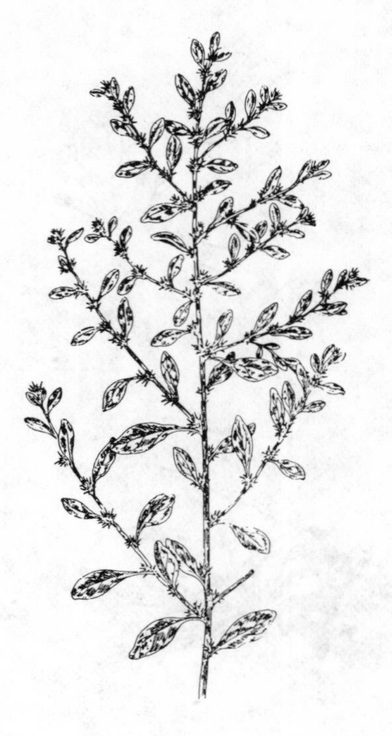

Tumbleweed, Tumble amaranth *(Amaranthus albus)*.

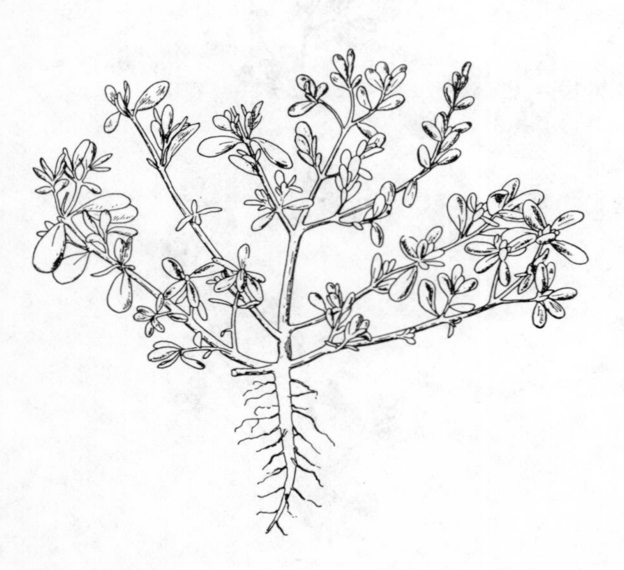

Purslane, Pusley *(Portulaca oleracea)*.

Russian Thistle, Common saltwort *(Salsola kali)*.

Yellow Rocket, Winter cress *(Barbarea vulgaris)*.

Lambsquarter *(Chenopodium album)*.

Bermuda Grass, Devilgrass *(Cynodon dactylon)*.

Yellow Foxtail, Yellow bristlegrass, Pigeon grass *(Setaria lutescens)*.

Goosegrass, Yardgrass, Silver crabgrass
(Eleusine indica).

Nimblewill *(Muhlenbergia schreberi)*.

Large Crabgrass, Large hairy crabgrass
(Digitaria sanguinalis).

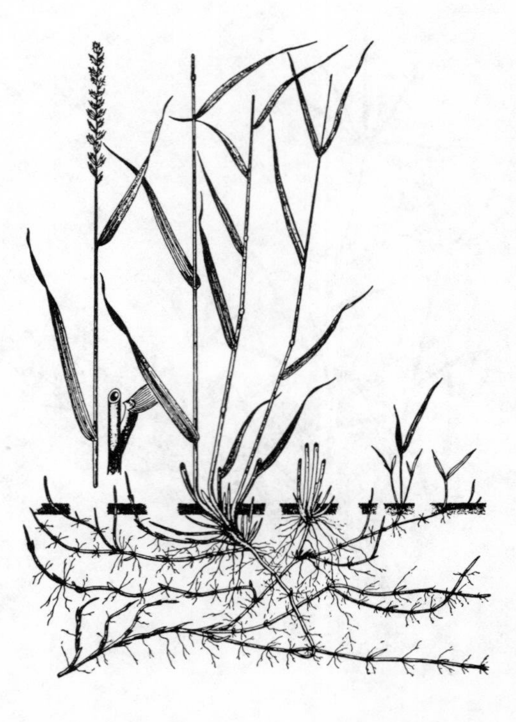

Quackgrass, Couchgrass *(Agropyron repens)*.

Yellow Nutgrass *(Cyperus esculentus)*.

Barnyard Grass *(Echinochloa crusgalli)*.

Fall Panicum, Spreading panicgrass
(Panicum dichotomiflorum).

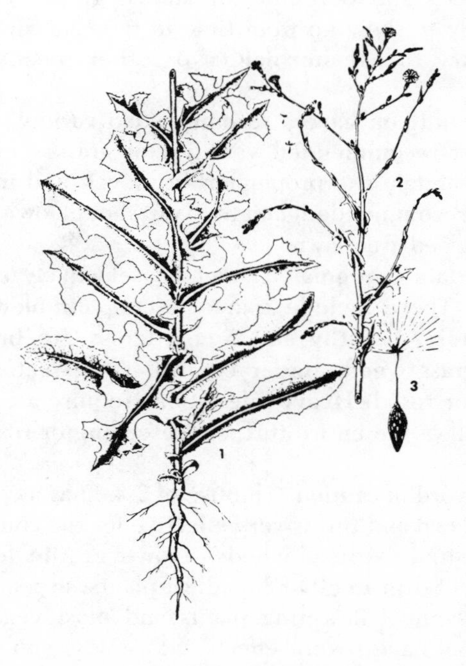

Prickly Lettuce
(Lactuca scariola)

Controlling Weeds

Weed control with herbicides is indeed both an art and a science. The art involves the delicate placement of the materials so as not to become a threat to non-target plants. The science requires the precise selection, mixing and timing of the herbicide. The art without the science, or vice versa, would result in total failure on every attempt.

Straight off, I do not recommend the use of herbicidal weed control in the home vegetable or flower garden. Most home gardeners are rank amateurs with respect to the art of herbicide employment. And, I am reluctant to recommend their use around woody ornamentals and trees. The benefit:risk ratio is too much out of proportion. Finally, I can suggest the use of herbicides for lawn and turf weed control. Very little art is required and the home gardener normally has a large enough target that a direct hit is generally inevitable. So, with that introduction, let's discuss briefly weed control in lawn and turf.

Several kinds of pests have a way of getting into lawn and turfgrass and weeds are no exception. Although a dense, healthy stand of grass is the most satisfactory method of controlling many weeds, it is not a fool-proof method. Broadleaf weeds and perennial or annual grass weeds are likely to show up from time to time, making it necessary to use herbicides or other means of control.

Should you have a recommended variety or a blend of recommended varieties of grasses for a lawn, maintain the proper fertility level, and mow at the recommended height, you should own an almost weed-free lawn.

Certain perennials cannot be selectively controlled. These include coarse fescue, nimblewill, quackgrass, timothy and orchardgrass. To bring these grass weeds under control, spot treat and reseed or resod. Heavy infestations require a complete kill of the entire turf and reseeding or resodding.

A word of caution. The use of 2,4-D as a spray for the lawn and turf is very effective for the control of certain broadleaf weeds. However the least amount of drift to other broadleaf plants, especially grapes, annual flowering plants and most vegetables, may have severe effects. So, unless you're a better artist than most home gardeners, don't spray this compound.

Weed Classification

The homeowner and gardener need to be able to recognize grass and broadleaf weeds because they differ in desirability as well as reaction to herbicides, culture, and various methods of control. For weed control purposes, plants are divided into three groups — grass, broadleaf, and woody.

Grass plants have one seed leaf. They generally have narrow, verticle, parallel-veined leaves and fibrous root systems. Broadleaved plants have two seed leaves, and they also generally have broad, net-veined leaves and tap roots, or coarse root systems. Woody plants include brush, shrubs, and trees. Brush and shrubs are regarded as woody plants that have several stems and are less than 10 feet tall.

Growth Patterns vs. Weed Control

Knowing the growing habits of annuals is important in planning how and when to control weeds. Annual plants complete their life cycles from seed in less than one year. Winter annuals germinate in the fall, survive the winter, mature, develop or set seed, and die in the spring or early summer when the weather grows warm and the days long. For best results with weed control, winter annuals should be controlled in the seedling stage of growth in the fall or early spring. Summer annual weeds germinate in the spring, grow, develop seed, and die before or during the fall. Summer annuals should be controlled soon after germination in the seedling stage. Some weeds are specifically winter or summer annuals, while other species can germinate and grow either in the fall or spring.

Biennial weeds complete their life cycles within two years, as their name implies. The first year the biennial weed forms basal or rosette leaves and a tap root. The second year it flowers, matures, develops seed and dies. Biennial weeds should be controlled in the first year of growth for best results.

Perennial plants live more than two years and may live indefinitely. Perennials reproduce by seed, and many are able to spread and reproduce vegetatively. They are difficult to control because of their persistent root system. Seedling perennials should not be allowed to become established. An alert gardener will adapt control of established perennials to the yearly growth cycle of the dominant weed species. Perennials should be controlled during the fast growth period prior to flowering or during the regrowth period after fruiting or cutting.

Simple perennials spread by seed, crown buds, and cut root segments. Most have large, fleshy tap roots.

Creeping perennials spread vegetatively as well as by seed. Grasses generally have a shallow root system compared to the deep root system of broadleaf plants.

Bulbous and tuberous perennials reproduce vegetatively from underground bulbs or tubers. Many also produce seed.

Brush, shrubs and trees may spread vegetatively as well as by seed. Woody plants can be controlled at any time of year.

When to Control with Herbicides

All weeds pass through four stages of growth: seedling, vegetative, flowering, and maturity. There is a best stage with each weed for its control. If control is not obtained at the best stage of growth, the control method may require changing.

Seedling growth is the same for annuals, biennials and perennials. All start from seed. Seedling weeds are small and succulent, requiring less effort for control than at any other stage. This holds true whether the effort is derived mechanically, chemically, or culturally.

For annuals, the vegetative stage is producing energy for production of stems, leaves, and roots, at this stage control is still feasible but more difficult than during the seedling stage.

As an annual weed continues to develop, a chemical messenger formed by the plant tells it to change from vegetative to flowering. At this time, most of the weed's energy goes to seed production. Chemical control now, for both broadleaf and grass, is not feasible because eliminating these older plants requires more effort than can be provided through herbicidal action.

Maturity and setting of seed in annuals complete the life cycles. Chemical control is no longer effective. The damage is done — viable seeds were produced, and the cycle is ready to be repeated.

Practically 100% of annual weeds can be controlled when herbicide application is made at the seedling stage. Applied later at the vegetative stage, control drops to 75%, and only 40% is achieved during the flowering stage. When the herbicide is applied at plant maturity, virtually no control is obtained.

The formula for weed control described above applies also to biennials. The only distinction is that biennials go through the same stages as annuals, but require two years.

Perennial weeds are a bit more complicated. The seedling stage of growth and its control are the same as for annuals and biennials. However, the stages of growth from vegetative through maturity are different. It should be pointed out here that shoots which emerge from established roots are not seedlings.

During the vegetative stage of perennials, part of the energy used in the production of stems and leaves is derived from energy stored in the underground roots and stems. Additional energy results from photosynthesis in the leaves. Chemical control is only mediocre at best during this stage of growth but improves as the weed approaches the bud stage.

The flowering stage in perennials, as with annuals, involves a messenger which directs the plant's energy into production of flowers and seed. Food storage in the roots is initiated and continues through maturity. Chemical control is most effective just prior to flowering, referred to as the bud stage.

With perennials, only the above ground portions die each year. The underground roots and stems remain very much alive through the winter and develop new plant growth the following spring. Control with most chemicals is not feasible at this stage, despite the temptation. However, a select few herbicides will give control when applied to mature plants. The herbicide label will provide this information if it applies.

Optimum herbicidal control of perennial weeds is obtained by treating perennials during the bud and regrowth stage, thus causing the greatest drain on the underground food reserves. Treatment at early flowering usually equals that obtained at the bud stage. However, when perennials reach full flowering, control declines. Application to regrowth following this is beneficial.

Fall applications of herbicides offer one important advantage — environmental safety. In the fall, desirable plants in gardens, lawns, and shrubs, have completed their growth and escape the effects of the herbicide.

Additionally, fall herbicide applications reach the underground plant parts through the natural translocation activity of the plant. In the fall, nutrients are moved from above ground parts in advance of the first killing frost and are stored over winter beneath the frost line. For this reason perennial weeds are most susceptible to herbicides in the fall. The herbicide moves with the above ground nutrients to the underground storage organs where control is achieved.

Biennials which develop from seed the first year and overwinter in a rosette, are also controlled by fall applications. Winter annuals which germinate in the fall are also controlled by fall applications.

Fall herbicide applications provide the target weed with three stresses: (1) herbicidal effects, (2) winter effects, and (3) effects of heavy nutrient demand caused by rapid spring growth. Fall application thus increases the chances of controlling the toughest of weeds, biennials and perennials.

Climatic Factors

Climatic factors influence weed control more than with any other form of pest control. These include temperature, humidity, precipitation, and wind.

As the temperature increases, the effect of herbicide activity increases. Weed control results are the same, regardless of temperature, but the higher the temperature the faster the kill. There are exceptions. Volatile herbicides should not be applied at the higher temperatures.

Low humidity reduces penetration of the herbicide into weed tissue. When a plant is growing under humid conditions, a foliar-applied herbicide will enter the leaf more easily and rapidly than at low humidity when penetration is slow. With high humidity the weed leaf is more succulent, has a thinner outer wax layer, and a thinner cuticle.

Rainfall, occurring after a foliar-applied herbicide treatment, may decrease its effectiveness. Soil-applied herbicides will be activated by rain. Heavy rains, however, may move the herbicide through and away from the target zone.

Wind can be hazardous to herbicide applications, by causing drift of the herbicide during spraying as well as movement of herbicide-laden dust particles. Wind and high temperature can also affect the weed. Hot, dry winds will cause plant stomata to close, leaf surfaces to thicken, and wax layers to harden, thus making herbicide penetration more difficult.

In closing, here are a few do's and don't's to follow to achieve the best results from your efforts with herbicides: Use as low pressure as possible in your weed sprayer; don't treat clear to the edge (leave an untreated edge); use less volatile formulations of the herbicide; spray when wind speed is near zero; spray when adjacent sensitive vegetation is either mature or not present and preferably upwind; and, finally, remember thy neighbor!

Mulches

The use of mulches in flower or vegetable gardens may well be your most valuable garden weed control practice. Good mulches reduce soil blowing and washing, prevent weed germination and growth, keep the soil moist and cool, and generously add to the organic matter in the soil.

Leaves, grass clippings, sawdust, bark chips, straw and compost make excellent mulches, and they are easy to apply. Simply spread a 3-6 inch layer of one of these organic materials on the soil surface around plants, making certain not to cover them. It is important to keep the layer deep enough to do the job. This means that it will be necessary to add more mulching material over the old layers to obtain all of the benefits of mulching.

Using lawn clippings as a mulch is a good way to dispose of a by-product that is usually hauled off with the trash. However, they may need to be mixed with other mulch materials to avoid compacting and preventing water from penetrating into the soil.

Sawdust and bark chips make better mulches if they are well rotted, or if 1 to 2 cups of ammonium sulfate or sodium nitrate are added to each bushel of fresh sawdust or chips before applying. Weed-free straw is excellent but loose straw can be a fire hazard. Sometimes compost is the best mulch, and it can be made from leftover plant materials from your garden.

Mulches prevent loss of moisture from the soil by evaporation. Moisture moves by capillary action to the surface and evaporates if the soil is not covered by a mulch. Sun and wind hasten this moisture loss.

You can reduce evaporation and control weeds by stirring the soil an inch or so deep, but plant roots cannot develop in this soil layer. A layer of organic material on the surface gives the same benefits and allows normal plant-root development.

The splattering action from falling raindrops is dissipated on a mulched soil. The result is less soil erosion and less soil compaction. Mulches suppress weeds, thus saving hours of back-breaking work. An occasional weed may poke through the mulch, but it is easily pulled because of the soft soil texture in which it is rooted.

Mulches prevent the soil from getting hot under intense sunlight. Many plants, including those in vegetable and flower gardens, need a cool soil surface.

Mulches, especially grass clippings and compost, add organic matter to the soil and furnish food for earthworms, which are valuable in aerating soil. The organic matter helps to keep the soil loose and easy to work. Farmers refer to this as good tilth. At the end of the growing season, the mulch can be worked into the soil to supply organic matter the following year. When mulches are used around perrennials in the winter, the mulches should be removed in the spring to allow the soil to thaw and warm up.

Many organic materials, such as straw and autumn leaves, are rich in carbohydrates and low in nitrogen. It is beneficial to add nitrogen fertilizer to the material before applying it as a mulch, which causes it to break down quicker and to avoid nitrogen deficiency. One to two cups of fertilizer high in nitrogen, such as ammonium sulfate or ammonium nitrate, should be allowed for each bushel of mulch. Direct contact of fertilizer with plants will cause burning.

To provide a continuous source of the best mulch, every gardener should have a compost bin, preferably two, for making compost from organic materials. Bins can be made by attaching ordinary wire fence or boards to solid posts or with open brickwork. Each bin should be 4 to 6 feet high, 3 to 5 feet wide, and any convenient length. One side of each should be removable for convenience in adding and removing the compost material. In late fall, a temporary piece of wire fence may be used to increase the height about 2 feet, which can be removed in the spring after the material settles.

Compost is not only the ideal mulch, but it is also a good fertilizer and soil conditioner when worked into the soil. Leaves, grass clippings, stems and stalks from harvested vegetables, corn husks, pea hulls, and fine twigs are good materials for composting. Always compost or shred leaves before using as a mulch. Raw leaves are flat and may prevent water from entering the soil. Avoid using any diseased plants, particularly the roots of plants having rootknot (nematodes) in the compost or mulch.

The ideal way to make compost is to use two bins. Fill one with alternate layers of organic material 6 to 12 inches thick and of garden soil about 1 inch thick. To each layer of organic material, add chemicals at the following rate: for each tightly-packed bushel of organic materials, ammonium sulfate, 1 cup; or ammonium nitrate, one-half cup,; or superphosphate, one-half cup and magnesium sulfate (Epsom salt) 1 tablespoon; or mixed fertilizer (5-10-5), 3 cups. Avoid the use of lime or ashes since their alkalinity causes loss of ammonia (NH_3) and is unneeded.

The organic material should be moistened thoroughly. Repeat this layering process until the bin is filled. Pack the material tightly around the edges but only lightly in the center so the center settles more than the edges and retains the water.

After 3 to 4 months of moderate to warm weather, commonly in June, begin turning the material by moving it from the first bin to the second. Before turning, it is best to move the material added the previous fall from the edges to the center because it has probably dried out.

In areas that have cool frosty winters, compost made from leaves in November and December can be turned the following May or June. Additional information can be obtained from your local County Agricultural Agent.

TABLE 28. Weed Control in the Home Garden With and Without Chemicals[1].

Garden Plants	Weeds To Control	Herbicide to Use	Non-Chemical Control
Seeded crops (beans, beets, carrots, corn, cucumber, lettuce, potatoes, spinach, etc.)	Annual weeds (pigweed, lambsquarter purslane crabgrass foxtail)	CDEC (Vegadex). Work soil thoroughly prior to planting. Treat soil with CDEC spray or granules immediately after planting. Rainfall or watering required within 24 hours.	Give garden a good, 6″ deep, cultivation before planting. Shallow cultivation, hoeing, hand pulling, and black plastic mulches are effective after garden has germinated.
		To control weeds between rows after crops have emerged, spray with undiluted paint thinner, diesel fuel or kerosene. Do not spray when windy. Drift is toxic to all green leaves.	
	Perennial and biennial weeds (quackgrass, thistles, bindweed, yellow rocket, curly dock)	Herbicide control is not recommended.	Black plastic mulch (1.5-4 mil is generally the most satisfactory.)
Transplanted crops (broccoli, cabbage, celery, melons, peppers, tomatoes)	Annual weeds (pigweed, lambsquarter, purslane, crabgrass, foxtail)	Stir soil thoroughly before transplanting. Treat soil as for seeded crops, but immediately after transplanting. Rainfall or watering required within 24 hours.	Give garden a good, 6″ deep, cultivation before planting. Shallow cultivation, hoeing, hand pulling, and black plastic mulches are effective after plants have begun growth.
	Perennial and biennial weed control and foliage weed treatments are the same as for seeded crops.	Herbicide control is not recommended.	Black plastic mulch.

[1] Refer to Table 37 for trade or proprietary names of recommended pesticides.

TABLE 29. Weed Control Suggestions for Lawn and Turf[1].

Lawn or Turf to Be Protected	Weeds	Herbicide to Use
Bentgrass (pre-emergence control)	Crabgrass Foxtail	Bensulide. Apply uniformly in spring before crabgrass emergence. Dacthal. Apply in early spring before crabgrass emergence. Do not use on Cohansey or Toronto varieties.
	Annual bluegrass	Bensulide. Apply uniformly in Aug.-Sept. before annual bluegrass emergence. Dacthal. Apply in late summer and early spring. Do not use on Cohansey or Toronto varieties.
Bermudagrass turf	Spurge and most annual weeds	Dacthal. Apply to soil in spring before weeds germinate. A second treatment in mid-summer may be needed. Water after application. Do not plant winter lawns where DCPA was used in summer or fall.
	Spurge and most broadleaf weeds.	Bromoxynil plus detergent. Apply to foliage when weeds are small. Repeat if new weeds emerge. May cause temporary yellowing of Bermudagrass.
	Crabgrass and most annual grass weeds.	Dacthal, Siduron, Trifluralin, Bensulide or Benefin. Apply to soil before crabgrass germinates (Feb.-April). Water after treatment.
	Annual bluegrass	Dacthal, Bensulide, Trifluralin, or Benefin. Apply to soil before bluegrass germinates in fall (Aug.-Sept.). Water after application. Do not use where winter lawns are to be planted.
	Crabgrass seedlings	DSMA or MSMA. Apply to crabgrass in June or July. Repeat if needed. May control other small seedlings.
	Established nutsedge, creeping chaffweed, woodsorrel and many perennial weeds	Glyphosate. Apply to weeds as spot treatment. Repeat in 8 weeks or when normal growth occurs. This treatment may injure desirable plants.
Bluegrass and Fescue (pre-emergence control)	Barnyardgrass Crabgrass Foxtail	Benefin. Apply uniformly in early spring before crabgrass emergence. Bensulide. Apply uniformly in early spring before crabgrass emergence. Dacthal. Apply before crabgrass emergence in early spring.
	Goosegrass	Bensulide or Dacthal. Same as for crabgrass. Goosegrass is more difficult to control than crabgrass, so don't expect more than 70-80% control.
Bluegrass and Fescue (post-emergence control)	Nutsedge	2,4-D Amine. Apply when actively growing. At least 3 applications at 10-14 day intervals are required to give control.
Seedling Turf	Broadleaf weeds	Bromoxynil. Follow label instructions. Apply after grasses emerge and before broadleaf weeds are past the 3-4 leaf stage.

[1] Refer to Table 37 for trade or proprietary names of recommended pesticides.

Lawn or Turf to Be Protected	Weeds	Herbicide to Use
Turf Conversion — Converting Bermudagrass to winter lawn of ryegrass.		Cacodylic acid plus detergent. Wet foliage thoroughly 2 to 6 weeks before frost. Wait 1-2 days for final seedbed preparation and planting.
Converting ryegrass to summer Bermudagrass.		Cacodylic acid plus detergent. Wet foliage thoroughly when night temperatures reach 60°F. Repeat treatment in 1 week if kill is incomplete.
Bermudagrass turf renovation.	Removal of Bermudagrass to establish new turf.	Glyphosate. Apply to foliage when Bermudagrass has headed out. After 2 weeks spade or renovate. New seeding or sprigging may be made immediately. Avoid drift onto shrubs or trees.
		Dalapon plus detergent. Apply to foliage when Bermudagrass is growing rapidly. Wait 7 days, spade or renovate and water. Wait 2-3 weeks before establishing new turf. Do not apply near trees and shrubs.
Dichondra turf	Annual grass, Wild celery, and other broadleaf weeds.	Diphenamid. Apply to soil in spring and fall, followed by thorough watering.
Gravel or stone yards and desert landscape	Most annual weeds	Dacthal or Trifluralin. Apply to soil before weed seed germinate, and water thoroughly.
	Annual weeds	Diquat or Cacodylic acid plus detergent. Wet foliage of small weeds thoroughly. Repeat when new weed seed germinate.
	Perennial weeds	Glyphosate. Apply to foliage when weeds are growing actively. Repeat in 8 to 12 weeks. Avoid drift onto shrubs and trees.
		Diquat or Cacodylic acid plus detergent. Wet foliage of weeds thoroughly, and repeat every 1-3 weeks as long as growth continues.
Non-selective Control	Fescue Nimblewill Orchardgrass Quackgrass Timothy	Dalapon, Amitrol or Dalapon + Amitrol. Wet foliage of actively growing grass in spring or summer. Repeat at 7 to 10-day intervals until complete control is obtained. Wait 30 days after using Dalapon and 21 days after Amitrol before reseeding or resodding.
Turf, General	Black medic	2,4-D Amine. Spray or granules.
	Chickweed	2,4-D Amine. Spray or granules.
	Clover	2,4-D Amine. Spray or granules.
	Crabgrass (pre-emergence)	Dacthal, Siduron, Bensulide, or Benefin. Follow label directions carefully.
	Dandelion	2,4-D Amine granules.
	Plantain (narrow and broad-leaved)	2,4-D Amine spray or granules.
	Wild onion	2,4-D Amine. Spray plants twice, 5 days apart.

TABLE 30. Weed Control in Shrubs and Woody Ornamentals[1].

Weeds to Control	Herbicide to Use
Annual weeds and perennial grass weeds from seed	Dacthal or Trifluralin. Apply to soil before weed seeds germinate, and water thoroughly.
Bermudagrass and nutsedge	Dichlobenil. Apply to soil and incorporate. Remove tops of weeds before or after treatment. Granular formulation is easier to use. Requires 2-3 applications per year for good control.
Nutsedge	Eptam. Apply to soil and incorporate. Remove tops before or after treatment. Granular formulation is easier to use. Requires 2-3 applications per year for good control.
All annual broad leaved and grass weeds	Kerosene, Paint Thinner, Diesel Fuel, or Gasoline. Apply with compressed air sprayer directly to weeds. These are generally very safe to use around delicate shrubs if applied during the quiet hours to avoid drift. When in doubt about the standard herbicides, these work like a charm and are fast, but will require retreatment in 10-14 days.

[1] Refer to Table 37 for trade or proprietary names of recommended pesticides.

RODENTS

Three blind mice,
see how they run!
They all ran after
the farmer's wife,
She cut off their tails with
a carving-knife,
Did you ever see
such a sight in your life,
As three blind mice?
Old English
Nursery Rhyme

CONTROLLING RATS, MICE AND OTHER MAMMALS

Rodents and several other small mammals damage man's dwellings, his stored products, and his cultivated crops. Among these are native rats and mice, squirrels, woodchucks, pocket gophers, hares, and rabbits. Rats are notorious freeloaders, and in some of the underprivileged countries where it is necessary to store grain in the open, as much as 20% may be consumed by rats before man has access to it.

The number of rodent species (order Rodentia) comprises about one-half of all mammalian species, and because they are so very highly productive and widespread, they are continuously competing with man for his food.

Rats and Mice

Of these, rats and mice are the most abundant, and consequently the most annoying and destructive. Rats and mice have accompanied man to probably all of the areas of the world in which he has settled, except perhaps the Arctic and Antarctic. Historically, they have been responsible for more human illnesses and deaths than any other group of mammals. Because of man's indifference and carelessness in handling food and trash-garbage, he has fostered populations of rats and mice in such close proximity to his home and place of work that they are referred to as domestic rodents.

These domestic rodents are the Norway rat, roof rat and house mouse, all members of the family Muridae, Order Rodentia (See Figure 20). The Muridae have a single pair of incisor biting teeth on each jaw and do not have canine teeth. (I really don't expect you to look to be certain!)

Norway rats are burrowing rodents and are the most common and largest of the rats. They are generally distributed throughout the U.S. and are known by several common names: wharf rat, brown rat, house rat, sewer rat and barn rat. The adults average one pound in weight, and have course, reddish brown fur, and a blunt nose. The droppings or fecal pellets are large, up to 3/4 inch long. They reach sexual maturity in 3 to 5 months, and have a gestation period averaging 22 days, each female averages about 20 young per year, born in 4 to 7 litters, with 8 to 12 young per litter. As in most rodents, there is considerable infant mortality.

The Norway rat lives an average of about one year. It prefers the outdoors and burrows in the ground under foundations and in trash dumps. Indoors it lives between floors and walls, in enclosed spaces of cabinets, shelving and appliances, in piles of trash, and in any place concealed from view. It ranges usually no more than 100 to 150 feet. It feeds on all foods, eating from 3/4 to 1 ounce of dry food and 1/2 to 1 ounce of water per day.

The roof rat is smaller than the Norway, is more agile, and lives mostly in the warmer areas of the U.S., namely the South and to the Pacific coast and Hawaii. It is rare or absent in the colder portions of the world. The adults are smaller than the Norway rat, weighing only 8 to 12 ounces. They are black in color and have a slender, pointed nose and a tail that is longer than the body and head combined. Its droppings are up to 1/2 inch long. They reach sexual maturity in 3 to 5 months, have a gestation period averaging 22 days, and average about 20 young per female per year, born in 4 to 6 litters, with 6 to 8 young per litter.

The roof rat also lives about a year. It appears to prefer above-ground dwellings, in attics, between walls, and in enclosed spaces of cabinets and shelving, or outdoors in trees and dense vines. It, too, ranges usually no more than 100 to 150 feet. It prefers fruits, vegetables and grains compared to the high protein preference of the Norway rat. It requires 1/2 to 1 ounce of dry food and up to 1 ounce of water per day.

The house mouse is the smallest of the household rodents and is found throughout the United States. It weights 1/2 to 3/4 ounce, is dull gray, and has a tail as long as its body and head combined. Its droppings are small, about 1/8 inch long. It reaches sexual maturity in 6 weeks, has a gestation period averaging 19 days, and weans 30 to 35 young per female per year, born in 6 to 8 litters, with 5 to 6 young per litter. They live about 1 year, and nest in any convenient, protected space. They range only up to about 30 feet. Mice will eat most anything, but prefer grain, nibble constantly, and require only 1/10 ounce of food and 3/10 ounce of water per day.

Control. Rat and mouse populations are controlled by storing of all food materials in rodent-proof containers, collecting and disposing of refuse, and the proper storage of usuable materials. Permanent removal of harborage and sources of food will eliminate existing rat and mouse populations.

Trapping is useful when poisoning fails or is too hazardous, or where the odor of unrecovered dead rodents would be a problem. If for some reason you wish not to kill these pests, live traps are available that permit their removal and relocation.

FIGURE 20.

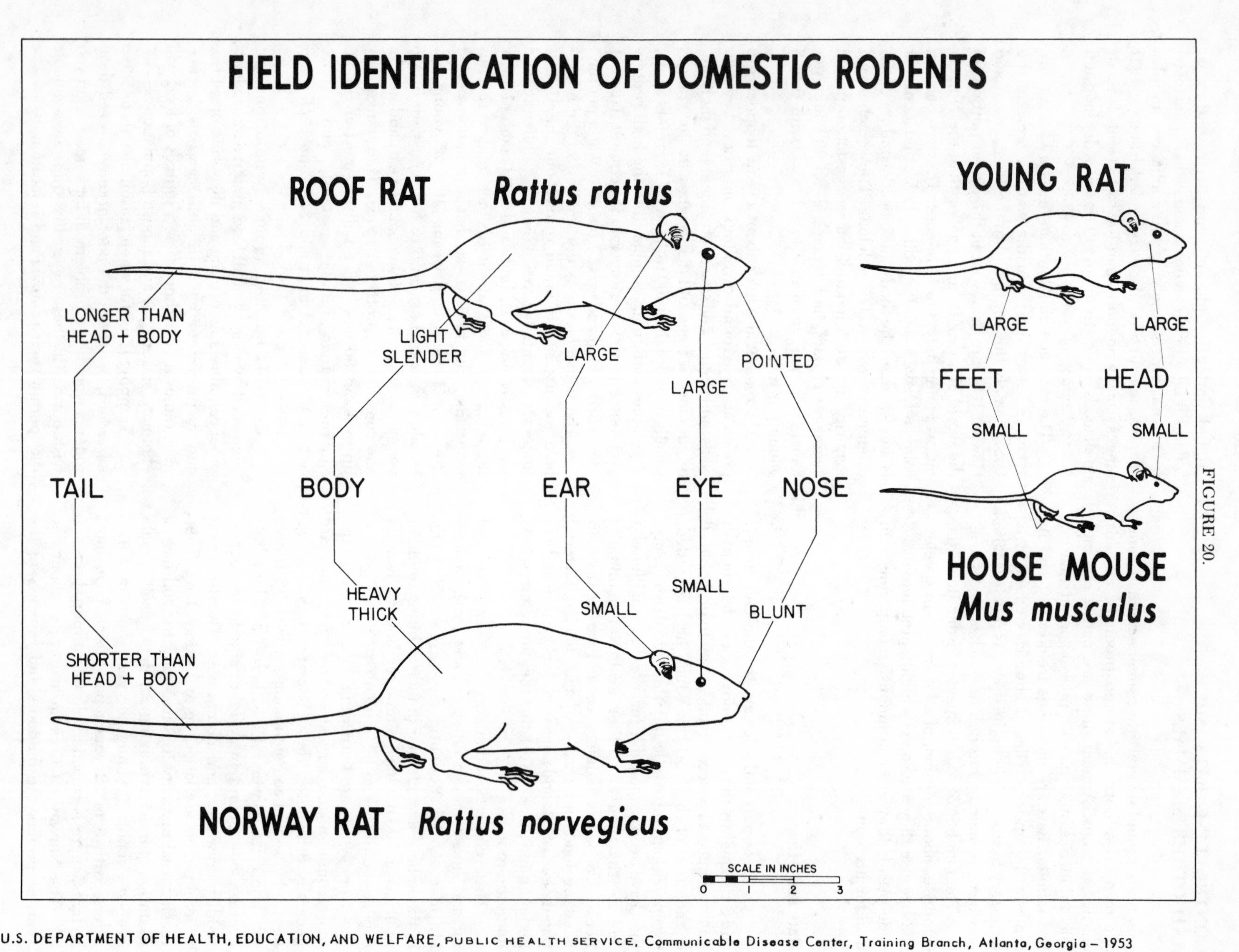

U.S. DEPARTMENT OF HEALTH, EDUCATION, AND WELFARE, PUBLIC HEALTH SERVICE, Communicable Disease Center, Training Branch, Atlanta, Georgia – 1953

FIGURE 21.

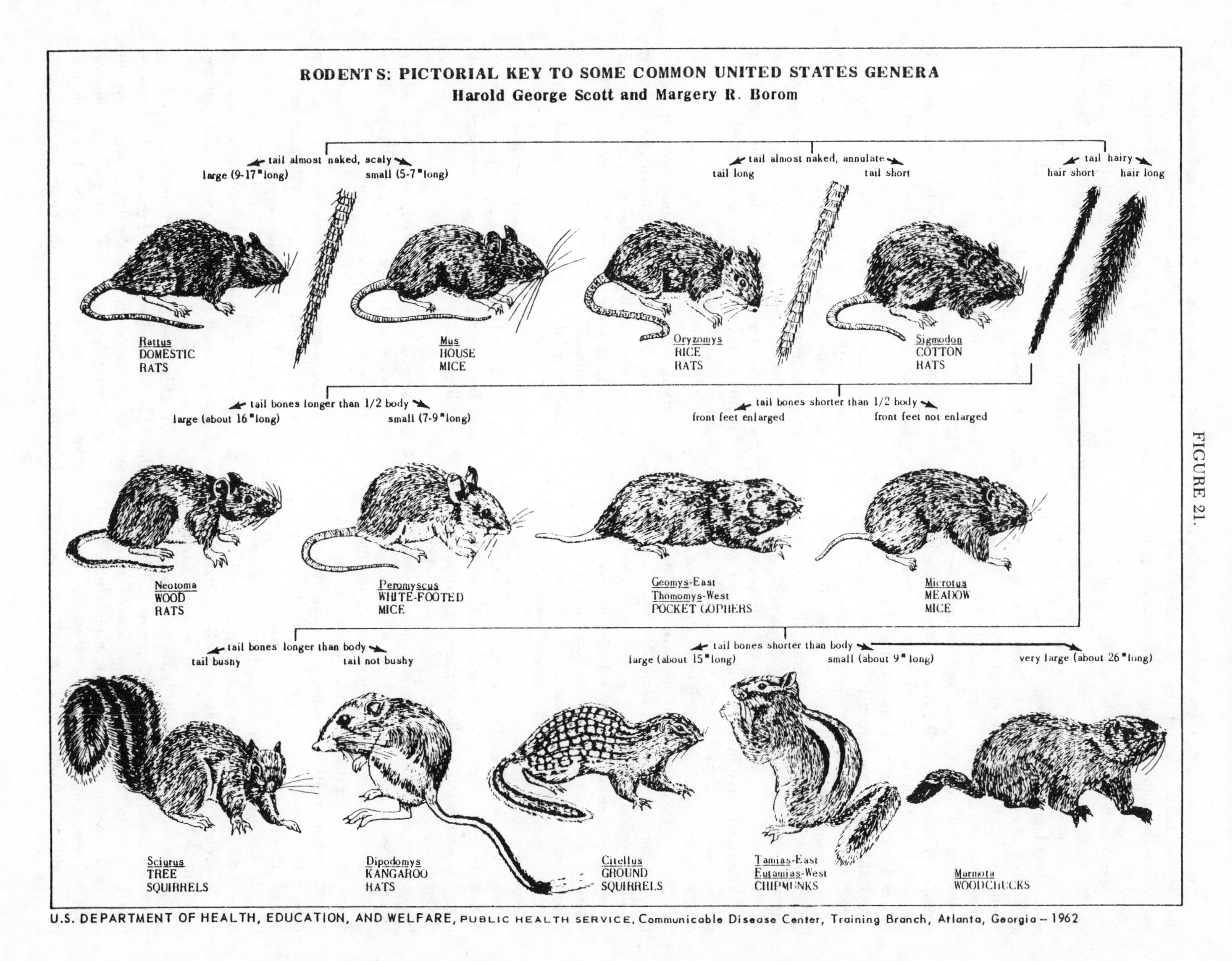
RODENTS: PICTORIAL KEY TO SOME COMMON UNITED STATES GENERA
Harold George Scott and Margery R. Borom
tail almost naked, scaly
large (9-17" long)
small (5-7" long)
tail almost naked, annulate
tail long
tail short
tail hairy
hair short
hair long
Rattus
DOMESTIC RATS
Mus
HOUSE MICE
Oryzomys
RICE RATS
Sigmodon
COTTON RATS
tail bones longer than 1/2 body
large (about 16" long)
small (7-9" long)
tail bones shorter than 1/2 body
front feet enlarged
front feet not enlarged
Neotoma
WOOD RATS
Peromyscus
WHITE-FOOTED MICE
Geomys-East
Thomomys-West
POCKET GOPHERS
Microtus
MEADOW MICE
tail bones longer than body
tail bushy
tail not bushy
tail bones shorter than body
large (about 15" long)
small (about 9" long)
very large (about 26" long)
Sciurus
TREE SQUIRRELS
Dipodomys
KANGAROO RATS
Citellus
GROUND SQUIRRELS
Tamias-East
Eutamias-West
CHIPMUNKS
Marmota
WOODCHUCKS
U.S. DEPARTMENT OF HEALTH, EDUCATION, AND WELFARE, PUBLIC HEALTH SERVICE, Communicable Disease Center, Training Branch, Atlanta, Georgia – 1962

Trapping is an art. There are many types of rat and mouse traps. One of the most effective and versatile is the snap trap, which is generally available in stores.

Many foods make good baits: peanut butter, nut meats, doughnuts, cake, fresh crisp-fried bacon, raisins, strawberry jam, and soft candies, particularly milk chocolate and gumdrops. Rats are attracted more to ground meat or fish. Surprisingly, recent studies indicate that mice aren't nearly as fond of cheese as tradition has it. Traps may also be baited by sprinkling rolled oats over and around the bait trigger. If possible, baits should be fastened by winding a short piece of thread or string around the bait and trigger. Where food is plentiful and nesting material scarce, good results can sometimes be obtained with cotton tied to the trigger.

Trap-shy individuals may be caught by hiding the entire trap under a layer of flour, dirt, sawdust, fine shavings or similar lightweight material.

The common wooden-base snap trap can be made more effective by enlarging the trigger with a piece of heavy cardboard or light screen wire. Cut the cardboard or screen in the shape of a square smaller then the limits of the guillotine wire and attach it firmly to the bait trigger. To bait the trap, smear a small dab of peanut butter in the center of the enlarged trigger or sprinkle rolled oats over the entire surface. This works well on both rat and mouse traps placed where the animals commonly run.

It is very important to place traps across the paths normally used by rats and mice. If their runways cannot be readily determined, sprinkle a light layer of talcum powder, flour, or similar material in foot-square patches in likely places. Place traps in the areas where tracks appear. Because rats and mice like to run close to walls, these spots should be checked first and traps set against walls.

Use boxes or other obstacles to force the rat or mouse to pass over the trigger. Two or more traps set close together produce good results where many are present or where trap-shy individuals are a problem. Use plenty of traps rather than rely on one or two for the job. Because mice travel such short distances, traps should be placed within 10-foot intervals.

To protect other animals or small children trap boxes should be constructed so that mice and rats are forced to enter them when the boxes are placed next to walls. Where the animals travel on rafters or pipes, nail the traps in place or set them on small nipples clamped to the pipes. Leave traps in place for a few days before moving them to other locations. Check traps regularly and adjust to a fine setting.

Rats and mice are accustomed to human odors, so it is not necessary to boil traps or handle them with gloves. Neither does the scent of dead individuals warn others away.

Rat- and mouse-proofing is vital in a complete rodent control program. This consists of changing structural details to prevent entry of rodents into buildings. Openings as small as a half-inch can be entered by young rats, and assuredly by mice.

There are 6 basic devices for keeping these rodents out of buildings:

1. The cuff and channel for wood doors to side and back entrances prevent rats from gnawing under or around the doors. The front doors of most establishments are less exposed to rats and are generally protected with a kick plate. Wooden door jambs can be flashed with sheet metal to protect them from rat gnawing. Because open doors provide ready entry for rodents, both the screen doors and wooden doors to food-handling establishments should be equipped with reliable self-closing devices.

2. Vents and windows can be made secure against rat entry by screening them with heavy wire mesh, preferably in a sheet-metal frame. If desired, fly screening can be incorporated into the frame also. Wooden surfaces exposed to gnawing must be covered by the frame.

3. Metal guards of suitable construction should be placed around or over wires and pipes to prevent rats from using them to gain entrance into a building.

4. Openings around pipes or conduits should either be covered with sheet metal patches or filled with concrete or brick and mortar.

5. The use of concrete for basement floors and for foundations not only prevents rat entry but also increases the value of the property.

6. Floor drains, transoms, letter drops, and fan openings must receive stoppage consideration.

In addition to sealing off entries, buildings should be planned or modified to avoid dead spaces such as double walls, double floors, and enclosed areas under stairways. Trash and garbage piles or other materials stacked against buildings should be removed. They provide the means by which rats and mice can bypass otherwise effective stoppage measures.

Repellents can be of value if used diligently against rats and mice. Of the chemical repellents mentioned later, only napthalene and paradichlorobenzene are useful in preventing rodents from entering protected spaces. The other compounds are intended to prevent eating of grains, seeds and other delectables.

Poisoning of rats and mice is by far the simplest method of control. And of the available poisons, I recommend the ready-prepared anticoagulant baits.

Warfarin, Pival, Fumarin and Diphacinone are effective in a variety of simple, inexpensive baits, and are safe to use around homes if label directions are followed. These corn-meal-like materials must be placed in bait stations, in the runs of the rodents, and fed upon by the target pests for 5 to 10 days or more. The effects are cumulative and the rodents finally die by internal hemorrhaging. They are slow, but home-safe and effective.

More recently a one-shot rat and mouse poison has become available, known as Vacor®. It works against Warfarin-resistant rats, and kills in a few hours after a single feeding. It is absolutely essential to use rodenticides according to the directions appearing on their labels.

Miscellaneous Animal Pests

The more common of these miscellaneous rodents are squirrels, chipmunks, rabbits, moles and bats (See Figures 21-23). They all have predictable needs and activities, and can be controlled utilizing these habits.

Squirrels and Chipmunks

Squirrels and chipmunks can be repelled with the use of moth balls or moth flakes (paradichlorobenzene or napthalene) placed in the runs and holes where they enter buildings. Or they can be trapped using one of the several kinds of live-traps and released in rural areas. The best baits for squirrels are walnut meats or similar nuts, while chipmunks are attracted to rolled oats, grains of corn, or peanut butter.

Rabbits

Rabbits can be live-trapped, using carrots or apples as bait, and released in more appropriate territory. Fruit and other trees can be protected from damage by making guards of 1/4 inch hardware cloth. These guards should be formed into cylinders about 2 inches larger than the diameter of the tree trunk, and long enough to protect the tree above the depth of the deepest snow expected. They should also be anchored in the soil at the base. Commercial repellents containing Thiram or Ziram fungicides used for plant diseases are also effective for rabbits, and should be painted on the areas of tree trunks to be protected. Other repellents include blood dust and nicotine. A recent and successful addition to the market is Repel® (Leffingwell Chemical Co., Brea, CA).

Moles

Moles in the lawns are frequently pests in all parts of the country. They are attracted to white grubs, the larvae of several large beetles, feeding on the roots of lawn grasses, and also to earthworms. Their elimination with soil insecticides will normally eliminate the moles simply by removing the food attraction. Trapping, using the old fashioned spring-activated trap is only partially effective, since they tend to occasionally bypass such devices. Leaving the burrows raised and adding paradichlorobenzene flakes or crystals or Thiram to the burrow floor every 6 feet will probably prevent these generally beneficial nuisances from returning.

Bats

Bats on occasion get into the eaves and attics, fouling the area with their odorous feces or guano and disturbing the occupants with their noise. As soon as they are detected openings in eaves and attic louvers should be sealed with 1/4 inch hardware cloth or screen, because their droppings attract new bat colonies after the original ones are broken up. Narrow cracks can be sealed with caulking compound. If a home is being completely bat-proofed, make sure all bats are outside before plugging the last openings. Normally, all bats leave at about the same time. If there are several openings leave one of them unplugged for several days, then close it in the evening after all bats have left the roost.

To discourage bats from roosting in attics, scatter 3 to 5 pounds of moth flakes over the floor, or hang them in mesh bags from the rafters. Floodlights in the attic or directed upon outside entrances for several nights will sometimes cause bats to leave a building. The only way to permanently control bats is to keep them out.

Skunks

Even though you may never have encountered a skunk before, you will immediately recognize both its appearance and odor. Skunks are kin to the weasels, and since they are not particularly disturbed by man's presence and activities, they occasionally move into the neighborhood.

Skunks are protected by law in most states and they are frequently found to be carriers of rabies. So, avoid handling at all costs.

Control is achieved by exclusion, that is, preventing them from returning to their sleeping or nesting quarters. Sprinkle a thin layer of flour around the hole or suspected entrance to form a tracking patch. Examine the area for skunk tracks soon after dark, and when the tracks lead out of the entrance, the opening can then be safely closed off with lumber, fencing, or concrete.

Skunks are known for their scent, which can be ejected for 6 to 10 feet. The persistence of the odor on anything touched by it is astonishing. A chemical known as Neutroleum-Alpha® is probably the most effective odor neutralizer available for countering that horrible odor. A tablespoonful in a water bath works well for dogs and humans unfortunate enough to be "hit". It is also quite effective for scrubbing floors, walls, garages, basements, outdoor furniture and the like, using 2 ounces to each gallon of water. It can also be sprayed on contaminated soil.

Substitutes for the above are chlorine laundry bleach or household vinegar, diluted 1 to 10 parts of water, with a little household detergent added to assist in wetting and saturating the area. Tomato juice is messy and won't work. Its supposed effectiveness is based on the fact that it contains a small quantity of the organic acid, aspartic acid. In large quantities this could be effective, but not so for tomato juice.

Deer

Most of you urbanites are thinking how romantic it would be to live in a rural atmosphere and have deer nibbling around the cabin. To those who have the problem, it isn't all that great, particularly when they prune leaves and limbs from fruit and ornamental trees up to 8 feet by standing on their hind legs. Materials registered as effective deer repellents include two fungicides, Thiram and Ziram, and bone oil. Spray trunks of trees and lower limbs with fairly generous portions of the above at 2-3 week intervals. Repel®, a recent addition to deer repellents is reported to be quite effective (See Rabbits).

Dogs and Cats

Probably nothing is more annoying than having someone's dog defecate in your front lawn, urinate on the tires of your station wagon parked in the driveway, or seeing a strange cat patrolling your backyard for unwary birds. Compared to some pest control problems, these are relatively simple.

Dogs and cats both respond to scolding, clapping of hands, broom-waving, hurled clods and stones, and repellents. The materials registered as dog and cat repellents are almost too numerous to describe.

Both forms of moth crystals (naphthalene and paradichlorobenzene) are very effective inside the home and out. For instance, a few crystals placed in the favorite winged-back chair will compel Fido or Kitty to seek another parking place. The same is true outside. Trees, fireplugs, and other urinating points for dogs, with the exception of the tires on family vehicles, can be equally unattractive with a few crystals sprinkled around the base. Cats will avoid walking or resting on walls with moth crystals sprinkled at regular intervals.

If the automobile tire episodes are really important to you, try one of the commercial dog repellent aerosols containing either citral, citronella, or creosote.

Other materials contained in commercial mixtures are allyl isothiocyanate, amyl acetate, anethole, bittrex, bone oil, capsaicin, citrus oil, cresylic acid, eucalyptus, geranium oil, lavender oil, lemongrass oil, menthol, methyl nonylketone, methyl salicylate, nicotine, pentanethiol, pyridine, sassafras oil and thymol. The formulated product should be first checked out with your own nose if it is to be used indoors, since some of the above are readily detectable and may be highly annoying. Better the dog or cat!

FIGURE 22.

*All measurements for adults.

U.S. DEPARTMENT OF HEALTH, EDUCATION, AND WELFARE
PUBLIC HEALTH SERVICE, Communicable Disease Center, Training Branch, Atlanta, Georgia – 1962

FIGURE 23.

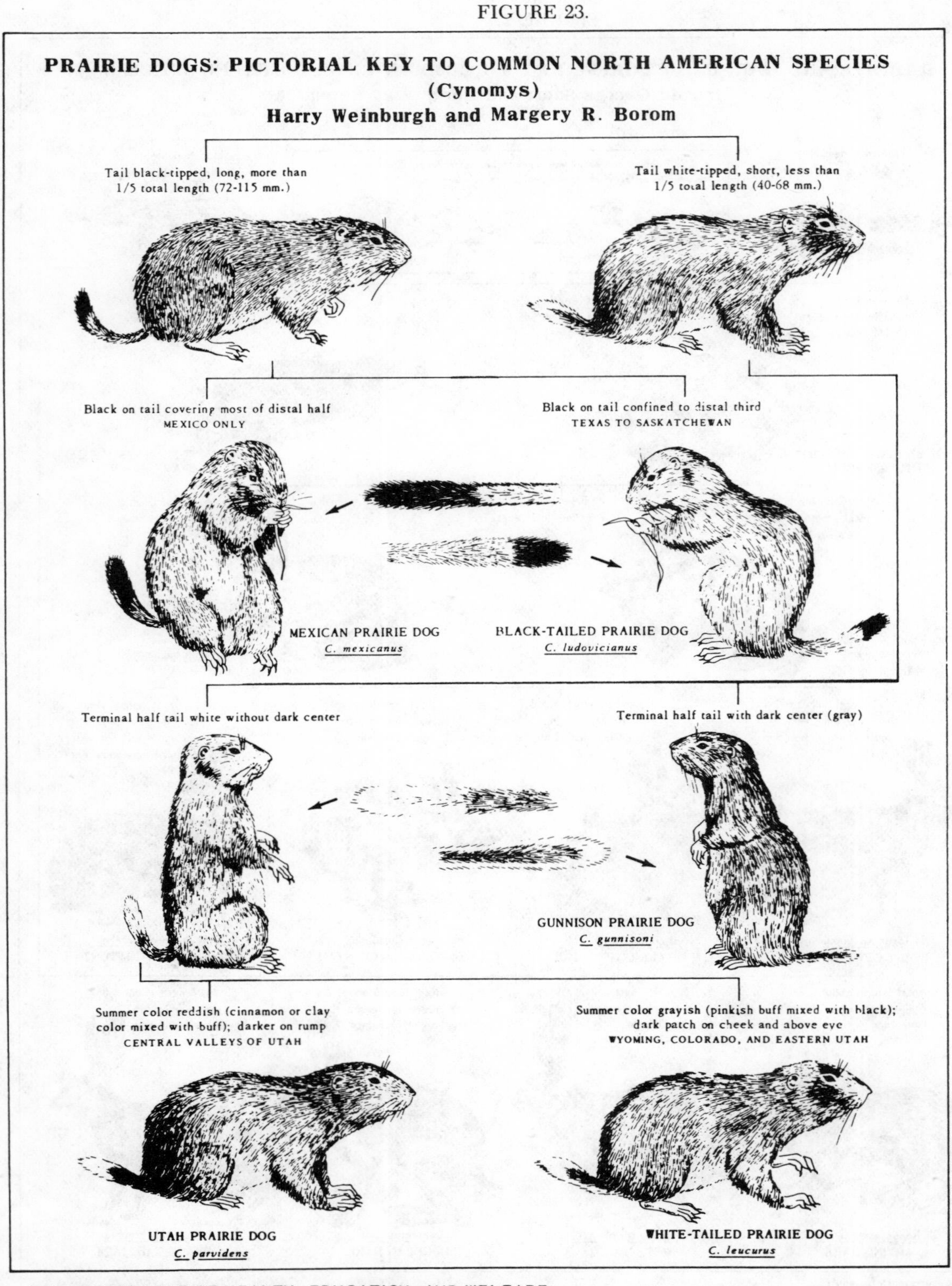

U.S. DEPARTMENT OF HEALTH, EDUCATION, AND WELFARE
PUBLIC HEALTH SERVICE, Communicable Disease Center, Training Branch, Atlanta, Georgia – 1964

BIRDS

A scarecrow in a garden of cucumbers keepeth nothing.

The Apocrypha,
Baruch VI, 70.

BIRDS

All birds can, in one way or another at times be beneficial to mankind. They provide enjoyment and wholesome recreation for most of us regardless of where we live. Despite the fact that wild bird populations are for the most part beneficial, there are occasions when individuals of certain species can seriously compete with the homeowner's interests. When these situations occur, some kinds of control measures are inevitable.

These beautiful winged creatures create pest problems singly or in small groups, but especially when in large aggregations. Most of the areas of conflict with man are: (1) destruction of agricultural foodstuffs and predation; (2) contamination of food-stuffs or defacing of buildings with their feces; (3) transmission of diseases, directly and indirectly, to man, poultry and dairy animals; (4) hazards at airports and freeways; and (5) being a general nuisance or affecting man's comfort, aesthetics, or sporting values.

Controlling Birds

It has been my experience that an alert and aggressive cat will play a big role in repelling pest birds — and song birds, as well, unfortunately. And, there are other ways of controlling birds, mainly by repelling. How do the professionals drive nuisance birds from their roosts and perches? With frightening devices (both visual and acoustical), chemical repellents including sticky pastes on ledges and roosts, mechanical barriers, trapping and shooting, toxic baits, soporifics (stupefacients), surfactants (feather-wetting agents), and biological control (nest or roost destruction, modification of habitat, and chemosterilants). However, the encouragement of natural predators or payment of bounties have not been generally successful.

You may be surprised to learn that the bird species commonly requiring control are as numerous or even more abundant today than they were years ago before controls became so widespread. The reason that the number of birds has not declined is that it is primarily the condition of food and cover that determine their density. The way man has altered his environment apparently has provided an improved habitat for many of the bird species. Our crops, and the landscaping of city and suburban homeowners, provide desirable food or nesting cover not available in the past to these species.

Whether a bird is a pest depends on how many there are, where they roost, and what they eat. Thus a list of pest species may or may not include your pest. Usually the avian pests would include: The English (house) sparrow, starling, common grackle, Brewer's blackbird, rusty blackbird, red-winged blackbird, feral (wild) pigeon, house finch, and cow birds (See Figure 24). Other birds that may become pests on occasion are the blue jay, Steller's jay, mourning and white wing dove, swallows, martins, mocking birds, sapsuckers, flickers, gulls and pelicans. Neither list is complete, much as a list of weeds would never be complete. For every bird could, under certain circumstances, be a pest to someone.

What can we do about birds that dig up the seed which we have planted in the garden, nibble on our strawberries, or attack our ripening fruit? You have about 3 choices: Repellents, stakes and flags, or continuous string flagging.

Repellents have been used for years in commercial plantings with only moderate success, and they can be used in the garden with about the same outcome. After the garden is planted, several handfuls of naphthalene granules or flakes scattered over the seed beds will do very satisfactorily. Or the seed may be treated prior to planting with one of the chemical repellents listed later.

Scare crows do not work. Stakes and flags do. Stakes or laths are fixed in the soil and strips of cloth or paper attached to their tops. The flag is usually tied to the end of the stake with a short string or sometimes it is tacked to the stake. For persistent birds in the garden space your stakes 15 or 20 feet apart in all directions.

The best of protective methods for garden seed and fruit eaters is the continuous string flagging. The needed materials are heavy stakes at least 4 feet long, strong cotton or sisal wrapping twine and paper, cloth or plastic streamers 2 to 2½ inches wide and 20 to 24 inches long. The stakes are driven firmly into the ground and may be 50 or more feet apart. The twine is stretched from stake to stake and streamers are fastened at 5-foot intervals, making 10 streamers to each 5-foot section between stakes. It is best to install continuous string flagging ahead of attack on that treasured garden. Spiral twirlers, shiny propellers, and other objects that flash in the sunlight or rustle and rattle as they spin are useful in small areas. These methods work for about a week. After that, birds grow accustomed to them and sit on them while they eat your berries.

For birds that assemble in your trees in large numbers during the fall and winter — these are probably grackles, blackbirds, or starlings — it is best to frighten them away on their first visit. Once the area is marked with their odorous white feces, it seems to become more attractive and populated.

Roost control or management now appears to be the ecologically sound method of preventing roosting by droves of birds. Recent work in Houston, Texas, by Drs. Heidi Good and Dan Johnson, indicate that removal of trees and tree trimming can prove so unattractive to blackbirds that they choose other roosts. With active roost trees, they were able to prevent roosting in particular groves for up to three years by trimming out only one-third of the canopy of all trees. Large branches are not removed unless necessary for the appearance of the tree. And, trimming only the sucker growth is ineffective. Birds prefer dense bushy trees to roost in presumably because they can afford them protection from foul weather and provide many small branches on which to roost.

Recommendations that resulted from their work were to trim or remove trees before the birds arrived. This prevents the roost from ever being initiated. Until the birds find another suitable roost to use each year, it will be necessary to keep the trees trimmed to retain control. Only enough trees should be removed or pruned to create an open space within the favored site. These birds like to cluster, and generally avoid roosting in isolated trees.

Carbide exploders have proven very successful, however, continued use seems to lose its effectiveness on the birds while gaining in effectiveness with the neighbors. Firecrackers can be just as effective, and less costly. Firecracker ropes can be made by inserting the fuses in the strands of cotton rope which serve as a slow-burning fuse. Silhouettes of large owls placed in the trees have been reported effective in preventing alighting. Hand-clapping in combination with yelling in view of the invaders is effective. Throwing empty #3 cans containing rocks, nails or other noise-producing objects into the occupied trees will also send them flying.

Placing a cheesecloth or other veil-like covering over fruit trees during ripening satisfactorily keeps birds out of fig, cherry, plum and other fruit trees.

For birds that rest on window sills, a single strand of wire stretched across the ledge and attached to both sides, about breast high on the bird, will prevent them from landing or roosting on these ledges. Also the application of a caulk-like sticky chemical to ledges, rain gutters, roof hips, tops and gables and other resting areas is also effective.

Poisoning is effective but not selective. Invariably several song and other protected species become inadvertent victims of attempts to control pest species, and consequently poisoning is not recommended. The chemical controls available to the home gardener are one new sterilizing chemical (chemosterilant) (Ornitrol®) and a few repellents, Captan, naphthalene, and PDB. (See Avicides).

FIGURE 24.

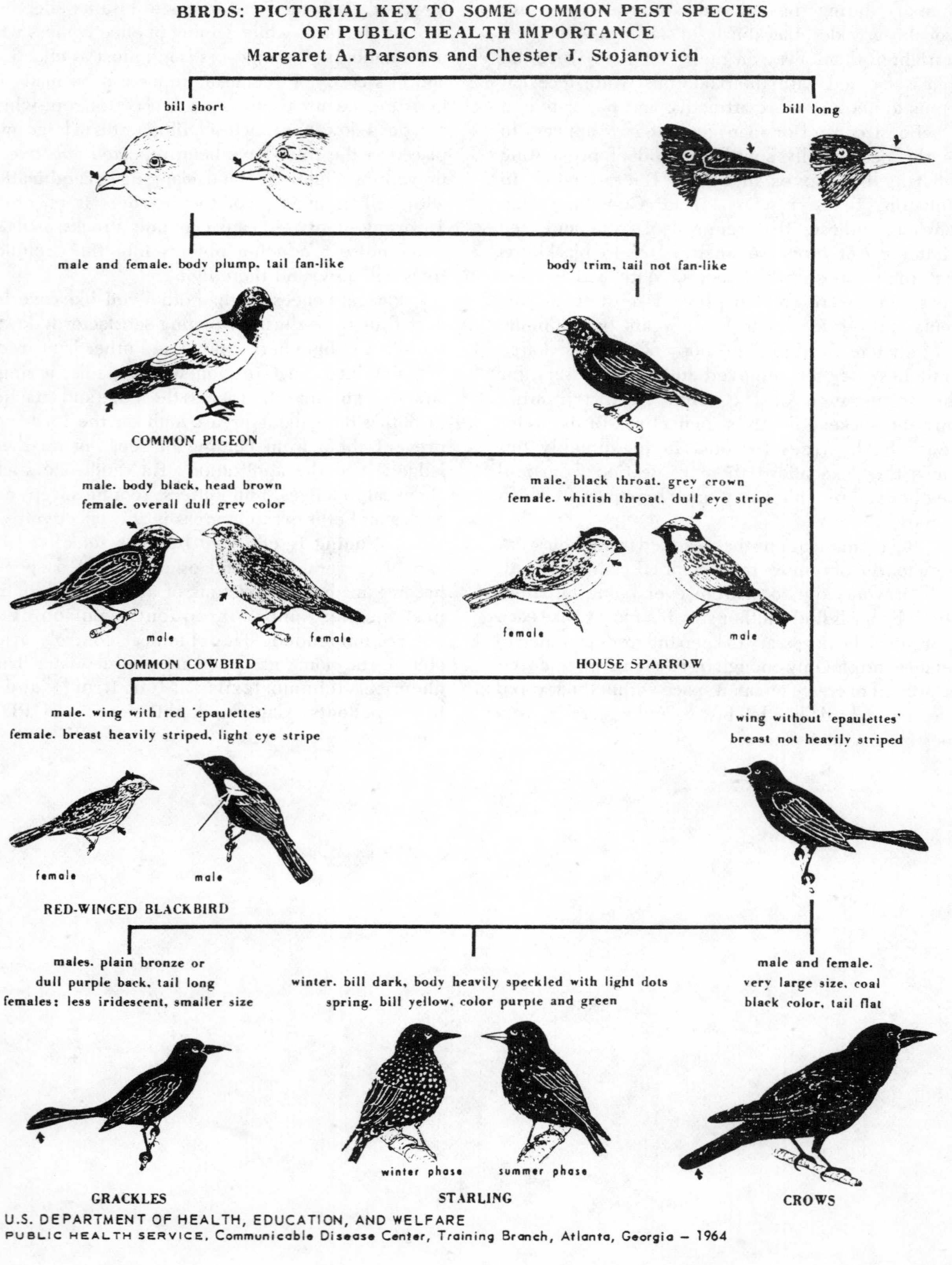
BIRDS: PICTORIAL KEY TO SOME COMMON PEST SPECIES
OF PUBLIC HEALTH IMPORTANCE
Margaret A. Parsons and Chester J. Stojanovich
bill short
bill long
male and female. body plump. tail fan-like
body trim, tail not fan-like
COMMON PIGEON
male. body black, head brown
female. overall dull grey color
male. black throat. grey crown
female. whitish throat. dull eye stripe
male
female
COMMON COWBIRD
female
male
HOUSE SPARROW
male. wing with red 'epaulettes'
female. breast heavily striped, light eye stripe
wing without 'epaulettes'
breast not heavily striped
female
male
RED-WINGED BLACKBIRD
males. plain bronze or
dull purple back, tail long
females: less iridescent, smaller size
winter. bill dark, body heavily speckled with light dots
spring. bill yellow, color purple and green
male and female.
very large size, coal
black color, tail flat
winter phase
summer phase
GRACKLES
STARLING
CROWS
U.S. DEPARTMENT OF HEALTH, EDUCATION, AND WELFARE
PUBLIC HEALTH SERVICE, Communicable Disease Center, Training Branch, Atlanta, Georgia – 1964

CHAPTER 6

It is better to be safe than sorry.
American Proverb.

THE SAFE HANDLING AND STORAGE OF PESTICIDES

Before any pesticide is applied in or around the home, READ THE LABEL! This is the first rule of safety in using any pesticide — read the label and follow its directions and precautions. All pesticides are safe to use, provided common-sense safety is practiced and provided they are used according to the label instructions. This especially means keeping them away from children, illiterate or mentally incompetent persons, and pets.

An old cliche of industrial and safety engineers is "Safety is a state of mind". Pesticide safety is more than a state of mind. It must become a habit with those who apply pesticides around the home, and certainly with those who have small children. You can control pests in your home and garden with absolute safety, if you use pesticides properly.

Pesticide Selection. When buying a pesticide, check the label. Make sure it lists the name of the pests you want to control. If in doubt, consult your County Agent or other authority. Select the pesticide that is recommended by competent authority and consider the effects it may have on nearby plants and animals. Make certain that the label on the container is intact and up-to-date; it should include directions and precautions. And finally, purchase only the quantity needed for the current season. Don't buy more than you need because it's a bargain.

Pesticide Mixing and Handling. If the pesticide is to be mixed before applying, read carefully the label directions and current recommendations from the County Agent's office when available. This information can be obtained easily by telephone. It's always a good idea to wear rubber gloves when mixing a pesticide and stand upwind of the mixing container. Handle the pesticide in a well-ventilated area. Avoid dusts and splashing when opening containers or pouring into the sprayer. Don't use or mix a pesticide on windy days. Measure the quantity of pesticide required accurately, using the proper equipment. Over-dosage is wasteful; it won't kill more pests; it may be injurious to plants and may leave an excess residue on fruits and vegetables. Don't mix a pesticide in areas where there is a chance that spills or overflows could get into any water supply. Clean up spills immediately. Wash the pesticide off skin promptly with plenty of soap and water, and change clothes immediately if they become contaminated.

Pesticide Application. Wear the appropriate protective clothing and equipment if the label calls for it. Make certain that equipment is calibrated correctly and is in satisfactory working condition. Apply only at the recommended rate, and to minimize drift, apply only on a calm day. Do not contaminate feed, food, or water supplies. This includes pet food and water bowls. Avoid damage to beneficial and pollinating insects by not spraying during periods when such insects are actively visiting flowering plants. Honey bees are usually inactive at dawn and dusk, which are good times for outdoor applications. Dusk is preferred.

Keep pesticides out of mouth, eyes, and nose. Don't use the mouth to blow out clogged hoses or nozzles. Observe precisely the waiting periods specified on the label between pesticide application and harvest of fruit and vegetables. Clean all equipment used in mixing and applying pesticides according to recommendations. Don't use the same sprayer for insecticide and herbicide applications.

Wash the sprayer, protective equipment and hands thoroughly after handling pesticides. If you should ever become ill after using pesticides and believe you have the symptoms of pesticide poisoning, call your physician and take the pesticide label with you. This situation is highly unlikely, but it's always good to know in an emergency.

Pesticide Storage. The rule of thumb around the home is lock up all pesticides. Lock the room, cabinet or shed where they are stored to discourage children. Don't store pesticides where food, feed, seed or water can be contaminated. Certainly don't store them beneath the kitchen sink. Store in a dry, well-ventilated place, away from sunlight, and at temperatures above freezing. If you should happen to have an operation larger than a typical homeowner's, mark all entrances to your storage area with signs bearing this caution: "PESTICIDES STORED HERE — KEEP OUT". This action would be in keeping with a commercial greenhouse or farming operation.

Pesticides should be kept only in original containers, closed tightly and labelled. Examine pesticide containers occasionally for leaks and tears. Dispose of leaking and torn containers, and clean up spilled or leaked material immediately. It's a good idea to date the container when purchased. This makes the disposal of outdated materials a simple matter. Because many pesticide spray formulations are flammable, take precautions against potential fire hazards.

Disposal of Empty Containers and Unused Pesticides. Analagous to the loaded gun, empty containers are never completely empty, so don't reuse for any purpose. Instead, break glass containers, rinse metal containers three times with water, punch holes in top and bottom and leave in your trash barrels for removal to the official landfill trash dump. Empty paper bags and cardboard boxes should be torn or smashed to make unusable, placed in a larger paper bag, rolled, and relegated to the trash barrel. In summary, don't leave anything tempting in the trash barrel or dump. That kid raiding your trash cans may be your own!

Unwanted Pesticides. Offer to give unwanted pesticides to a responsible person in need of the materials. If this is not practicable, bury dry pesticides at a depth of at least 18 inches in a safe disposal site. Pour liquid pesticides into a pit dug in sandy soil away from trees and shrubs. Do not take unwanted pesticides to an incinerator.

Hundreds of "do" and "don't" rules for handling pesticides have generated over the past two decades with the increased awareness of pesticide hazards. The above gleanings are from many sources and may be useful to the reader in his own home situation, around the commercial greenhouse, on the farm, in preparation for talks on safety, for inspections of schools and public buildings for safe playing and working conditions, or just for your own confident reference.

The Label

The pesticide label on the container is the single most important tool to the homeowner in using pesticides safely. The FEDERAL ENVIRONMENTAL PESTICIDE CONTROL ACT (FEPCA), which is discussed in the chapter on pesticide laws, contains 3 very important points concerning the pesticide label which should be further emphasized. They pertain to reading the label, understanding the label directions, and following carefully these instructions.

The first two provisions of FEPCA are that the use of any pesticide inconsistent with the label is prohibited, and deliberate violations by growers, applicators, or dealers can result in heavy fines or imprisonment or both. The third provision is found in the general standards for certification of commercial applicators, which in essence will license them to use restricted-use pesticides, the area of label and labeling comprehension. For certification applicators are to be tested on (a) the general format and terminology of pesticide labels and labeling, (b) the understanding of instructions, warning, terms, symbols, and other information commonly appearing on pesticide labels, (c) classification of the product (general or restricted use), and (d) the necessity for use consistent with the label.

In Figure 23 is shown the format label for general-use pesticides as required by EPA to appear on all containers beginning in October of 1977. This label is keyed as follows:

1. Product name
2. Company name and address
3. Net contents
4. EPA pesticide registration number
5. EPA formulator manufacturer establishment number

6A. Ingredients statement
6B. Pounds/gallon statement (if liquid)

7. Front panel precautionary statements
7A. Child hazard warning, "Keep Out of Reach of Children"
7B. Signal word — CAUTION
7D. Statement of practical treatment
7E. Referral statement
8. Side/back panel precautionary statements
8A. Hazards to humans and domestic animals
8B. Environmental hazards
8C. Physical or chemical hazards
9B. Statement of pesticide classification
9C. Misuse statement
10A. Re-entry statement
10C. Storage and Disposal block
10D. Directions for use

Emergencies Involving Pesticides

All pesticides can be used safely, provided common-sense safety is practiced and provided they are used according to the label instructions; this includes keeping them away from children and illiterate or mentally incompetent persons. Despite the most thorough precautions, accidents will occur. Below are given two important sources of information in the event of any kind of serious pesticide accident.

The first and most important source of information is for human-poisoning cases: the nearest Poison Control Center. Look it up in the telephone directory under POISON CONTROL CENTERS, or ask the telephone operator for assistance. Poison Control Centers are usually located in the larger hospitals of most cities and can provide emergency treatment information on all types of human poisoning, including pesticides. The telephone number of the nearest Poison Control Center should be kept as a ready reference by parents of small children, or employers of persons who work with pesticides and other potentially hazardous materials.

The second and one you will not likely use, is the CHEMTREC telephone number. From this toll-free long-distance number can be obtained emergency information on all pesticide accidents, pesticide-poisoning cases, pesticide spills, and pesticide spill cleanup teams. This telephone service is available twenty-four hours a day. The toll-free number is:

CHEMTREC 800-424-9300

Specific pesticide poisoning information can also be obtained in writing or by telephone from:

National Clearing House for
Poison Control Centers
Food & Drug Administration
Bureau of Drugs
5401 Westbard Avenue
Bethesda, Maryland 20016

FIGURE 23. Format for the General Use pesticide label.

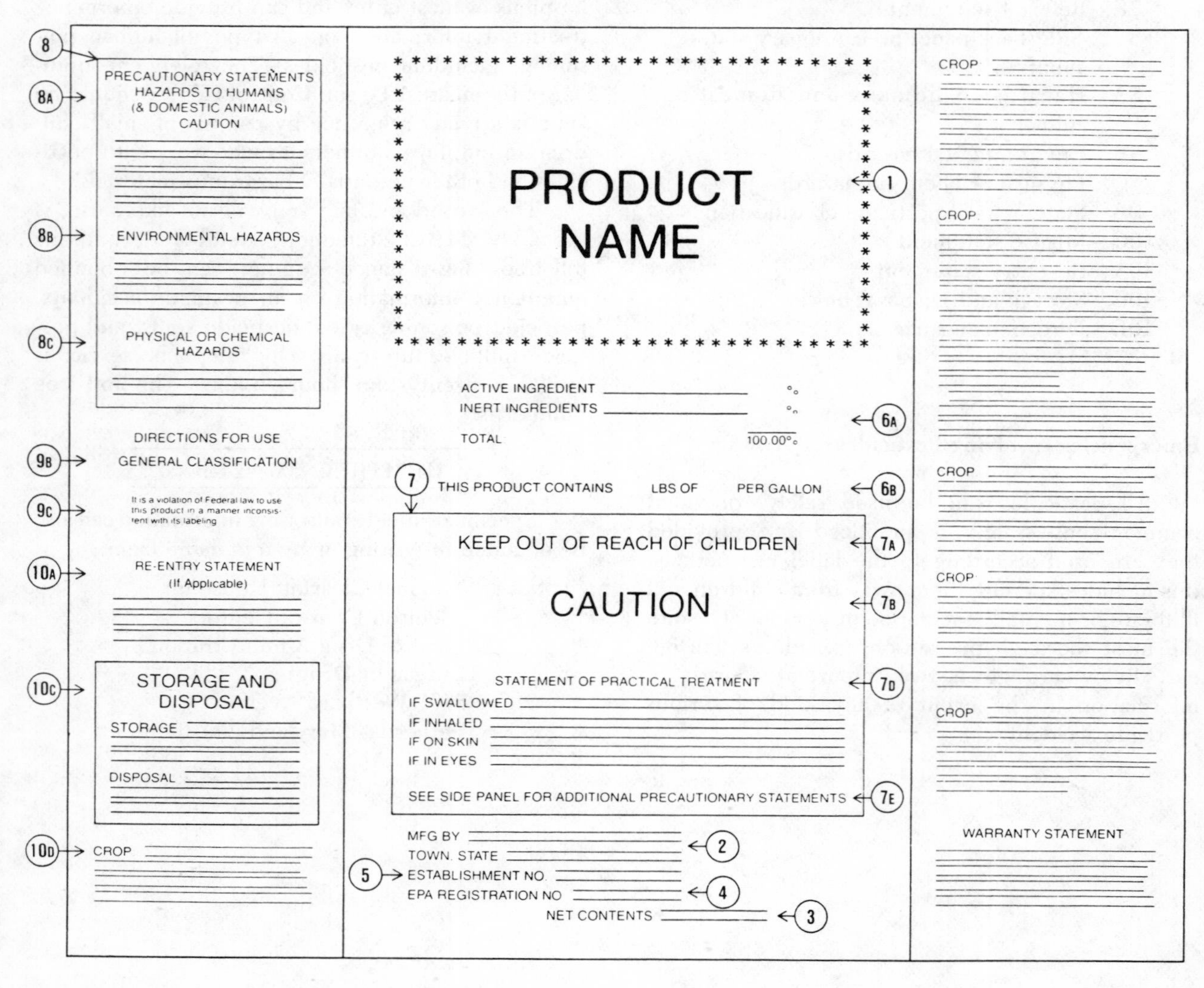

CHAPTER 7

Good laws lead to the making of better ones; bad ones bring about worse.
Jean Jacques Rousseau,
Contrat Social, Book III,
Chap. 15.

THE PESTICIDE LAWS

PESTICIDES AND THE LAW

Pesticide use is controlled by both federal and state, and occasionally county or city laws. Federal laws have protected the user of pesticides, his pets and domestic animals, his neighbor, and the consumer of treated products for several decades. Nothing is left unprotected, and after reading briefly about the laws themselves, the reader will be convinced of their omnipotence.

The first federal law, the Federal Food, Drug and Cosmetic Act of 1906, known as the Pure Food Law, required that food (fresh, canned and frozen) shipped in interstate commerce be pure and wholesome. There was not a single word in the law that pertained to pesticides.

The Federal Insecticide Act of 1910, which covered only insecticides and fungicides, was signed into law by President William Howard Taft. The act was the first to control pesticides and was designed mainly to protect the farmer from substandard or fraudulent products, for they were plentiful at the turn of the century. This was probably one of our earliest consumer protection laws.

The Pure Food Law of 1906 was amended in 1938 to include pesticides on foods, primarily the arsenicals such as lead arsenate and Paris green. It also required the adding of color to white insecticides, including sodium fluoride and lead arsenate, to prevent their use as flour or other look-alike cooking materials. This was the first federal effort toward protecting the consumer from pesticide-contaminated food, by providing tolerances for pesticide residues, namely arsenic and lead, in foods where these materials were necessary for the production of a food supply.

In 1947 the Federal Insecticide, Fungicide, and Rodenticide Act (FIFRA) became law. It superseded the 1910 Federal Insecticide Act and extended the coverage to include herbicides and rodenticides and required that any of these products must be registered with the U.S. Department of Agriculture before they could be marketed in interstate commerce. Basically, the law was one requiring good and useful labeling, making the product safe to use if label instructions were followed. The label was required to contain the manufacturer's name and address, name, brand and trademark of the product, its net contents, an ingredient statement, an appropriate warning statement to prevent injury to man, animals, plants, and useful invertebrates, and directions for use adequate to protect the user and the public.

In 1945 the Miller amendment to the Food, Drug and Cosmetic Act (1906, 1938) was passed. It provided that any raw agricultural commodity may be condemned as adulterated if it contains any pesticide chemical whose safety has not been formally cleared or that is present in excessive amounts (above tolerances). In essence, this clearly set tolerances on all pesticides in food products, for example, 7.0 ppm DDT in lettuce (though it is no longer in use) or 1.0 ppm ethyl parathion on string beans.

Two laws, the Federal Insecticide, Fungicide and Rodenticide Act (FIFRA) and the Miller Amendment to the Food, Drug, and Cosmetic Act, supplement each other and are interrelated by law in practical operation. Today they serve as the basic elements of protection for the applicator, the consumer of treated products, and the environment, as modified by the following amendments.

The Food Additives Amendment to the Food, Drug and Cosmetic Act (1906, 1938, 1954) was passed in 1948. It extended the same philosophy to all types of food additives that has been applied to pesticide residues on raw agricultural commodities by the 1954 amendment. However, this also controls pesticide residues in processed foods that had not previously fitted into the 1954 designation of raw agricultural commodities. Of greater importance, however, was the inclusion of the Delaney clause, which states that any chemical found to cause cancer (a carcinogen) in laboratory animals when fed at any dosage may not appear in foods consumed by man. This has become the most controversial segment of the entire spectrum of federal laws applying to pesticides, mainly with regard to the dosage found to produce cancer in experimental animals.

The various statutes mentioned so far apply only to commodities shipped in interstate commerce. In 1959, FIFRA (1947) was amended to include nematicides, plant regulators, defoliants, and desiccants as economic poisons (pesticides). (Poisons and repellents used against amphibians, reptiles, birds, fish, mammals, and invertebrates have since been included as economic poisons.) Because FIFRA and the Food, Drug and Cosmetics Act are allied, these additional economic poisons were also controlled as they pertain to residues in raw agricultural commodities.

In 1964, FIFRA (1947, 1959) was again amended to require that all pesticide labels contain the Federal Registration Number. It also required caution words such as WARNING, DANGER, CAUTION, and KEEP OUT OF REACH OF CHILDREN to be included on the front label of all poisonous pesticides. Manufacturers also had to remove safety claims from all labels.

Until December, 1970, the administration of FIFRA was the responsibility of the Pesticides Regulation Division of the U.S. Department of Agriculture. At that time the responsibility was transferred to the newly-designated U.S. Environmental Protection Agency (EPA). Simultaneously, the authority to establish pesticide tolerances was transferred from the Food and Drug Administration (FDA) to EPA. The enforcement of tolerances remains the responsibility of the FDA.

In 1972, FIFRA (1947, 1959, 1964) was revised by the most important pesticide legislation of this century. THE FEDERAL ENVIRONMENTAL PESTICIDE CONTROL ACT (FEPCA), sometimes referred to as 1972 FIFRA amendment. Some of the provisions of FEPCA are abstracted as follows:

1. Use of any pesticide inconsistent with the label is prohibited.
2. Deliberate violations of FEPCA by growers, applicators or dealers can result in heavy fines and/or imprisonment.
3. All pesticides will be classified into (a) General Use or (b) Restricted Use categories by October 1977.
4. Anyone applying Restricted Use pesticides must be certified by the state in which he lives.
5. Pesticide manufacturing plants must be registered and inspected by EPA.
6. States may register pesticides on a limited basis when intended for special local needs.
7. All pesticide products must be registered by EPA, whether shipped in interstate or intrastate commerce.
8. For a product to be registered the manufacturer is required to provide scientific evidence that the product, when used as directed, will (1) effectively control the pests listed on label, (2) not injure humans, crops, livestock, wildlife, or damage the total environment, and (3) not result in illegal residues in food or feed.

In 1978, President Jimmy Carter signed into law a series of amendments to FIFRA as amended in 1972. These provisions will have a great influence on the way pesticides are registered and used.

Many of the changes in that legislation were designed to improve the registration process, which was slowed significantly by regulations resulting from the 1972 Act. The more important points are as follows:

1. Efficacy data can be waived. The EPA has the option of setting aside requirements for proving the efficacy of a pesticide before registration. This leaves the manufacturer to decide whether a pesticide is effective enough to market, and final proof will depend on product performance.
2. Generic standards will be set for the active ingredients rather than for each product. This change permits EPA to make safety and health decisions for the active ingredient in a pesticide, instead of treating each product on an individual basis. (It is easy to see how this provision will speed registration, considering that there are only about 1,000 active ingredients in the 35,000 formulations currently on the market.) This document will specify the information submitted in the past to register an active ingredient and identify the data now needed under reregistration.

3. Reregistration of all older products will be required. Under reregistration, all compounds are currently being reexamined to make certain the supporting data for a registered pesticide satisfies today's requirements for registration, in light of new knowledge concerning human health and environmental safety. Many pesticides were registered under the old data requirements (prior to August 1975), and registrants must submit new information to carry the product through the reregistration process.

4. Pesticides can now be given conditional registration. EPA may now grant a conditional registration for a pesticide even though certain supporting data have not been completed. That information will still be required, but it may be deferred to a later date. Conditional registration can be granted by EPA if:

   a. The uses are identical or greatly similar to those which exist on labels for already registered products with the same active ingredient;

   b. New uses are being added, providing a notice of Rebuttable Presumption Against Registration (RPAR) has not been issued on the product, or in the case of food or feed use, there is no other available or effective alternative;

   c. New pesticides have had additional data requirements imposed since the date of the original submission.

5. The use of data from one registrant can be used by other manufacturers or formulators if paid for. This new law spells out how data submitted by one registrant for an active ingredient may be used by other applicants. All data provided from 1970 on can be used for a 15-year period by other registrants, if they offer to pay "reasonable compensation" for this use. In the future, registrants will have 10 years of exclusive use of data submitted for a new pesticide active ingredient. During that time, other applicants may request and be granted permission to use the information but must obtain approval.

6. Trade secrets will be protected. A clear and detailed outline of what information is considered confidential is described. EPA may reveal data on most pesticide effects (including human, animal, and plant hazard evaluation), efficacy, and environmental chemistry. Four categories of data are generally to be kept confidential but may be released under certain circumstances:

   a. Manufacturing and quality control processes;

   b. Methods of testing, detecting, or measuring deliberately added inert ingredients;

   c. Identity or quality of deliberately added inerts;

   d. Production, distribution, sale, and inventories of pesticides.

7. The state now has primary enforcement responsibility. Under the 1978 law, the primary authority for use enforcement under federal law will be assigned to the states. Any suspected misuse of pesticides will be investigated by and acted upon by the state regulatory boards. Before the enforcement authority can be legally transferred by EPA, the state must indicate that their regulatory methods will meet or exceed the federal requirements. If a state does not take appropriate action within 30 days of alleged misuse, EPA can act. In addition, enforcement authority can be taken away from any state which consistently fails to take proper action.

8. States can register pesticides for Special Local Needs (SLN). The state authority to register materials for Special Local Needs is increased, and EPA's role is decreased. All states are given the authority automatically to register products for use within a state for special situations. Registrations can be made for new products, using already registered active ingredients. Existing product labels can be amended for new uses, including chemicals which have been subject to cancellation or suspension in the past. Only those specific uses which are cancelled or suspended may not be registered by the state for SLN's.

9. Uses inconsistent with the labeling are defined. The phrase, "to use any registered pesticide in a manner inconsistent with its labeling" is defined. Certain use practices, which previously were covered by Pesticide Enforcement Policy Statements (PEPS), are now permitted. Persons who derive income from the sale or distribution of pesticides may *not* make recommendations which call for uses inconsistent with labeling. Users and applicators may now:

a. Use a pesticide for control of a target pest not named on the label.

b. Apply the pesticide using any method not specifically prohibited on the label.

c. Mix one or more pesticides with other pesticides or fertilizers, provided the current labeling does not actually prohibit this practice;

d. Use a pesticide at less than labeled dosage, providing the total amount applied does not exceed that currently allowed on the labeling.

These are only the most important aspects of FEPCA that you, the interested novice, need be acquainted with.

Beyond the federal laws providing rather strict control over the use of insecticides, each state usually has two or three similar laws controlling the application of pesticides and the sales and use of pesticides. These may or may not involve the licensing of aerial and ground applicators as one group, and the structural applicator or pest control applicator as another. Only the latter is of concern to the homeowner, who should insist that anyone selling him pest control services be both licensed within that state and certified to apply Restricted Use pesticides.

CHAPTER 8

> The single factor that determines the degree of harmfulness of a compound is the dose . . .
>
> Ted A. Loomis,
> *Essentials of Toxicology*, 1974

THE HAZARDS OF PESTICIDES

All pesticides are toxic, but their use is not necessarily a hazard. Let's get it straight at the outset. There is a marked distinction between *toxicity* and *hazard*. These two terms are not synonymous. *Toxicity* refers to the inherent toxicity of a compound. In other words, how toxic is it to animals under experimental conditions. *Hazard* is the risk or danger of poisoning when a chemical is used or applied, sometimes referred to as *use hazard*. The factor with which the user of a pesticide is really concerned is the use hazard and not the inherent toxicity of the material. Hazard depends not only upon toxicity but also upon the chance of exposure to toxic amounts of the material.

Webster defines the word poison as "any substance which introduced into an organism in relatively small amounts, acts chemically upon the tissues to produce serious injury or death." One can immediately spot several flaws in this definition. The "relatively small amount" statement is open to wide interpretation. For instance, many chemical agents to which man is exposed regularly could be termed poisons under this definition. An oral dose of 400 milligrams of sodium chloride per kilogram (mg/kg) of body weight, ordinary table salt, will make a person violently ill. A standard aspirin tablet contains about 5 grains of aspirin, chemically known as acetylsalicylic acid. A fatal dose of aspirin to man is in the range of 75 to 225 grains or 15 to 45 tablets. Approximately 85 deaths occur every year (about one-third are children) as a result of overdoses of aspirin. To take a third example, let us consider nicotine. A fatal oral dose of this naturally occurring alkaloid to man is about 50 milligrams (mg), or approximately the amount of nicotine contained in two unfiltered cigarettes. In smoking, however, most of the nicotine is decomposed by burning, and, thus, it is not absorbed by the smoker.

Here man is not exposed during ordinary use to amounts of salt, aspirin, and nicotine which cause toxicity problems. Therefore, it is obvious that the *hazard* from normal exposure is very slight even though the compounds themselves would be toxic under other circumstances.

There's a better definition for the term "poison": "A chemical substance which exerts an injurious effect in the majority of cases in which it comes into contact with living organisms during normal use." The compounds mentioned above would obviously be excluded by such a definition and so would the majority of pesticides.

By necessity, pesticides are poisons, but the toxic hazards of different compounds vary greatly. As far as the possible risks associated with the use of pesticides are concerned, we can distinguish between two types: First, acute poisoning, resulting from the handling and application of toxic materials; and second, chronic risks from long-term exposure to small quantities of materials or from ingestion of them. The question of acute toxicity is obviously of paramount interest to people engaged in manufacturing and formulating pesticides and to those responsible for their application. Supposed chronic risks, however, are of much greater public interest because of their potential effect on the consumer of agricultural products.

Fatal human poisoning by pesticides is uncommon in the United States and is due to accident, ignorance, suicide, or crime. Fatalities represent only a small fraction of all recorded cases of poisoning, as demonstrated by these recent United States statistics, Table 31. It will be noted that in 1968, 2.8% of deaths from accidental poisoning were from pesticides, while in 1977 this cause had dropped to 1.0%. When the children's part of these statistics are viewed, it becomes a much more serious matter.

TABLE 31. Total Deaths from Accidental Poisonings by Solids and Liquids.

POISONING CLASSIFICATION	1968 (base)	1974	1975	1976	1977
Antibiotics & other anti-infectives	15	9	27	20	23
Hormones & synthetic substitutes	8	28	32	26	20
Systemic & hematologic agents	27	53	61	58	58
Analgesics & antipyretics	390	967	1275	1067	526
(Salicylate & congeners)	(120)	(83)	(98)	(79)	(58)
Sedatives & hypnotics	827	620	557	508	363
(Barbiturates)	(321)	(330)	(266)	(224)	(208)
Autonomic nervous system & psychotherapeutic drugs	99	165	178	190	193
(Tranquilizers)	(68)	(106)	(103)	(92)	(88)
Central nervous system depressants and stimulants	25	40	37	36	39
(Amphetamines)	(13)	(18)	(11)	(11)	(10)
Cardiovascular drugs	46	143	158	138	157
Gastro-intestinal drugs	7	6	3	4	4
Other & unspecified drugs & medicants	248	711	804	792	831
TOTAL DRUGS	1692	2742	3132	2839	2214
Alcohol	182	370	391	337	337
Cleaning & polishing agents	23	13	12	14	10
Disinfectants	9	8	6	6	5
Paints & varnishes	2	1	5	0	1
Petroleum products and other solvents	70	54	54	45	52
Pesticides, fertilizers or plant foods	72	35	30	31	34
Heavy metals (and their fumes)	53	23	13	15	21
Corrosives and caustics	28	13	16	22	17
Noxious foodstuffs and poisonous plants	10	5	6	6	7
Other & unspecified solid & liquid substances	442	752	1029	846	676
TOTAL NON-DRUG SOLID AND LIQUID SUBSTANCES	891	1274	1562	1322	1160
TOTAL ALL SOLIDS AND LIQUIDS	2583	4016	4694	4161	3374

Source: Mortality Statistics — Special Reports
Division of Vital Statistics
National Center for Health Statistics, Health Resources Administration
Public Health Service, Food and Drug Administration

In 1968, 11% of all accidental poisoning deaths were children under 5 years of age, while in 1977 that figure had dropped to 2.8%, a remarkable improvement. Still another good statistic is that of the less-than-5-year-olds poisoned in 1968, 11% again were from pesticides, while in 1977 that percentage had dropped to 7.4%. On the grimmer side is the statement that of all deaths attributed to accidental poisoning by pesticides in 1977, 20% were children under 5 years of age (Table 32).

Let's examine another type of data for 1973. Of 117,589 reported accidental ingestions among children under 5 years of age, pesticides were responsible for only about one half the number attributed to cosmetics or two-thirds of those attributed to aspirin (Table 33).

TABLE 32. Deaths from Accidental Poisoning by Solids and Liquids in 1977.

POISONING CLASSIFICATION	Under 5 Years of Age	5 Years and Over*	Total Deaths All ages
Antibiotics & other anti-infectives	1	22	23
Hormones & synthetic substitutes	0	20	20
Systemic & hematologic agents	8	50	58
Analgesics & antipyretics	16	510	526
(Salicylate & congeners)	(11)	(47)	(58)
Sedatives & hypnotics	4	359	363
(Barbiturates)	(3)	(205)	(208)
Autonomic nervous system & psychotherapeutic drugs	9	184	193
(Tranquilizers)	(2)	(86)	(88)
Central nervous system depressants and stimulants	1	38	39
(Amphetamines)	(0)	(10)	(10)
Cardiovascular drugs	5	152	157
Gastro-intestinal drugs	3	1	4
Other & unspecified drugs & medicines	10	821	831
TOTAL DRUGS	57	2157	2214
Alcohol	0	337	337
Cleaning & polishing agents	2	8	10
Disinfectants	3	2	5
Paints & varnishes	0	1	1
Petroleum products and other solvents	12	40	52
Pesticides, fertilizers or plant foods	7	27	34
Heavy metals (and their fumes)	3	18	21
Corrosives and caustics	6	11	17
Noxious foodstuffs and poisonous plants	0	7	7
Other & unspecified solid & liquid substances	4	672	676
TOTAL NON-DRUG SOLID AND LIQUID SUBSTANCES	37	1123	1160
TOTAL ALL SOLIDS & LIQUIDS	94	3280	3374

* Includes Unspecified Age

Source: Mortality Statistics — Special Reports
Division of Vital Statistics
National Center for Health Statistics, Health Resources Administration
Public Health Service, Food and Drug Administration

TABLE 33. Accidental Ingestions Among Children Under 5 Years of Age

	1973	
Type of Substance	No.	%
Medicines	52,113	44.3
*Internal	42,215	35.9
*Aspirin	7,763	6.6
*Other	34,452	29.3
*External	9,898	8.4
Cleaning and Polishing Agents	19,132	16.3
Petroleum Products	4,974	4.2
Cosmetics	10,362	8.8
Pesticides	5,591	4.8
Gases and Vapors	140	0.1
Plants	7,032	6.0
Turpentine, Paints, Etc.	6,988	5.9
Miscellaneous	10,517	9.0
Not Specified	740	0.6
Total	117,589	100.0

Source: Individual case reports submitted to the National Clearinghouse for Poison Control Centers; 1973, from 517 centers in 45 states.

Bull. Natl. Clearinghouse for Poison Control Centers. U.S. Food and Drug Admin. Bur. Drugs, H.E.W., May-June 1974.

Regardless of your attitude toward pesticides and their presumed hazard, they have an excellent safety track record, and it grows better each year, mainly through education and labeling of containers.

PESTICIDE EFFECTS ON MAN

Pesticides were developed to kill unwanted organisms, and are toxic materials which produce their effects by several different mechanisms. Under certain conditions they may be toxic to man, and an understanding of the basic principles of toxicity and the differences between toxicity and hazard is essential. As you already know, some pesticides are much more toxic than others, and severe illness may result when only a small amount of a certain chemical has been ingested, while with other compounds no serious effects would result even after ingesting large quantities. Some of the factors that influence this are related to (1) the toxicity of the chemical, (2) the dose of the chemical, especially concentration, (3) length of exposure, and (4) the route of entry or absorption by the body.

Early in the development of a pesticide for further experiments and exploration, toxicity data are collected on the pure toxicant as required by the Environmental Protection Agency. These tests are conducted on test animals that are easy to work with and whose physiology, in some instances, is like that of man, for example, dog. Test animals include white mice, white rats, white rabbits, guinea pigs, and beagle dogs. For instance, intravenous tests are determined usually on mice and rats, whereas dermal tests are conducted on shaved rabbits and guinea pigs. Acute oral toxicity determinations are most commonly made in rats and dogs, with the test substance being introduced directly into the stomach by tube. Chronic studies are conducted on the same two species for extended periods, and the compound is usually incorporated in the animal's daily ration. Inhalation studies may involve any of the test animals, but rats, guinea pigs, and rabbits are most commonly used.

These procedures are necessary to determine the overall toxic properties of the compound to various animals. From this information, toxicity to man can generally be extrapolated, and eventually some micro-level portion of the pesticide may be permitted in his food as a residue, which is expressed in parts per million, that is parts of a chemical per million parts of a particular food, by weight.

Toxicologists use rather simple animal toxicity tests to rank pesticides according to their toxicity. Long before pesticides are registered with the Environmental Protection Agency and eventually released for public use the manufacturer must declare the toxicity of their pesticide to the white rat under laboratory conditions. This toxicity is defined by the LD_{50}, expressed as milligrams (mg) of toxicant per kilogram (kg) of body weight, the dose which kills 50% of the test animals to which it is administered under experimental conditions.

This toxicity value, or LD_{50}, is measured in terms of oral (fed to, or placed directly in the stomachs of rats), dermal (applied to the skin of rats or rabbits), and respiratory toxicity (inhaled). Using two of these tests, oral and dermal, a toxicologic ranking is shown for the organophosphate and organochlorine insecticides in Fig. 26-27. The materials on the top of the list are the most toxic and those at the bottom the least. The size of the dose is the most important single item in determining the safety of a given chemical, and actual statistics of human poisonings correlate reasonably well with these toxicity ratings.

ESTIMATING PESTICIDE TOXICITY TO MAN

In addition to toxicity, the dose, length of exposure, and route of absorption are the other important variables. The amount of pesticide required to kill a man can be correlated with the LD_{50} of the material to rats in the laboratory. In Table 34 below, for example, the acute oral LD_{50} expressed as mg/kg dose of the technical material, is translated into the amount needed to kill a 170 lb. man. Dermal LD_{50}s are included for a better understanding of the relationship of expressed animal toxicity to human toxicity.

In a manner of generalizing, oral ingestions are more toxic than respiratory inhalations which are more toxic than dermal absorption. Additionally, there are physical and chemical differences between pesticides which make them more likely to produce poisoning. For instance, parathion changes to the more toxic metabolite paraoxon under certain conditions of humidity and temperature. Parathion is more toxic than methyl parathion to field workers, yet there is not that great a difference in their oral toxicities. Workers' exposure is usually dermal, which explains why many more illnesses are reported in workers exposed to parathion than those exposed to methyl parathion. So, we see that toxicity, route of absorption, dose, length of exposure and the physical and chemical properties of the pesticide contribute to its relative hazard. Hazard, then, is an expression of the potential of a pesticide to produce human poisoning.

TABLE 34. Combined Tabulation of Pesticide Toxicity Classes.

	Routes of Absorption		
Toxicity Rating*	LD_{50} Single ORAL Dose Rats mg/kg	LD_{50} Single DERMAL Dose Rabbits mg/kg	Probable Lethal Oral Dose for Man
6 - Super toxic	less than 5 mg	20 mg or less	A taste, a grain
5 - Extremely toxic	5 - 50 mg	20 - 200	A pinch - 1 teaspoon
4 - Very toxic	50 - 500	200 - 1,000	1 teaspoonful - 2 tablespoons
3 - Moderately toxic	500 - 5,000	1,000 - 2,000	1 ounce - 1 pint
2 - Slightly toxic	5,000 - 15,000	2,000 - 20,000	1 pint - 1 quart
1 - Practically nontoxic	>15,000	>20,000	>1 quart

* Modified from: Clinical Toxicology of Commercial Products, 2nd Edition 1963. Gleason, M. N., Gosselin, R. E., and Hodge, H. C. The Williams and Wilkins Company, Baltimore, Maryland.

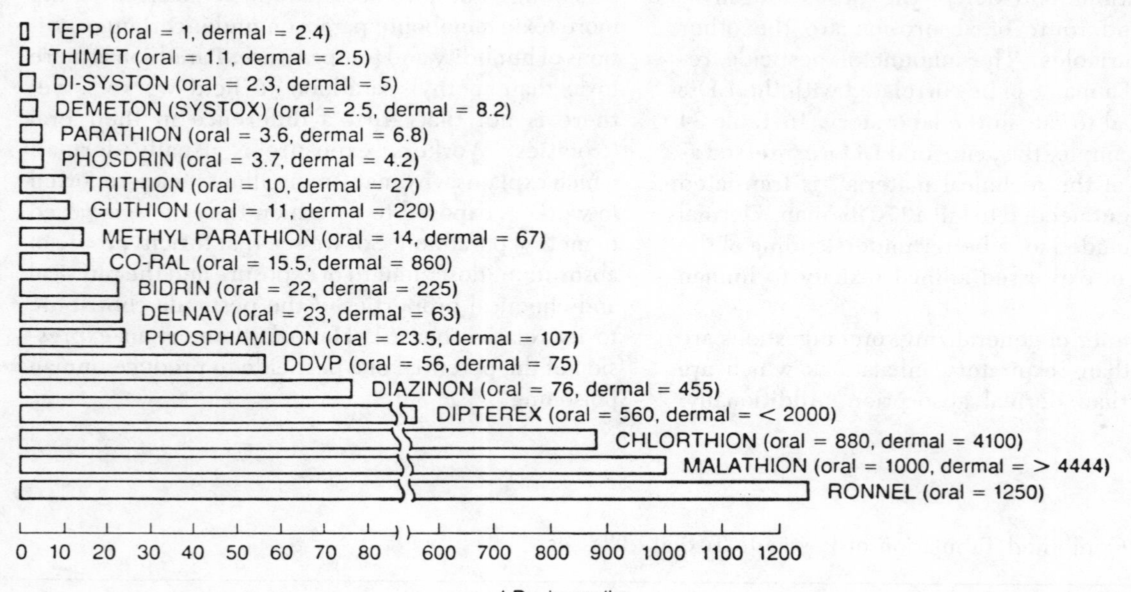

FIGURE 26. Acute Oral and Dermal Toxicity Values to Rats for Some Organophosphate Pesticides.[1]

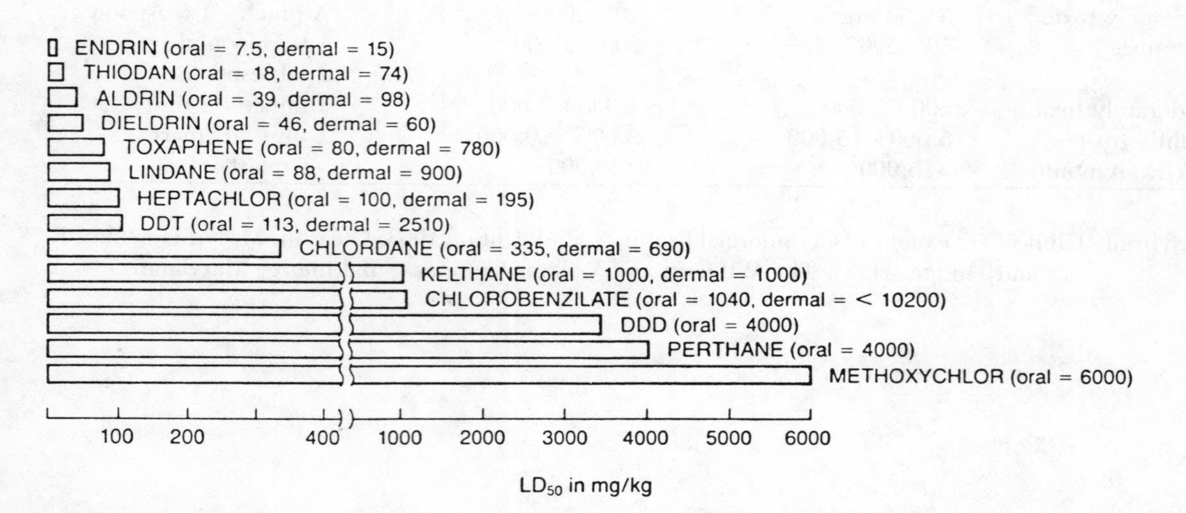

FIGURE 27. Acute Oral and Dermal Toxicity Values to Rats for Some Chlorinated Hydrocarbon Pesticides.[1]

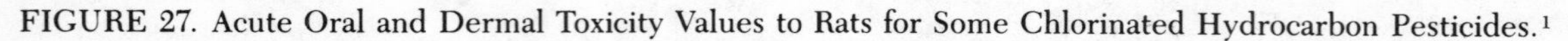

[1] Source: Unpublished chart prepared by the Bureau of Occupational Health, State of California Department of Public Health. Reproduced by permission.

FIRST AID FOR PESTICIDE POISONING

The symptoms of poisoning may appear almost immediately after exposure or may be delayed for several hours depending on the chemical, dose, length of exposure and the individual. These symptoms may include, but are not restricted to headache, giddiness, nervousness, blurred vision, cramps, diarrhea, a feeling of general numbness, or abnormal size of eye pupils. In some instances there is excessive sweating, tearing, or mouth secretions. Severe cases of poisoning may be followed by nausea and vomiting, fluid in the lungs, changes in heart rate, muscle weakness, breathing difficulty, confusion, convulsions, coma, or death. However, pesticide poisoning may mimic brain hemorrhage, heat stroke, heat exhaustion, hypoglycemia (low blood sugar), gastroenteritis (intestinal infection), pneumonia, asthma, or other severe respiratory infection.

No matter how trivial the exposure may seem, if poisoning is present or suspected, obtain medical advice at once. If a physician is not immediately available by phone, take the person directly to the emergency ward of the nearest hospital and take along the pesticide label and telephone number of the nearest Poison Control Center.

First aid treatment is extremely important, regardless of the time that may elapse before medical treatment is available. The first aid treatment received during the first 2-3 minutes following a poisoning accident may very well spell the difference between life and death.

FIRST AID FOR CHEMICAL POISONING

If You Are Alone With the Victim

First — See that the victim is breathing; if not, give artificial respiration.

Second — Decontaminate him immediately; i.e. wash him off thoroughly. Speed is essential!

Third — Call your physician.

Note: Do *not* substitute first aid for professional treatment. First aid is only to relieve the patient before medical help is reached.

If Another Person Is With You and the Victim

Speed is essential; one person should begin first aid treatment while the other calls a physician.

The physician will give you instructions. He will very likely tell you to get the victim to the emergency room of a hospital. The equipment needed for proper treatment is there. Only if this is impossible should the physician be called to the site of the accident.

General

1. Give mouth-to-mouth artificial respiration if breathing has stopped or is labored.
2. Stop exposure to the poison and if poison is on skin cleanse the person, including hair and fingernails. If swallowed, induce vomiting as directed below.
3. Save the pesticide container and material in it if any remains; get readable label or name of chemical(s) for the physician. If the poison is not known, save a sample of the vomitus.

Specific

Poison on Skin

Drench skin and clothing with water (shower, hose, faucet).

Remove clothing.

Cleanse skin and hair thoroughly with soap and water; speed in washing is most important in reducing extent of injury.

Dry and wrap patient in blanket.

Poison in Eye

Hold eyelids open, wash eyes with gentle stream of clean running water immediately. Use copious amounts. Delay of a few seconds greatly increases extent of injury.

Continue washing for 15 minutes or more.

Do *not* use chemicals or drugs in wash water. They may increase the extent of injury.

Inhaled Poisons (Dusts, Vapors, Gases)

If victim is in enclosed space, do not go in after him without air-supplied respirator.

Carry patient (do not let him walk) to fresh air immediately.

Open all doors and windows, if any.

Loosen all tight clothing.

Apply artificial respiration if breathing has stopped or is irregular.

Call a physician.

Prevent chilling (wrap patient in blankets but don't overheat him).

Keep patient as quiet as possible.

If patient is convulsing, watch his breathing and protect him from falling and striking his head on the floor or wall. Keep his chin up so his air passage will remain free for breathing.

Do not give alcohol in any form.

Swallowed poisons

CALL A PHYSICIAN IMMEDIATELY

Do not induce vomiting if:

Patient is in a coma or unconscious.

Patient is in convulsions.

Patient has swallowed petroleum products (that is, kerosene, gasoline, lighter fluid).

Patient has swallowed a corrosive poison (strong acid or alkaline products); symptoms: severe pain, burning sensation in mouth and throat.

If the patient can swallow after ingesting a corrosive poison, give the following substances by mouth. A corrosive substance is any material which in contact with living tissue will cause destruction of tissue by chemical action such as lye, acids, Lysol, etc.

For acids: Milk, water, or milk of magnesia (1 tablespoon to 1 cup of water);

For alkali: Milk or water; for patients 1-5 years old, 1 to 2 cups; for patients 5 years and older, up to 1 quart.

Universal: Condensed canned milk, as much as the victim can consume.

Induce Vomiting (If Possible) When Non-Corrosive Substance has been Swallowed

Give milk or water (for patient 1-5 years old — 1 to 2 cups; for patients over 5 years — up to 1 quart).

Induce vomiting by placing the blunt end of a spoon, not the handle, or your finger at the back of the patient's throat, or by use of this emetic — 2 tablespoons of salt in an 8-10 oz glass of warm water.

When retching and vomiting begin, place patient face down with head lowered, thus preventing vomitus from entering the lungs and causing further damage. Do not let him lie on his back.

Do not waste excessive time in inducing vomiting if the hospital is a long distance away. It is better to spend the time getting the patient to the hospital where drugs can be administered to induce vomiting and/or stomach pumps are available.

Clean vomitus from person. Collect some in case physician needs it for chemical tests.

Chemical Burns of Skin

Wash with large quantities of running water.

Remove contaminated clothing.

Immediately cover with loosely applied clean cloth, any kind will do, depending on the size of the area burned.

Avoid use of ointments, greases, powders, and other drugs in first aid treatment of burns.

Treat shock by keeping patient flat, keeping him warm, and reassuring him until arrival of physician.

APPENDIXES

APPENDIX A

TABLE 35. Common, Trade and Chemical Names of Insecticides and Acaricides and Their Oral and Dermal LD_{50}s to Laboratory Animals.

Common Name and Trade Name	Chemical Name	Oral LD_{50}	Dermal LD_{50}	Manufacturer
Acaraban (see chlorobenzylate)				
acephate Orthene	*O,S*-dimethyl acetylphos = phoroamidothioate	866	2000	Chevron
Bacillus popillae Doom, Japidemic	Microbial insecticide for Japanese beetle grubs	non-toxic		Fairfax
Bacillus thuringiensis Dipel, Thuricide Bactur, BT	Microbial insecticide for caterpillars	non-toxic		Abbot Sandoz Thompson-Hayward
Bactur (see *Bacillus thuringiensis)*				
Baygon (see propoxur)				
γ-BHC (see Lindane)				
Bidrin (see dicrotophos)				
BT (see *Bacillus thuringiensis)*				
Butacide (see piperonyl butoxide)				
carbaryl Sevin	1-naphthyl methylcarbamate	307	2000	Union Carbide
chlordane Octachlor	1,2,4,5,6,7,8,8-Octachloro-3a,4,7,7a-tetrahydro-4,7-methanoindan	283	580	Velsicol
chlorobenzilate[1] Acaraben	ethyl 4,4′-dichlorobenzilate	700	10200	Ciba-Geigy
chlorpyrifos Dursban	*O,O*-diethyl *O*-(3,5,6-tri = chloro-2-pyridyl) phos = phorothioate	97	2000	Dow
cryolite, Kryocide	sodium alumino-fluoride	>10000	Very low toxicity	Pennwalt
cubé (see rotenone)				
Cygon (see dimethoate)				
Cythion (see malathion)				
DDT Topocide (lousicide)	1,1,1-trichloro-2,2-bis (*p*-chlorophenyl) ethane	87	1931	Montrose
DDVP (see dichlorvos)				

[1] No longer approved by the Environmental Protection Agency.

Common Name and Trade Name	Chemical Name	Oral LD_{50}	Dermal LD_{50}	Manufacturer
derris (see rotenone)				
diazinon Spectracide	*O,O*-diethyl-*O*-(2-isopropyl-6-methyl-4-pyrimidinyl) phosphorothioate	66	379	Ciba-Geigy
Dibrom (see naled)				
p-dichlorobenzene PDB	*p*-dichlorobenzene	500	2000	Dow Monsanto PPG Industries
dichlorvos Vapona, DDVP	2,2-dichlorovinyl dimethyl phosphate	25	59	Shell
dicofol Kelthane	4,4′-dichloro-α-trichloro = methyl) benzhydrol	575	4000	Rohm & Haas
dicrotophos Bidrin	dimethyl phosphate ester with (*E*)-3-hydroxy-*N,N*-dimethylcrotonamide	22	225	Shell Ciba-Geigy
dieldrin	1,2,3,4,10,10-hexachloro-6,7-epoxy 1,4,4a,5,6,7,8,8a-octahydro-1,4-endo-exo-5,8-dimethanophthalene	40	65	Shell Intl.
dimethoate Cygon	*O,O*-dimethyl *S*-(*N*-methyl = carbamoylmethyl) phosphoro = dithioate	250	150	American Cyanamid
Dipel (see *Bacillus thuringiensis*)				
disulfoton Di-Syston	*O,O*-diethyl *S*-[2-(ethylthio) ethyl] phosphorodithioate	2	20	Mobay
Di-Syston (see disulfoton)				
Doom (See *Bacillus popillae*)				
Dormant oil (see Superior oil)				
Dursban (see chlorpyrifos)				
Dylox (see trichlorfon)				
Elcar	*Heliothis* spp. virus	non-toxic		Sandoz
endosulfan Thiodan	6,7,8,9,10,10-hexachloro-1,5,5a,6,9,9a-hexahydro-6,9-methano-2,4,3-benzodioxa = thiepin 3-oxide	18	74	FMC Hooker Velsicol
ethion Nialate	*O,O,O′O′*-tetraethyl *S,S′*-methylene bis (phosphoro = dithioate)	27	62	FMC
Heliothis virus (see Elcar)				
heptachlor	1,4,5,6,7,8,8-heptachloro-3a,4,7,7a,tetrahydro-4,7-methanoindene	40	119	Velsicol

Common Name and Trade Name	Chemical Name	Oral LD_{50}	Dermal LD_{50}	Manufacturer
Japidemic (see *Bacillus popillae)*				
Kelthane (see dicofol)				
Korlan (see ronnel)				
Kryocide (see cryolite)				
Kwell (see Lindane)				
Lindane γ-BHC Kwell	1,2,3,4,5,6-hexachloro = cyclohexane, γ isomer of not less than 99% purity	76	500	Hooker
malathion Cythion	diethyl mercaptosuccinate *S*-ester with *O,O*-dimethyl phosphorodithioate	885	4000	American Cyanamid
Marlate (see methoxychlor)				
Meta Systox-R (see oxydemetonmethyl)				
methoxychlor Marlate	1,1,1-trichloro-2,2-bis = (p,-methoxyphenyl) ethane	5000	2820	Chemical Formulators Dupont Prentiss
Morestan (see oxythioquinox)				
naled Dibrom	1,2-dibromo-2,2-dichloro = ethyl dimethyl phosphate	430	1100	Chevron
Nialate (see ethion)				
Naphthalene	naphthalene	Low toxicity	Very low toxicity	Several
nicotine nicotine sulfate	1-3-(1-methyl-2-pyrrolidyl) pyridine	50	140	Chemical Formulators Prentiss
Octachlor (see chlordane)				
Orthene (see acephate)				
oxydemetonmethyl Meta Systox-R	*S*-[2-(ethylsulfinyl) ethyl] = *O,O*-dimethyl phosphorothioate	65	100	Mobay
oxythioquinox Morestan	cyclic S,S-(6-methyl-2,3-quinoxalinediyl) dithiocarbonate	2500	2000	Mobay
paradichlorobenzene (see *p*-dichlorobenzene)				
PDB (see *p*-dichlorobenzene)				
piperonyl butoxide Butacide	α-[2-(2-butoxy) ethoxy]-4,5-(methylenedioxy)-2-propyltoluene	7500	7500	Prentiss Alpha Laboratories Fairfield American McLaughlin Gormley King
propoxur Baygon	*o*-isopropoxyphenyl methylcarbamate	95	1000	Mobay

Common Name and Trade Name	Chemical Name	Oral LD50	Dermal LD50	Manufacturer
pyrethrins	mixture of pyrethrins and cinerins	200	1800	Fairfield American Penick Prentiss McLaughlin Gormley King
ronnel Korlan	*O,O*-dimethyl *O*-(2,4,5-trichlorophenyl) phos = phorothioate	906	1000	Dow
rotenone cubé, derris	1,2,12,12a-tetrahydro-2-isopropenyl-8,9-dimethoxy [1]benzopyrano[2,4-*b*] furo[2,3-*b*][1]benzopyran-6 (6a*H*)-one	60	>1000	Blue Spruce Fairfield American Penick Prentiss
ryania	*Ryania speciosa*	750	4000	manufacture discontinued
sabadilla	*Schoenocaulon officinale*	4000	—	Prentiss
sesamex Sesoxane	2-(2-ethoxyethoxy) ethyl. 3,4-(methylenedioxy) phenyl acetal of acetaldehyde	2000	11000	Shulton
Sesoxane (see sesamex)				
Sevin (see carbaryl)				
Spectracide (see diazinon)				
sulfoxide Sulfox-Cide	1,2-(methylenedioxy)-4-[2-(octylsulfinyl) propyl] = benzene	2000	9000	manufacture discontinued
Superior oil (dormant oil)		non-toxic		Sun Oil FMC Chevron Exxon
Tedion (see tetradifon)				
tetradifon Tedion	*p*-chlorophenyl 2,4,5-trichlorophenyl sulfone	5000	1000	Philips-Duphar B.V. (Netherlands)
Thiodan (see endosulfan)				
Thuracide (see *Bacillus thuringiensis*)				
Topocide (see DDT)				
trichlorfon Dylox	dimethyl (2,2,2-tri = chloro-1-hydroxy-ethyl) phosphonate	450	2000	Mobay
Vapona (see dichlorvos)				
virus, *Heliothis* (see Elcar)				

APPENDIX B

TABLE 36. Common, Trade and Chemical Names of Fungicides and Their Oral and Dermal LD_{50}s to Laboratory Animals.

Common Name and Trade Name	Chemical Name	Oral LD_{50}	Dermal LD_{50}	Manufacturer
Agrimycin (see streptomycin)				
Agri-Strep (see streptomycin)				
anilazine Dyrene	2,4-dichloro-6-(*o*-chloro = anilino)-s-triazine	2710	—	Mobay
Arasan (see thiram)				
basic copper sulfate	basic copper sulfate	nontoxic	—	Cities Service Kocide Chem. Phelps Dodge
Benlate (see benomyl)				
benomyl Benlate	methyl-1-(butylcarbamoyl)-2-benzamidazole carbamate	>10000	—	DuPont
Bordeaux	mixture of copper sulfate and calcium hydroxide forming basic copper sulfates	nontoxic	—	Chemical Formulators
Bravo (see chlorothalonil)				
captan Orthocide	*N*-(trichloromethylthio)-4-cyclohexene-1,2-dicarboximide	>10000	—	Chevron Stauffer
Carbamate (see ferbam)				
chlorothalonil Bravo, Daconil 2787	tetrachloro isophthalonitrile	>10000	10000	Diamond Shamrock
copper oxychloride	basic copper chloride	low	—	many
copper oxychloride sulfate	mixture of basic copper chloride and basic copper sulfate	nontoxic	—	many
Corozote (see ziram)				
Cuman (see ziram)				
Cyprex (see dodine)				
Daconil 2787 (see chlorothalonil)				
Dexon (see fenaminosulf)				
dinocap Karathane, Mildex	mixture of 2,4-dinitro-6-octylphenyl crotonate and 2,6-dinitro-4-octylphenyl crotonate	980	>4700	Rohm & Haas
Dithane D-14 (see nabam)				

Common Name and Trade Name	Chemical Name	Oral LD_{50}	Dermal LD_{50}	Manufacturer
Dithane M-22 (see maneb)				
Dithane M-45 (see mancozeb)				
Dithane Z-78 (see zineb)				
dodine Cyprex, Melprex	n-dodecylguanidine acetate	1000	>1500	American Cyanamid
Dyrene (see anilazine)				
fenaminosulf Dexon, Lexan	sodium [4-(dimethylamino) = phenyl] diazene sulfonate	64	100	Mobay
ferbam Carbamate	ferric dimethyldithiocarbamate	>17000	—	FMC
folpet Phaltan	*N*-(trichloromethylthio)-phthalimide	>10000	—	Chevron Stauffer
Fore (see mancozeb)				
Fungo 50 (see methyl thiophanate)				
Karathane (see dinocap)				
Lesan (see fenaminosulf)				
lime sulfur	calcium polysulfide	skin irritant	—	many
mancozeb Fore, Dithane M-45	coordination product of zinc ion and manganese ethylene bisdithiocarbamate	8000		
maneb Dithane M-22, Manzate, Manzeb, Tersan	manganese ethylenebisdithiocarbamate	6750	—	many
Manzate (see maneb)				
Manzeb (see maneb)				
Melprex (see dodine)				
metam-sodium Vapam	sodium methyldithiocarbamate	820	—	Stauffer
methyl thiophanate Fungo 50	dimethyl 4,4-o-phenylenebis = (3-thioallophanate)	9700	>8000	Mallinckrodt
metiram Polyram-Combi	mixture of ammoniates of ethylenebis (dithiocarbamate)-zinc and ethylenebisdithio = carbamic acid cyclic anhydro = sulfides and disulfides	>10000	—	BASF (Germany)
Mildex (see dinocap)				
Morestan (see oxythioquinox)				
nabam Dithane D-14	disodium ethylenebisdithio = carbamate	395	—	Chemical Formulators Rohm & Haas

Common Name and Trade Name	Chemical Name	Oral LD_{50}	Dermal LD_{50}	Manufacturer
Orthocide (see captan)				
oxythioquinox Morestan	6-methyl-1-3-dithiolo- (4,5-b) quinoxalin-2-one	2500	>1000	Mobay
PCNB Terraclor	pentachloronitrobenzene	1700	—	Olin
Phaltan (see folpet)				
Polyram-Combi (see metiram)				
streptomycin (sulfate or nitrate) Agri-Strep Agrimycin	2,4-diguanidino-3,5,6-tri = hydroxycyclohexyl-5-deoxy-2- *O*-2-deoxy-2-methylamino-*a*- glucopyranosyl)-3-formyl pentanofuranoside	9000	—	Charles Pfizer Merck
sulfur	elemental sulfur in many forms	nontoxic	—	many
Terraclor (see PCNB)				
Terra-Coat (see Terrazole)				
Terrazole Terra-Coat	5-ethoxy-3-trichloro = methyl-1,2,4-thiadiazole	1077	—	Olin
Tersan (see maneb)				
thiophanate Topsin E	1,2-bis (3-ethoxycarbonyl- 2-thioureido) benzene	>15000	—	Pennwalt
thiophanate methyl (see methyl thiophanate)				
thiram Arasan, TMTD	tetramethylthiuramidisulfide	780	—	W.A. Cleary DuPont Pennwalt Vineland
TMTD (see thiram)				
Topsin E (see thiophanate)				
Vapam (see metam-sodium)				
zineb Dithane Z-78	zinc ethylenebis (dithiocarbamate)	5200	—	FMC Rohm & Haas Crystal
Z-C Spray (see ziram)				
Zerlate (see ziram)				
ziram Corozote, Cuman Zerlate, Z-C Spray	zinc dimethyldithiocarbamate	1400	—	Pennwalt FMC

APPENDIX C

TABLE 37. Common, Trade and Chemical Names of Herbicides and Their Oral and Dermal LD_{50}s to Laboratory Animals.

Common Name or Designation	Chemical Name	Oral LD_{50}	Dermal LD_{50}	Manufacturer
Aminotriazole (see amitrole)				
amitrole Aminotriazole	3-amino-1,2,4-triazole	1100	>10000	Amchem American Cyanamid
Balan (see benefin)				
benefin Balan	*N*-butyl-*N*-ethyl-α,α,α-trifluoro-2,6-dinitro-*p*-toluidine	>10000	—	Elanco
bensulide Betasan, Prefar	*O,O*-diisopropyl phos = phorodithioate *S*-ester with N-(2-mercaptoethyl) = benzenesulfonamide	1082	3950	Stauffer
Betasan (see bensulide)				
Brominal (see bromoxynil)				
bromoxynil Brominal	3,5-dibromo-4-hydroxybenzonitrile	190	—	Amchem Rhodia
cacodylic acid Dilic, Phytar 138	hydroxydimethylarsine oxide	700	—	Crystal Vineland
Casoron (see dichlobenil)				
CDEC Vegedex	2-chlorallyl diethyldi = thiocarbamate	850	2200	Monsanto
Check-Mate (see MSMA)				
Dacthal (see DCPA)				
dalapon Dowpon	2,2-dichloropropionic acid	970	—	Dow
DCPA Dacthal	dimethyl tetrachloro = terephthalate	3000	10000	Diamond Shamrock
Ded-weed (see silvex)				
dichlobenil Casoron	2,6-dichlorobenzonitrile	3160	1350	Thompson-Hayward
Dilic (see cacodylic acid)				
diphenamid Dymid, Enide	*N,N*-dimethyl-2,2-diphenyl = acetamide	1000	—	Elanco Upjohn
diquat	6,7-dihydrodipyrido [1,2-α:2′,1′-*c*] pyrazinediium ion	400	>400	Chevron

Common Name or Designation	Chemical Name	Oral LD_{50}	Dermal LD_{50}	Manufacturer
Diumate (see MSMA)				
Dowpon (see dalapon)				
Dymid (see diphenamid)				
DSMA	disodium methanearsonate	1000	—	many
Enide (see diphenamid)				
Eptam (see EPTC)				
EPTC Eptam	*S*-ethyldipropylthiocarbamate	1630	10000	Stauffer
glyphosate Roundup	N-(phosphonomethyl) glycine	4300	>5010	Monsanto
Kuron (see silvex)				
Mesamate (see MSMA)				
MSMA Check-Mate Diumate Mesamate Weed-E-Rad	monosodium methanearsonate	700	—	many
paraquat	1,1′-dimethyl′4,4′-bipyridinium ion	150	>480	Chevron
Phytar 138 (see cacodylic acid)				
Prefar (see bensulide)				
Princep (see simazine)				
Roundup (see glyphosate)				
siduron Tupersan	1-(2-methylcyclohexyl)-3-phenylurea	>7500	—	DuPont
silvex[1] Kuron, Ded-weed	2-(2,4,5-trichlorophenoxy) = propionic acid	375	—	Dow Thompson-Hayward
simazine Princep	2-chloro-4,6-bis (ethyl = amino)-*s*-triazine	>5000	—	Ciba-Geigy
Treflan (see trifluralin)				
trifluralin Treflan	α,α,α-trifluoro-2,6-dinitro-*N*,*N*-dipropyl-*p*-toluidine	3700	—	Elanco
Tupersan (see siduron)				
Vegedex (see CDEC)				
Weed-E-Rad (see MSMA)				
2,4-D	(2,4-dichlorophenoxy) acetic acid	375	—	many
2,4,5-T[1]	(2,4,5-trichlorophenoxy) acetic acid	300	—	many

[1] No longer approved by the Environmental Protection Agency.

APPENDIX D

TABLE 38. Common, Trade and Chemical Names and Uses of Miscellaneous Recommended Pesticides and Their Oral LD_{50}s to Laboratory Animals.

Common Name and Trade Name	Use	Chemical Name	Oral LD_{50}
alkyldimethylbenzyl-ammonium chloride	algicide, disinfectant	alkyldimethylbenzyl-ammonium chloride	445
anthraquinone Corbit, Morkit	bird repellent	9,10-anthraquinone	>5000
Antimilace (see metaldehyde)			
Arasan (see thiram)			
Azochloramide	disinfectant	azochloramide	<500
Biomet 12	animal repellent	tri-n-butyltin chloride	10000(D)
calcium hypochlorite	algicide, disinfectant	calcium hypochlorite	corrosive
captan Merpan, Orthocide Vondcaptan	fungicide, bird repellent	cis-N-[(trichloro = methyl) thio]-4-cyclo = hexane-1,2-dicarboximide	9000
carbaryl Sevin	insecticide, molluscicide	1-napthyl N-methyl carbamate	500-850
Chloralose	bird repellent	gluchloralose	<400
Chloramine-T	disinfectant	sodium *p*-toluenesulfon-chloramine	<500
Corbit (see anthraquinone)			
coumafuryl Fumarin	rodenticide	3-(*a*-acetonylfurfuryl) -4-hydroxy coumarin	25
DBCP[1] Fumazone, Nemagon	nematicide, insecticide	1,2-dibromo-3-chloropropane	170-300
d-Con (see Warfarin)			
D-D Telone, Vidden-D	nematicide, insecticide	1,3-dichloropropene and isomers	140
deet Delphene, Detamide Metadelphene, Off	personal repellent	N,N-diethyl-*m*-toluamide	1950-2000
Delphene (see deet)			
Detamide (see deet)			

D = Dermal LD_{50}
[1] No longer approved by the EPA.

Common Name and Trade Name	Use	Chemical Name	Oral LD_{50}
dichlone Phygon	fungicide, algicide	2,3-dichloro-1,4- naphthoquinone	1300
diphacinone Diphacin, Promar, Ramik	rodenticide	2-diphenylacetyl-1,3- indandione	3
Diphacin (see diphacinone)			
ethyl hexanediol (see Rutgers 6-12)			
formaldehyde	fungicide, disinfectant	methanal	<500
Fumarin (see coumafuryl)			
Fumazone (see DBCP)			
glutaraldehyde	disinfectant	glutaraldehyde	<600
Halazone	disinfectant	*p*-sulfone dichloro- amidobenzoic acid	<500
lithium hypochlorite	algicide, disinfectant	lithium hypochlorite	corrosive
mercaptodimethur Mesurol	insecticide, molluscicide	3,5-dimethyl-4- (methylthio) phenol methylcarbamate	87-130
Merpan (see captan)			
Mesurol (see mercaptodimethur)			
Metadelphene (see deet)			
metaldehyde Antimilace, Namekil	molluscicide	metacetaldehyde	630
Metam-sodium (see SMDC)			
mexacarbate Zectran	insecticide, molluscicide	4-dimethylamino-3,5- xylyl N-methylcarbamate	19
Morkit (see anthraquinone)			
Namekil (see metaldehyde)			
naphthalene	insecticide, animal repellent	naphthalene	slightly toxic
Nemagon (see DBCP)			
nicotine	insecticide, animal repellent	*l*-3-(1-methyl-2- pyrrolidyl) pyridine	50
Off (see deet)			
Ornitrol (SC-12937)	bird chemosterilant	20,25-diazacholesterol- dihydrochloride	60
Orthocide (see captan)			
Paracide (see PDB)			
Paradow (see PDB)			
PDB Paracide, Paradow	insecticide, animal repellent	p-dichlorobenzene	1000-4000

Common Name and Trade Name	Use	Chemical Name	Oral LD_{50}
Phygon (see dichlone)			
pindone Pival	rodenticide	2-pivaloylindane-1,3-dione	280
Pival (see pindone)			
potassium dichloro-isocyanurate	algicide	potassium dichloro= isocyanurate	<750
Promar (see diphacinone)			
pyriminil Vacor, RH-787	rodenticide	N-3-pyridylmethyl N′-nitrophenyl urea	4.8 to 18.0
R-55 Repellent	animal repellent	tertiary butylsulfenyl dimethyl-dithio carbamate	>15000
Ramik (see diphacinone)			
RH-787 (see pyriminil)			
Rutgers 6-12 ethyl hexanediol	personal repellent	2-ethyl-1,3-hexanediol	6500
SC-12937 (see Ornitrol)			
Sevin (see carbaryl)			
simazine	herbicide, algicide	2-chloro-4,6-bis (ethyl = amino)-*s*-triazine	5000
SMDC Metam-sodium, Vapam	insecticide, nematicide, herbicide, fungicide	sodium N-methyldithio = carbamate	820
sodium chlorite	algicide, disinfectant	sodium chlorite	1200
sodium dichloro-isocyanurate	algicide	sodium dichloroisocyanurate	<750
sodium hypochlorite	algicide, disinfectant	sodium hypochlorite	corrosive
strychnine	rodenticide, avicide	alkaloid from tree, *Strychnos nux-vomica*	1 to 30
succinchlorimide	disinfectant	succinchlorimide	<500
Telone (see D-D)			
thiram Arasan, TMTD	fungicide, animal repellent	tetramethylthiuramidisulfide	780
TMTD (see thiram)			
Vapam (see SMDC)			
Vidden-D (see D-D)			

Common Name and Trade Name	Use	Chemical Name	Oral LD_{50}
Vondcaptan (see captan)			
Warf (see warfarin)			
warfarin d-Con, Warf	rodenticide	3-(a-acetonylbenzyl)- 4-hydroxycoumarin	25
Zectran (see mexacarbate)			
ziram (many names)	fungicide, animal repellent	zinc dimethyldithiocarbamate	1400

APPENDIX E — PESTICIDES THAT CAN BE MADE AT HOME

INSECTICIDES

Nicotine extract. Soak one or two shredded, cheap cigars, or a standard plug of chewing tobacco in one gallon of water overnight at room temperature. Remove tobacco parts and add one teaspoon of household detergent. Used as a foliar spray against aphids and other small insects it is usually as effective as sprays made commercially. CAUTION. This nicotine extract is very toxic to all living animals, including man. It may also transmit tobacco virus to tomatoes and other sensitive plants.

Lime-sulfur concentrate. Use one pound of unslaked lime (quicklime), two pounds of sulfur and one gallon of water for each batch. Make a paste of the sulfur with some of the water. In a well-ventilated room or out-of-doors heat the water and lime until the lime starts to dissolve, then add the sulfur paste. Bring to a boil and simmer for about 20-30 minutes or until the material becomes dark amber. Add water as necessary to maintain original level. When the free sulfur has all disappeared the reaction is complete, and the mixture should be strained and stored in jars. This is the *concentrate*. To use as a dormant spray, dilute 1 part of concentrate to 9 parts of water. As a summer foliage spray, dilute 1 part to 49 parts of water. Lime sulfur sprays will stain the skin and clothing. Do not apply to vegetables to be canned. Jars are known to explode which contain trace amounts of elemental sulfur.

Dormant oil spray concentrate. Heat 1 quart of good lubricating oil (10 wt) free of additives, white oil such as mineral oil is preferred, with 1 pint of water and 4 ounces of potash fish-oil soap. Yellow kitchen bar soap can be used as a substitute for fish-oil soap. Heat until the boiling point is reached. Pour back and forth until blended. For deciduous trees mix 5 teaspoons of concentrate with 1 quart of water. For evergreens use 2 teaspoons per quart. For leftover concentrate, reheat and mix again before using.

Kerosene emulsion concentrate. Dissolve 2 cubic inches of kitchen grade bar soap in 1 pint of water. Add 1 quart of kerosene and beat with egg beater or electric mixer until a thick cream results. For dormant deciduous trees dilute 1 part of concentrate with 7 parts of water. For ordinary mid-summer growth, dilute with 10 parts water, and for a weak spray, dilute with 15 parts water.

Bordeaux mixture. Add 3 ounces of copper sulfate (bluestone) to 3 gallons of water and dissolve thoroughly. Add 5 ounces of hydrated lime and mix completely. Use without further dilution. This fungicide-insecticide can be stored but is corrosive to metal containers. While primarily a fungicide, it is also very repellent to many insects such as flea beetles, leafhoppers, and potato psyllid, when sprayed over the leaves of plants.

Wormwood. Use dry or fresh wormwood leaves. Cover with water and bring to a boil. Dilute with 4 parts of water and spray on plants immediately. Wormwood sprays are purported to kill slugs, aphids and crickets.

Quassia. Soak 2 ounces of quassia wood chips in 1 gallon water for 2-3 days. Simmer for 2-3 hours over low heat. Remove chips and mix with 2 ounces of soft soap. It's not the best but will control aphids.

Soaps. Soap dilutions have been used for control of soft-bodied insects, such as aphids, since 1787, when this control method first appeared in writing. Undoubtedly it had been used long before that. Most often these soaps were derived from either plants (cotton seed, olive, palm, or coconuts) or from animal fat, such as lard, whale oil, or fish oil. Vegetable or plant-derived soaps are more effective than those derived from petroleum. Commercial soaps today vary greatly in composition and purity, therefore vary widely in effectiveness.

If in doubt, you should try true soap suds from a known brand of inexpensive laundry soap against aphids on only a few plants. If this proves successful, the practice could be expanded. Old-fashioned homemade soap may be prepared using inexpensive waste lard or tallow, lye, water, and borax (optional). Six pounds of fat and a can of lye will make 7 pounds of soap, or 12 to 15 pounds of soft soap.

To make, strain 3 quarts of heated-to-melting fat. Dissolve a can of commercial lye in one quart of cold water. When this is dissolved and the water is still warm, stir into it the warm fat, and add one cup of ammonia and two tablespoons of borax which have been dissolved in a half-cup of warm water. If fragrance is desired add a teaspoon of citronella oil. Stir thoroughly to mix, and allow to cool. The soap will harden and can then be cut up into bars.

If soft soap is preferred, add 3 quarts of hot water where the oil of citronella is called for. This of course will prevent hardening and can be spooned or poured after cooling.

HERBICIDES

Petroleum oils. A non-selective, all-purpose weed killer can be made of 1 part old crankcase oil and 20 parts of kerosene, gasoline, paint thinner, or diesel fuel. This is an effective contact herbicide for all vegetation and should be used with caution. Though only temporary in effect, it is fast-acting and probably the safest of the materials to use around the home. Especially good for driveways, alleys, and paths. Caution: Leaves oily residue on concrete and stones.

Copper sulfate (bluestone). Copper in concentrated form is lethal to all plant life. To make an all-purpose weed killer that will also sterilize the soil on which it falls for 2-3 years, dissolve 8 ounces of copper sulfate in 1 gallon of water. This is corrosive to metal, and sprayers should be rinsed thoroughly after use.

FUNGICIDES

Bordeaux mixture. See instructions for making under "Insecticides". It is particularly effective against downy mildew and several diseases of deciduous fruits when applied as a dormant spray.

Sulfur. Dusting sulfur, flotation or colloidal sulfur, and wettable sulfur are all available for fungicidal use. Sulfur is quite effective against powdery mildews, and most effective when temperature is above 70°F. Caution: Do not use when temperatures exceed 90°F., or on susceptible plants. Do not apply to vegetables to be canned. Jars are known to explode which contain trace amounts of elemental sulfur.

Lime-Sulfur. Primarily a dormant fungicide for deciduous fruit trees. (See "Insecticides" for recipe.)

ALGICIDES

Copper sulfate. (Bluestone) Copper is lethal to all algae. Where other plants are not involved and there is no need to control bacteria, copper sulfate can be effectively used to inhibit algae growth at the rate of 1 to 2 ounces per 100 gallons of water.

INSECT REPELLENTS FOR PLANTS

Garlic and onion. In a blender place onion, garlic, horseradish, peppers, mint or any other aromatic herbs. Add enough water to blend into a frappe. Dilute with an equal volume of water and let stand overnight at room temperature. Filter through an old stocking, add 1 teaspoonful of detergent and spray on plants to be protected. This is only a repellent to a few species and will not kill any insect.

Horseradish. See "Garlic and Onion".

Peppers. See "Garlic and Onion".
Mint. See "Garlic and Onion".
None of these are very good and are not recommended.

INSECT REPELLENTS FOR HUMANS

Anise oil. A fragrant oil with slight repellency to gnats, biting flies, and mosquitoes.

Oil of citronella. One of the better natural repellents against mosquitoes and biting flies, with a pleasant fragrance.

Asefetida. A terrible smelling herb that has absolutely no insect repellent qualities. (It probably keeps other humans away!)

Oil of Pennyroyal. A pleasant, short-life repellent for mosquitoes and biting flies. Pennyroyal herb rubbed fresh on skin is reported also to be effective for a short period. The fragrance resembles mint.

Camphor. Dissolve 2 squares of camphor in 1 pint of olive oil.

INSECT REPELLENTS FOR ANIMALS

Tansy. Dried tansy leaves rubbed into a pet's fur are reputed to drive off fleas.

Oil of lavender. Purported to drive fleas off dogs when rubbed into fur. Dogs probably won't be any happier for your efforts, however!

INSECT REPELLENTS FOR THE HOME

Tansy. Dried tansy leaves are supposed to keep ants off kitchen shelves and protect stored woolens and furs from moths by its repellent action.

Camphor. Camphor crystals have long been used for moth protection of woolens and furs in closed containers as chests and drawers.

Cedar. Still an old favorite is the pungent fragrance of cedar-lined chests and closets which are only moderately effective in repelling clothes moths from furs and woolens stored therein.

Oil of lavendar. Supposed to drive fleas and sand fleas from houses in the summer by sprinkling on carpets and floors. Of doubtful effectiveness.

Diatomaceous earth. When sprinkled heavily on dung heaps (chicken, cow, horse and dog) diatomaceous earth apparently repels flies so that egg laying is prevented, thus the usual reproduction is prevented.

Oil of Pennyroyal. The oil or the fresh herb rubbed on kitchen shelves will act as a short-lived ant repellent.

ANIMAL REPELLENTS

Mole and gopher repellent. Blend 2 ounces of castor oil with 1 ounce of liquid detergent. Add an equal volume of water and blend again. Fill a 2-quart sprinkling can with warm water and add 2 tablespoons of the mix, stir, and sprinkle over infested areas. Puncture mole tunnels and treat sparingly.

Deer. Deer can be repelled from fruit trees and areas by dusting the subject targets with blood meal or animal blood procured from a slaughter house.

Dogs and shrubs. Red pepper powder or napthalene flakes sprinkled on the subject urinating targets work equally well.

Dogs and cats on furniture. Dogs and cats can be prevented from sleeping in your best arm chairs or other locations by placing one or two moth balls or a few moth crystals in their normal resting place, or by giving the fabric (not wood surfaces) a very light spray of nicotine.

Dogs and rabbits in shrubs and flowers. Tobacco dust, lightly sprinkled over the area to be protected every 5-7 days acts as a good repellent.

SOURCES OF PEST CONTROL

APPENDIX F

Beneficial Insects With Abbreviations Used

Fly Pupae Parasites (FP)
Ladybeetles (LB)
Lacewings (LW)
Predatory Mites (MI)
Praying Mantids (PM)
Trichogramma Wasps (TW)

Beneficial Insectary
2544 First Avenue
San Bernardino, CA 92405
(FP)

Bio-Control Company
10180 Ladybird Drive
Auburn, CA 95603
(LB, PM)

Beneficial Biosystems
1525 63rd Street
Emeryville, CA 94608
(FP)

Burpee Seed Company
Warminster, PA 18991
(LB, PM)

Biotactics, Inc.
22412 Pico Street
Colton, CA 92324
(MI)

ORCON Organic Control, Inc.
5132 Venice Boulevard
Los Angeles, CA 90019
(LB, PM)

Pyramid Nursery
Box 22245
Sacramento, CA 95822
(LB, PM)

Mincemoyer Nursery
R.D. 5, Box 379
Jackson, NJ 08527
(PM)

John Staples
389 Rock Beach Road
Rochester, NY 14617
(PM)

King Entomological Labs
P.O. Box 69
Limerick, PA 19468
(LB, PM, TW)

Robert D. Robbins
424 N. Courtland
E. Stroudsburg, PA 18301
(PM)

Biogenesis, Inc.
Rt. 1, Box 36
Mathis, TX 78368
(LW, TW)

Gothard, Inc.
P.O. Box 370
Canutillo, TX 79835
(TW)

Fossil Flower Natural Bug Controls, Inc.
463 Woodbine Avenue
Toronto, Ontario M4E 2H5
CANADA
(LB, PM, TW)

Lakeland Nurseries Sales
Hanover, PA 17331
(LB, PM)

Montgomery Ward, Dept. B-4
618 W. Chicago Ave., Montgomery Ward Plaza
Chicago, IL 60671
(LB)

Bo-Biotrol, Inc.
54 S. Bear Creek Drive
Merced, CA 95340
(FP, LW)

Natural Pest Controls
9397 Premier Way
Sacramento, CA 95826
(LB, PM)

Gurney Seed & Nursery Co.
Yankton, SD 57078
(PM)

Beneficial Insect Co.
P.O. Box 323
Brownville, CA 95919
(LB)

Beneficial Insects, Ltd.
P.O. Box 154
Banta, CA 95304
(FP, LW, TW)

Rincon-Vitova Insectary, Inc.
P.O. Box 95
Oak View, CA 93022
(LB, LW, TW)

W. Atlee Burpee Co.
P.O. Box 748
Riverside, CA 92502
(LB, PM)

World Garden Products
2 First Street
E. Norwalk, CT 06855
(LB, LW, PM, TW)

Insect Disease Pathogens

Milky Spore Disease for Japanese beetle grub control (Doom)

Bacillus thuringiensis for control of various caterpillars (Thuricide, Dipel, Bactur, and BT.)

Virus for Tobacco Budworm and Cotton Bollworm Control (Elcar)

Fairfax Biological Laboratories
Clinton Corners, NY 12514
(Doom)

Hanna Biologicals
199 Derrick Avenue
Uniontown, PA 15401
(Doom)

Mellinger's, Inc.
2334 N. Range Road
North Lima, OH 44452
(Doom)

Reuter Laboratories
2400 James Madison Highway
Haymarket, VA 22069
(Doom)

Galt Research, Inc.
Box 245-A, RR 1
Trafalger, IN 46181
(Dipel)

Burpee Seed Co.
Warminster, PA 18991
(Thuricide)

ORCON Organic Control, Inc.
5132 Venice Boulevard
Los Angeles, CA 90019
(BT)

Sandoz, Inc.
480 Camino del Rio South
San Diego, CA 92108
(Thuricide, Elcar)

Thompson-Hayward Chemical Co.
P.O. Box 2383
Kansas City, KS 66110
(Bactur)

Nichols Garden Nursery
1190 N. Pacific Highway
Albany, OR 97321
(Thuricide)

International Minerals & Chemical Corp.
Crop Aids Dept.
5401 Old Orchard Road
Skokie, IL 60076
(Thuricide)

Hopkins Agricultural Chemical Co.
P.O. Box 7532
Madison, WI 53707
(BT)

Abbott Laboratories
Agricultural & Chemical Products Div.
Dept. 95M
North Chicago, IL 60064
(Dipel)

Botanical Insecticides

Nicotine sulfate, pyrethrins, rotenone, ryania, and sabadilla. (Also available through seed and garden catalogs and garden centers.)

Sunnybrook Farms Nursery
9448 Mayfield Road
Chesterland, OH 44026
(Nicotine Sulfate)

Burpee Seed Co.
Warminster, PA 18991
(Rotenone)

ORCON Organic Control, Inc.
5132 Venice Boulevard
Los Angeles, CA 90019
(Pyrethrins, rotenone, ryania)

George W. Park Seed Co., Inc.
Greenwood, SC 29647
(Rotenone)

S. B. Penick & Co.
1050 Wall Street
Lyndhurst, NJ 07071
(Rotenone)

Prentiss Drug & Chemical Co., Inc.
363 Seventh Ave.
New York, NY 10001
(Sabadilla)

Hopkins Agricultural Chemical Co.
P.O. Box 7532
Madison, WI 53707
(Rotenone, ryania)

Cryolite

Pennwalt Corp.
Agchem Div.
1630 E. Shaw Ave.
Fresno, CA 93710
(Kryocide)

Diatomaceous Earth

Flow Laboratories
1601 W. Orange
Orange, CA 92668
(Perma-Guard)

Dormant Oils

B. G. Pratt Co.
204 21st Street
Paterson, NJ 07503
(Scaleside)

Adhesive for Crawling Insects

The Tanglefoot Co.
314 Straight Avenue, S.W.
Grand Rapids, MI 49504
(Tree Tanglefoot)

Animal Repellants, Inc.
P.O. Box 999
Griffin, GA 30224
(Tack Trap)

Insect Traps

Burpee Seed Co.
Warminster, PA 18991
(Blacklight, Japanese beetle)

Ellisco Inc.
American and Luzerne Streets
Philadelphia, PA 19140
(Blacklight)

Zoecon Corporation
975 California Avenue
Palo Alto, CA 94304
(Gypsy moth lure)

Beneficial Biosystems
1525 63rd Street
Emeryville, CA 94608
(Fly traps)

Animal Repellents, Inc.
P.O. Box 999, Griffin, GA 30224
(Cockroach traps)

Agrilite Systems, Inc.
404 Barringer Building
Columbia, SC 29201
(Blacklight)

D-Vac Co.
P.O. Box 2095
Riverside, CA 92506
(Blacklight)

Flypaper

Aeroxon Products, Inc.
10 Cottage Place
New Rochelle, NY 10802

Bird Netting

Animal Repellents, Inc.
P.O. Box 999
Griffin, GA 30224

French Textiles Co.
835 Bloomfield Avenue
Clifton, NJ 07012

Ross Daniels, Inc.
P.O. Box 430
West Des Moines, IA 50265

Burpee Seed Co.
Warminster, PA 18991

Bird Repellents

Animal Repellents, Inc.
P.O. Box 999
Griffin, GA 30224

The Tanglefoot Co.
314 Straight Ave., S.W.
Grand Rapids, MI 49504

Farmer Seed and Nursery Co.
Faribault, MN 55021
(Scarecrows, plastic owls)

Animal Repellents

Animal Repellents, Inc.
P.O. Box 999
Griffin, GA 30224
(Dog, cat, rabbit, deer)

Burpee Seed Co.
Warminster, PA 18991
(Dog, cat)

George W. Park Seed Co., Inc.
Greenwood, SC 29647
(Dog, cat)

Lakeland Nurseries Sales
Hanover, PA 17331
(Mole, mechanical)

The Tanglefoot Co.
314 Straight Ave., S.W.
Grand Rapids, MI 49504
(Squirrel)

Animal Traps

Burpee Seed Co.
Warminster, PA 18991
(Mole, Havahart® for rat, weasel, chipmunk, squirrel, mink, rabbit, woodchuck, oppossum.)

Earthworms

Burpee Seed Co.
Warminster, PA 18991

Lakeland Nurseries Sales
Hanover, PA 18991

Brazos Worm Farms
P.O. Box 4185
Waco, TX 76705

Pyramid Nursery
P.O. Box 22245
Sacramento, CA 95822

Frogs

Nu-Tex Frog Farm
P.O. Box 4029
Corpus Christi, TX 78400

Deodorants

Fritzsche, Dodge and Olcolt, Inc.
76 9th Avenue
New York, NY 10011
(Neutroleum Alpha)

Cline-Buckner, Inc.
Cerritos, CA 90701
(Super Hydrasol)

Prentiss Drug & Chemical Co., Inc.
363 — 7th Avenue
New York, NY 10001
(Meelium)

DIRECTORY OF STATE COOPERATIVE EXTENSION SERVICES

APPENDIX G

Your first and best source of sound pest control information is your local County Agent located in nearly every county of the nation. He can be found by looking in the white pages of the telephone directory for Cooperative Extension Service listed under the local County Government. If there is no County Agent near you or you need additional information, call or write the Cooperative Extension Service at your State Land Grant University listed below.

School of Agriculture
Auburn University
Auburn, Alabama 36830

School of Agriculture and Land
Resources Management
University of Alaska
Fairbanks, Alaska 99701

College of Agriculture
University of Arizona
Tucson, Arizona 85721

College of Agriculture
University of Arkansas
Fayetteville, Arkansas 72701

College of Agricultural Sciences
University of California
Berkeley, California 94720

College of Agricultural Sciences
Colorado State University
Fort Collins, Colorado 80523

College of Agriculture
and Natural Resources
University of Connecticut
Storrs, Connecticut 06268

College of Agricultural Sciences
University of Delaware
Newark, Delaware 18711

Institute of Food
and Agricultural Sciences
University of Florida
Gainsville, Florida 32611

College of Agriculture
University of Georgia
Athens, Georgia 30602

College of Tropical Agriculture
University of Hawaii
Honolulu, Hawaii 96822

College of Agriculture
and Life Sciences
University of Guam
Agana, Guam 96910

College of Agriculture
University of Idaho
Moscow, Idaho 83843

College of Agriculture
University of Illinois
Urbana, Illinois 61801

School of Agriculture
Purdue University
West Lafayette, Indiana 47907

College of Agriculture
Iowa State University
Ames, Iowa 50011

College of Agriculture
Kansas State University
Manhattan, Kansas 66506

College of Agriculture
University of Kentucky
Lexington, Kentucky 40506

College of Agriculture
Louisiana State University
Baton Rouge, Louisiana 70893

College of Life Sciences
and Agriculture
University of Maine
Orono, Maine 04473

College of Agriculture
University of Maryland
College Park, Maryland 20742

College of Food and
Natural Resources
University of Massachusetts
Amherst, Massachusetts 01002

College of Agriculture and
Natural Resources
Michigan State University
East Lansing, Michigan 48824

College of Agriculture
University of Minnesota
St. Paul, Minnesota 55101

College of Agriculture
Mississippi State University
Mississippi State, Mississippi 39762

College of Agriculture
University of Missouri
Columbia, Missouri 65201

College of Agriculture
Montana State University
Bozeman, Montana 59717

Institute of Agriculture
and Natural Resources
University of Nebraska
Lincoln, Nebraska 68503

College of Agriculture
University of Nevada
Reno, Nevada 89507

College of Life Sciences
and Agriculture
University of New Hampshire
Durham, New Hampshire 03824

Cook College
Rutgers State University
New Brunswick, New Jersey 08903

College of Agriculture
and Home Economics
New Mexico State University
Las Cruces, New Mexico 88003

College of Agriculture
and Life Sciences
Cornell University
Ithaca, New York 14853

School of Agriculture
and Life Sciences
North Carolina State University
Raleigh, North Carolina 27607

College of Agriculture
North Dakota State University
Fargo, North Dakota 58102

College of Agriculture
and Home Economics
Ohio State University
Columbus, Ohio 43210

College of Agriculture
Oklahoma State University
Stillwater, Oklahoma 74074

School of Agriculture
Oregon State University
Corvallis, Oregon 97331

College of Agriculture
Pennsylvania State University
University Park, Pennsylvania 16802

College of Agricultural Sciences
University of Puerto Rico
Mayaguez, Puerto Rico 00708

College of Resource Development
University of Rhode Island
Kingston, Rhode Island 02881

College of Agricultural Sciences
Clemson University
Clemson, South Carolina 29631

College of Agriculture
and Biological Sciences
South Dakota State University
Brookings, South Dakota 57007

College of Agriculture
University of Tennessee
P.O. Box 1071
Knoxville, Tennessee 37901

College of Agriculture
Texas A & M University
College Station, Texas 77843

College of Agriculture
Utah State University
Logan, Utah 84322

College of Agriculture
University of Vermont
Burlington, Vermont 05401

College of the Virgin Islands
P.O. Box L
Kingshill, St. Croix
Virgin Islands 00850

College of Agriculture
and Life Sciences
Virginia Polytechnic Institute
Blacksburg, Virginia 24061

College of Agriculture
Washington State University
Pullman, Washington 99164

College of Agriculture
and Forestry
West Virginia University
Morgantown, West Virginia 26506

College of Agricultural
and Life Sciences
University of Wisconsin
Madison, Wisconsin 53706

College of Agriculture
University of Wyoming
Laramie, Wyoming 82071

APPENDIX H — CONVERSION OF U.S. MEASUREMENTS TO THE METRIC SYSTEM[1]

The metric system originated in France in 1790 and spread throughout Europe, Latin America, and the East during the 19th century. With the exception of the United States (now considering the system seriously) and a few former British associated areas, the metric system is the official language of measurements.

Simply described, the metric system is a decimal system of weights and measures in which the gram (.0022046 pound), the meter (39.37 inches) and the liter (61.025 cubic inches) are the basic units of weight, length, and volume respectively. Most names of the various other units are formed by the addition of the following prefixes to these three terms, namely:

- milli—(one thousandth), as 1 millimeter = 1/1000 meter
- centi—(one hundredth), as 1 centimeter = 1/100 meter
- deci—(one tenth), as 1 decimeter = 1/10 meter
- deca or deka—(ten), as 1 decameter = 10 meters
- hecto—(one hundred), as 1 hectometer = 100 meters
- kilo—(one thousand), as 1 kilometer = 1,000 meters

Distance:

1 inch = 2.54 cm.; 12 inches = 1 foot; 3 feet = 1 yard; 1 yard = .91 (meter) M.

1,760 yards (5,280 feet) = 1 mile; 1 mile = 1,619 M.

[1] Courtesy of George W. Ware, Sr.

Therefore, 1 (kilometer) Km = .62 miles or 1 mile = 1.62 Km.

Accordingly, for example, 50 Km. = 31 miles (50 miles × .62 miles per Km. = 31).

Area:

1 acre (43,560 square feet) = 4,047 square meters = 0.405 (hectare) Ha.

1 square mile (a section of land) = 640 acres or 259 Ha.

1 Ha. = 2.47 acres. To convert acres into Ha., simply multiply acres by 0.405.

Accordingly, for example, 200 acres = 81 Ha. (200 acres × 0.405 Ha. per acre = 81).

Volume:

1 quart (58 cubic inches) = .9463 liter. One liter = 1.0567 quarts.

4 quarts = a liquid gallon = 231 cubic inches or 3.78 (liter) L.

1 cubic yard (36 × 36 × 36 inches) = .76 (cubic meter) M^3.

1 bushel (dry measure) = 2,150 cubic inches = 35.2 liters.

Weight:

1 ounce = 28.35 grams; 16 ounces = 1 pound (#); 100# = 45.35 Kg; 2,000# = 1 short ton; 1 short ton = .907 metric ton.

1 Kg = 2.205 pounds; 1,000 Kg. = 2,204.6 pounds or metric ton; 1 Cwt (England) = 112#; 20 Cwt = 2240# = long ton.

Temperature:

Temperature in the U.S. is measured in Fahrenheit (F), while under the metric system in Centigrade (C). Freezing is 32°F and boiling 212°F — a difference of 180 degrees F., compared with freezing at 0 Centigrade and boiling at 100 C — a difference of 100 degrees (See comparison below). Accordingly, 1 degree F = 5/9 degree C, and 9 degrees F = 5 degrees C.

Approximate F and C comparisons are: 50°F = 10°C; 60°F = 16°C; 70°F = 21°C; 80°F = 27°C; 90°F = 32°C; 100°F = 38°C, etc.

System	Freezing Point	Boiling Point	Difference Between Freezing & Boiling
(F) Fahrenheit	32°F	212°F	180 degrees F
(C) Centigrade	0°C	100°C	100 degrees C

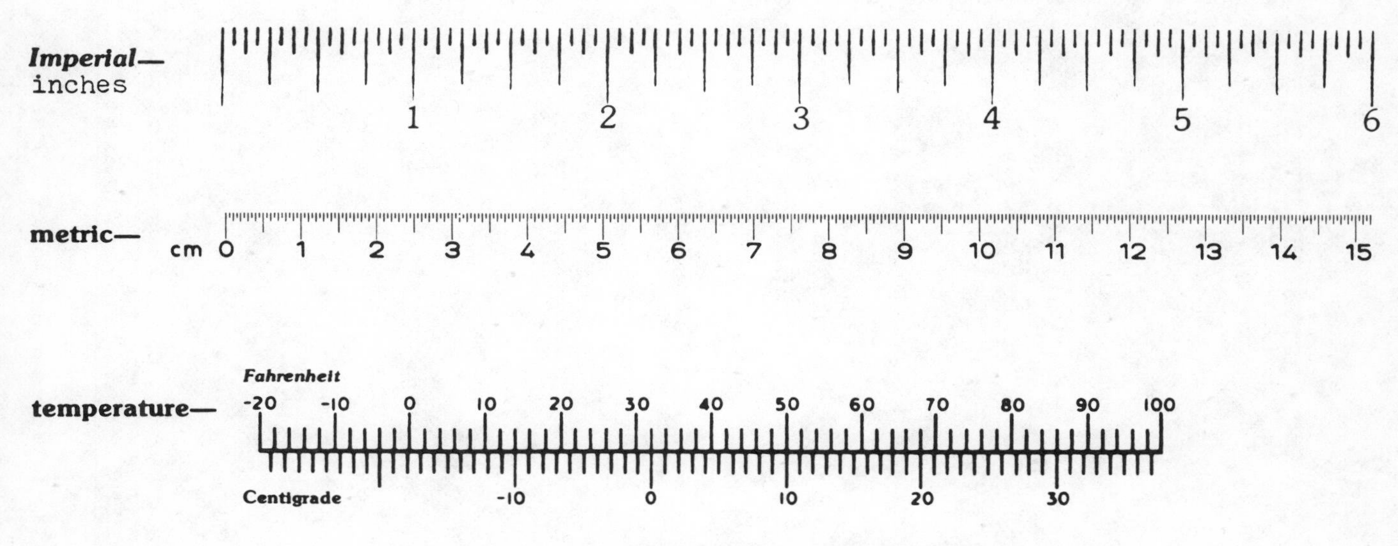

Visual Scales for Converting Simple Distances
and Temperatures to The Metric System

Abscission — Process by which a leaf or other part is separated from the plant.

Absorption — Process by which pesticides are taken into tissues, namely plants, by roots or foliage (stomata, cuticle, etc.).

Acaricide (miticide) — An agent that destroys mites and ticks.

Acetylcholine (ACh) — Chemical transmitter of nerve and nerve-muscle impulses in animals.

Activator — Material added to a fungicide to increase toxicity.

Active ingredient (a.i.) — Chemicals in a product that are responsible for the pesticidal effect.

Acute toxicity — The toxicity of a material determined at the end of 24 hours to cause injury or death from a single dose or exposure.

Adjuvant — An ingredient that improves the properties of a pesticide formulation. Includes wetting agents, spreaders, emulsifiers, dispersing agents, foam suppressants, penetrants, and correctives.

Adsorption — Chemical and/or physical attraction of a substance to a surface. Refers to gases, dissolved substances, or liquids on the surface of solids or liquids.

Adulterated pesticide — A pesticide that does not conform to the professed standard or quality as documented on its label or labeling.

Aerosol — Colloidal suspension of solids or liquids in air.

Algicide — Chemical used to control algae and aquatic weeds.

Annual — Plant that completes its life cycle in one year, i.e., germinates from seed, produces seed, and dies in the same season.

Antagonism — (Opposite of synergism) decreased activity arising from the effect of one chemical on another.

Antibiotic — Chemical substance produced by a microorganism and that is toxic to other microorganisms.

Anticoagulant — A chemical which prevents normal bloodclotting. The active ingredient in some rodenticides.

Antidote — A practical treatment, including first aid, used in the treatment of pesticide poisoning or some other poison in the body.

Anti-transpirant — A chemical applied directly to a plant which reduces the rate of transpiration or water loss by the plant.

Apiculture — Pertaining to the care and culture of bees.

Aromatics — Solvents containing benzene, or compounds derived from benzene.

Atropine (atropine sulfate) — An antidote used to treat organophosphate and carbamate poisoning.

Attractant, insect — A substance that lures insects to trap or poison-bait stations. Usually classed as food, oviposition, and sex attractants.

Auxin — Substance found in plants, which stimulates cell growth in plant tissues.

Avicide — Lethal agent used to destroy birds, but also refers to materials used for repelling birds.

Bactericide — Any bacteria-killing chemical.

Bacteriostat — Material used to prevent growth or multiplication of bacteria.

Biennial — Plant that completes its growth in 2 years. First year it produces leaves and stores food; the second year it produces fruit and seeds.

Biological control agent — Any biological agent that adversely affects pest species.

Biomagnification — The increase in concentration of a pollutant in animals as related to their position in a food chain, usually referring to the persistent, organochlorine insecticides and their metabolites.

Biota — Animals and plants of a given habitat.

Biotic insecticide — Usually microorganisms known as insect pathogens that are applied in the same manner as conventional insecticides to control pest species.

Biotype — Subgroup within a species differing in some respect from the species such as a subgroup that is capable of reproducing on a resistant variety.

Blight — Common name for a number of different diseases on plants, especially when collapse is sudden — e.g. leaf blight, blossom blight, shoot blight.

Botanical pesticide — A pesticide produced from naturally-occurring chemicals found in some plants. Examples are nicotine, pyrethrum, strychnine and rotenone.

Brand — The name, number, or designation of a pesticide.

Broadcast application — Application over an entire area rather than only on rows, beds, or middles.

Broad-spectrum insecticides — Nonselective, having about the same toxicity to most insects.

Calibrate, calibration — To determine the amount of pesticide that will be applied to the target area.

Canker — A lesion on a stem.

Carbamate insecticide — One of a class of insecticides derived from carbamic acid.

Carcinogen — A substance that causes cancer in living animal tissue.

Carrier — An inert material that serves as a diluent or vehicle for the active ingredient or toxicant.

Causal organism — The organism (pathogen) that produces a given disease.

Certified applicator — Commercial or private applicator qualified to apply Restricted Use Pesticides as defined by the EPA.

Chelating agent — Certain organic chemicals (i.e. ethylenediaminetetraacetic acid) that combine with metal to form soluble chelates and prevent conversion to insoluble compounds.

Chemical name — Scientific name of the active ingredient(s) found in the formulated pesticide. The name is derived from the chemical structure of the active ingredient.

Chemotherapy — Treatment of a diseased organism, usually plants, with chemicals to destroy or inactivate a pathogen without seriously affecting the host.

Chemtrec — A toll-free, long-distance, telephone service that provides 24-hour emergency pesticide information (800-424-9300).

Chlorosis — Loss of green color in foliage.

Cholinesterase (ChE) — An enzyme of the body necessary for proper nerve function that is inhibited or damaged by organophosphate or carbamate insecticides taken into the body by any route.

Chronic toxicity — The toxicity of a material determined usually after several weeks of continuous exposure.

Common pesticide name — A common chemical name given to a pesticide by a recognized committee on pesticide nomenclature. Many pesticides are known by a number of trade or brand names but have only one recognized common name. Example: The common name for Sevin insecticide is carbaryl.

Compatible (Compatibility) — When two materials can be mixed together with neither affecting the action of the other.

Concentration — Content of a pesticide in a liquid or dust, for example lbs/gallon or percent by weight.

Contact Herbicide — Phytotoxic by contact with plant tissue rather than as a result of translocation.

Contamination — The presence of an unwanted pesticide or other material in or on a plant, animal, or their by-products; soil; water; air; structure; etc. (See residue).

Cumulative pesticides — Those chemicals which tend to accumulate or build up in the tissues of animals or in the environment (soil, water).

Curative pesticide — A pesticide which can inhibit or eradicate a disease-causing organism after it has become established in the plant or animal.

Cutaneous toxicity — Same as dermal toxicity.

Cuticle — Outer covering of insects or leaves. Chemically they are quite different.

Days-to-Harvest — The least number of days between the last pesticide application and the harvest date, as set by law. Same as "harvest intervals".

Deciduous — Plants that lose their leaves during the winter.

Decontaminate — The removal or breakdown of any pesticide chemical from any surface or piece of equipment.

Deflocculating agent — Material added to a spray preparation to prevent aggregation or sedimentation of the solid particles.

Defoliant — A chemical that initiates abscission.

Deposit — Quantity of a pesticide deposited on a unit area.

Dermal toxicity — Toxicity of a material as tested on the skin, usually on the shaved belly of a rabbit; the property of a pesticide to poison an animal or human when absorbed through the skin.

Desiccant — A chemical that induces rapid desiccation of a leaf or plant part.

Desiccation — Accelerated drying of plant or plant parts.

Detoxify — To make an active ingredient in a pesticide or other poisonous chemical harmless and incapable of being toxic to plants and animals.

Diatomaceous earth — A whitish powder prepared from deposits formed by the silicified skeletons of diatoms. Used as diluent in dust formulations.

Diluent — Component of a dust or spray that dilutes the active ingredient.

Disinfectant — A chemical or other agent that kills or inactivates disease producing microorganisms in animals, seeds, or other plant parts. Also commonly referred to chemicals used to clean or surface sterilize inanimate objects.

DNA — Deoxyribonucleic acid.

Dormant spray — Chemical applied in winter or very early spring before treated plants have started active growth.

Dose, dosage — Same as rate. The amount of toxicant given or applied per unit of plant, animal, or surface.

Drift, spray — Movement of airborne spray droplets from the spray nozzle beyond the intended contact area.

EC_{50} — The median effective concentration (ppm or ppb) of the toxicant in the environment (usually water) which produces a designated effect in 50% of the test organisms exposed.

Ecology — Derived from the Greek *oikos*, house or place to live. A branch of biology concerned with organisms and their relation to the environment.

Ecosystem — The interacting system of all the living organisms of an area and their nonliving environment.

Ectoparasite — A parasite feeding on a host from the exterior or outside.

ED_{50} — The median effective dose, expressed as mg/kg of body weight, which produces a designated effect in 50% of the test organisms exposed.

Emulsible (Emulsifiable) Concentrate — Concentrated pesticide formulation containing organic solvent and emulsifier to facilitate emulsification with water.

Emulsifier — Surface active substances used to stabilize suspensions of one liquid in another, for example, oil in water.

Emulsion — Suspension of miniscule droplets of one liquid in another.

Endoparasite — A parasite that enters host tissue and feeds from within.

Environment — All the organic and inorganic features that surround and affect a particular organism or group of organisms.

Environmental Protection Agency (EPA) — The Federal agency responsible for pesticide rules and regulations, and all pesticide registrations.

EPA — The U.S. Environmental Protection Agency.

EPA Establishment Number — A number assigned to each pesticide production plant by EPA. The number indicates the plant at which the pesticide product was produced and must appear on all labels of that product.

EPA Registration Number — A number assigned to a pesticide product by EPA when the product is registered by the manufacturer or his designated agent. The number must appear on all labels for a particular product.

Eradicant — Applies to fungicides in which a chemical is used to eliminate a pathogen from its host or environment.

Exterminate — Often used to imply the complete extinction of a species over a large continuous area such as an island or a continent.

FEPCA — The Federal Environmental Pesticide Control Act of 1972.

FIFRA — The Federal Insecticide, Fungicide and Rodenticide Act of 1947.

Filler — Diluent in powder form.

Fixed coppers — Insoluble copper fungicides where the copper is in a combined form. Usually finely divided, relatively insoluble powders.

Flowable — A type of pesticide formulation in which a very finely ground solid particle is mixed in a liquid carrier.

Foaming agent — A chemical which causes a pesticide preparation to produce a thick foam. This aids in reducing drift.

Fog treatment — The application of a pesticide as a fine mist for the control of pests.

Foliar treatment — Application of the pesticide to the foliage of plants.

Food chain — Sequence of species within a community, each member of which serves as food for the species next higher in the chain.

Formulation — Way in which basic pesticide is prepared for practical use. Includes preparation as wettable powder, granular, emulsifiable concentrate, etc.

Full coverage spray — Applied thoroughly over the crop to a point of runoff or drip.

Fumigant — A volatile material that forms vapors which destroy insects, pathogens, and other pests.

Fungicide — A chemical that kills fungi.

Fungistatic — Action of a chemical that inhibits the germination of fungus spores while in contact.

Gallonage — Number of gallons of finished spray mix applied per 1000 square feet, acre, tree, hectare, square mile, or other unit.

General Use pesticide — A pesticide which can be purchased and used by the general public without undue hazard to the applicator and environment as long as the instructions on the label are followed carefully. (See Restricted Use pesticide).

Germicide — A substance that kills germs (microorganisms). (Antiquated term).

Growth regulator — Organic substance effective in minute amounts for controlling or modifying (plant or insect) growth processes.

Harvest intervals — Period between last application of a pesticide to a crop and the harvest as permitted by law.

Hormone — A product of living cells that circulates in the animal or plant fluids and that produces a specific effect on cell activity remote from its point of origin.

Host — Any plant or animal attacked by a parasite.

Hydrolysis — Chemical process of (in this case) pesticide breakdown or decomposition involving a splitting of the molecule and addition of a water molecule.

Hyperplasia — Abnormal increase in the number of cells of a tissue.

Hypertrophy — Abnormal increase in the size of cells of a tissue.

Incompatible — Two or more materials which cannot be mixed or used together.

Inert ingredients — The inactive materials in a pesticide formulation, which would not prevent damage or destroy pests if used alone.

Ingest — To eat or swallow.

Ingredient statement — That portion of the label on a pesticide container which gives the name and amount of each active ingredient and the total amount of inert ingredients in the formulation.

Inhalation — Exposure of test animals either to vapor or dust for a predetermined time.

Inhalation toxicity — To be poisonous to man or animals when breathed into the lungs.

Insect growth regulator (IGR) — Chemical substance which disrupts the action of insect hormones controlling molting, maturity from pupal stage to adult, and others.

Insect pest management — The practical manipulation of insect (or mite) pest populations using any or all control methods in a sound ecological manner.

Integrated control — The integration of the chemical and biological methods of pest control.

Integrated pest management — A management system that uses all suitable techniques and methods in as compatible a manner as possible to maintain pest populations at levels below those causing economic injury.

Intramuscular — Injected into the muscle.

Intraperitoneal — Injected into the viscera, but not into the organs.

Intravenous — Injected into the vein.

Invert emulsion — One in which the water is dispersed in oil rather than oil in water. Usually a thick, salad-dressing-like mixture results.

Kg or kilogram — A unit of weight in the metric system equal to 2.2 pounds.

Label — All printed material attached to or part of the pesticide container.

Labeling — Supplemental pesticide information which complements the information on the label, but is not necessarily attached to or part of the container.

LC_{50} — The median lethal concentration, the concentration which kills 50% of the test organisms, expressed as milligrams (mg), or cubic centimeters (cc), if liquid, per animal. It is also the concentration expressed as parts per million (ppm) or parts per billion (ppb) in the environment (usually water) which kills 50% of the test organisms exposed.

LD_{50} — A lethal dose for 50% of the test organisms. The dose of toxicant producing 50% mortality in a population. A value used in presenting mammalian toxicity, usually oral toxicity, expressed as milligrams of toxicant per kilogram of body weight (mg/kg).

Leaching — The movement of a pesticide chemical or other substance downward through soil as a result of water movement.

Low volume spray — Concentrate spray, applied to uniformly cover the crop, but not as a full coverage to the point of runoff.

mg/kg (milligrams per kilogram) — Used to designate the amount of toxicant required per kilogram of body weight of test organism to produce a designated effect, usually the amount necessary to kill 50% of the test animals.

Microbial insecticide — A microorganism applied in the same way as conventional insecticides to control an existing pest population.

Mildew — Fungus growth on a surface.

Miscible liquids — Two or more liquids capable of being mixed in any proportions, and that will remain mixed under normal conditions.

M.L.D. — Median lethal dose, or the LD_{50}.

Molluscicide — A chemical used to kill or control snails and slugs.

Mosaic — Leaf pattern of yellow and green or light green and dark green produced by certain virus infections.

Mutagen — Substance causing genes in an organism to mutate or change.

Mycoplasma — A microorganism intermediate in size between viruses and bacteria possessing many virus-like properties and not visible with a light microscope.

Necrosis — Death of tissue, plant or animal.

Negligible residue — A tolerance which is set on a food or feed crop permitting an ultra-small amount of pesticide at harvest as a result of indirect contact with the chemical.

Nematicide — Chemical used to kill nematodes.

Oncogenic — The property to produce tumors (not necessarily cancerous) in living tissues. (See carcinogenic.)

Oral toxicity — Toxicity of a compound when given by mouth. Usually expressed as number of milligrams of chemical per kilogram of body weight of animal (white rat) when given orally in a single dose that kills 50% of the animals. The smaller the number, the greater the toxicity.

Organochlorine insecticide — One of the many chlorinated insecticides, e.g. DDT, dieldrin, chlordane, BHC, lindane, etc.

Organophosphate — Class of insecticides (also one or two herbicides and fungicides) derived from phosphoric acid esters, e.g. malathion, diazinon, etc.

Ovicide — A chemical that destroys an organism's eggs.

Pathogen — Any disease-producing organism or virus.

Penetrant — An additive or adjuvant which aids the pesticide in moving through the outer surface of plant tissues.

Perennial — Plant that continues to live from year to year. Plants may be herbaceous or woody.

Persistence — The quality of an insecticide to persist as an effective residue due to its low volatility and chemical stability, e.g. certain organochlorine insecticides.

Pesticide — An "economic poison" defined in most state and federal laws as any substance used for controlling, preventing, destroying, repelling, or mitigating any pest. Includes fungicides, herbicides, insecticides, nematicides, rodenticides, desiccants, defoliants, plant growth regulators, etc.

Pheromones — Highly potent insect sex attractants produced by the insects. For some species laboratory-synthesized pheromones have been developed for trapping purposes.

Physical selectivity — Refers to the use of broad-spectrum insecticides in such ways as to obtain selective action. This may be accomplished by timing, dosage, formulation, etc.

Physiological selectivity — Refers to insecticides which are inherently more toxic to some insects than to others.

Phytotoxic — Injurious to plants.

Piscicide — Chemical used to kill fish.

Plant regulator (Growth regulator) — A chemical which increases, decreases, or changes the normal growth or reproduction of a plant.

Poison — Any chemical or agent that can cause illness or death when eaten, absorbed through the skin, or inhaled by man or animals.

Poison Control Center — Information source for human poisoning cases, including pesticides, usually located at major hospitals.

Post-emergence — After emergence of the specified weed or crop.

ppb — Parts per billion (parts in 10^9 parts) is the number of parts of toxicant per billion parts of the substance in question.

ppm — Parts per million (parts in 10^6 parts) is the number of parts of toxicant per million parts of the substance in question. They may include residues in soil, water or whole animals.

Predacide — Chemical used to poison predators.

Pre-planting treatment — Made before the crop is planted.

Propellant — An inert ingredient in self-pressurized products that produces the force necessary to dispense the active ingredient from the container. (See aerosol.)

Protectant — Fungicide applied to plant surface before pathogen attack to prevent penetration and subsequent infection.

Protective clothing — Clothing to be worn in pesticide-treated fields under certain conditions as required by federal law, e.g. reentry intervals.

Protopam chloride (2-PAM) — An antidote for certain organophosphate pesticide poisoning, but not for carbamate poisoning.

Rate — Refers to the amount of active ingredient applied to a unit area regardless of percentage of chemical in the carrier (dilution).

Raw agricultural commodity — Any food in its raw and natural state, including fruits, vegetables, nuts, eggs, raw milk, and meats.

Reentry intervals — Waiting interval required by federal law between application of certain hazardous pesticides to crops and the entrance of workers into those crops without protective clothing.

Registered pesticides — Pesticide products which have been approved by the Environmental Protection Agency for the uses listed on the label.

Repellent (insects) — Substance used to repel ticks, chiggers, gnats, flies, mosquitoes and fleas.

Residual — Having a continued killing effect over a period of time.

Residue — Trace of a pesticide and its metabolites remaining on and in a crop, soil, or water.

Resistance (insecticide) — Natural or genetic ability of an organism to tolerate the poisonous effects of a toxicant.

Restricted Use pesticide — One of several pesticides designated by the EPA that can be applied only by certified applicators, because of their inherent toxicity or potential hazard to the environment.

RNA — Ribonucleic acid.

Rodenticide — Pesticide applied as a bait, dust, or fumigant to destroy or repel rodents and other animals, such as moles and rabbits.

Rust — A disease with symptoms that usually include reddish-brown or black pustules; a group of fungi in the Basidiomycetes.

Safener — Chemical that reduces the phytotoxicity of another chemical.

Scientific name — The one name of a plant or animal used throughout the world by scientists, and based on Latin and Greek.

Secondary pest — A pest which usually does little if any damage but can become a serious pest under certain conditions, e.g. when insecticide applications destroy its predators and parasites.

Selective insecticide — One which kills selected insects, but spares many or most of the other organisms, including beneficial species, either through differential toxic action or the manner in which insecticide is used.

Selective pesticide — One which, while killing the pest individuals, spares much or most of the other fauna or flora, including beneficial species, either through differential toxic action or through the manner in which the pesticide is used (formulation, dosage, timing, placement, etc.)

Senescence — Process or state of growing old.

Sex lure — Synthetic chemical which acts as the natural lure (pheromone) for one sex of an insect species.

Signal word — A required word which appears on every pesticide label to denote the relative toxicity of the product. The signal words are either "Danger — poison" for highly toxic compounds, "Warning" for moderately toxic, or "Caution" for slightly toxic.

Slimicide — Chemical used to prevent slimy growth, as in wood pulping processes for manufacture of paper and paperboard.

Slurry — Thin, watery mixture, such as liquid mud, cement, etc. Fungicides and some insecticides are applied to seeds as slurries to produce thick coating and reduce dustiness.

Smut — A fungus with sooty spore masses; a group of fungi in the Basidiomycetes.

Soil application — Application of pesticide made primarily to soil surface rather than to vegetation.

Soil persistence — Length of time that a pesticide application on or in soil remains effective.

Soluble powder — A finely ground, solid material which will dissolve in water or some other liquid carrier.

Spore — A single to many-celled reproductive body in the fungi that can develop a new fungus colony.

Spot treatment — Application to localized or restricted areas as differentiated from overall, broadcast, or complete coverage.

Spreader — Ingredient added to spray mixture to improve contact between pesticide and plant surface.

Sterilize — To treat with a chemical or other agent to kill every living thing in a certain area.

Sticker — Ingredient added to spray or dust to improve its adherence to plants.

Stomach poison — A pesticide that must be eaten by an insect or other animal in order to kill or control the animal.

Structural pests — Pests which attack and destroy buildings and other structures, clothing, stored food, and manufactured and processed goods. Examples: Termites, cockroaches, clothes moths, rats, dry-rot fungi.

Stupefacient or soporific — Drug used as a pesticide to cause birds to enter a state of stupor so they can be captured and removed or to frighten other birds away from the area.

Subcutaneous toxicity — The toxicity determined following its injection just below the skin.

Surfactant — Ingredient that aids or enhances the surface-modifying properties of a pesticide formulation (wetting agent, emulsifier, spreader).

Suspension — Finely divided solid particles or droplets dispersed in a liquid.

Swath — The width of the area covered by a sprayer or duster making one sweep.

Synergism — Increased activity resulting from the effect of one chemical on another.

Synthesize — Production of a compound by joining various elements or simpler compounds.

Systemic — Compound that is absorbed and translocated throughout the plant or animal.

Tank mix — Mixture of two or more pesticides in the spray tank at time of application: Such mixture must be cleared by EPA.

Target — The plants, animals, structures, areas, or pests to be treated with a pesticide application.

Temporary tolerance — A tolerance established on an agricultural commodity by EPA to permit a pesticide manufacturer or his agent time, usually one year, to collect additional residue data to support a petition for a permanent tolerance; in essence, an experimental tolerance. (See tolerance.)

Teratogenic — Substance which causes physical birth defects in the offspring following exposure of the pregnant female.

Tolerance — Amount of pesticide residue permitted by federal regulation to remain on or in a crop. Expressed as parts per million (ppm).

Tolerant — Capable of withstanding effects.

Topical application — Treatment of a localized surface site such as a single leaf blade, on an insect, etc., as opposed to oral application.

Toxic — Poisonous to living organisms.

Toxicant — A poisonous substance such as the active ingredient in pesticide formulations that can injure or kill plants, animals, or microorganisms.

Toxin — A naturally occurring poison produced by plants, animals, or microorganisms. Examples: The poison produced by the black widow spider, the venom produced by snakes, the botulism toxin.

Trade name (Trademark name, proprietary name, brand name) — Name given a product by its manufacturer or formulator, distinguishing it as being produced or sold exclusively by that company.

Translocation — Transfer of food or other materials such as herbicides from one plant part to another.

Trivial name — Name in general or common-place usage; for example, nicotine.

Ultra low volume (ULV) — Sprays that are applied at 0.5 gallon or less per acre or sprays applied as the undiluted formulation.

Vector — An organism, as an insect, that transmits pathogens to plants or animals.

Vermin — Pests, usually rats, mice, or insects.

Viricide — A substance that inactivates a virus completely and permanently.

Virustatic — Prevents the multiplication of a virus.

Volatilize — To vaporize.

Weed — Plant growing where it is not desired.

Wettable powder — Pesticide formulation of toxicant mixed with inert dust and a wetting agent which mixes readily with water and forms a short-term suspension (requires tank agitation).

Wetting agent — Compound that causes spray solutions to contact plant surfaces more thoroughly.

BIBLIOGRAPHY

Ashton, Floyd M. and Alden S. Crafts. 1973. Mode of Action of Herbicides. John Wiley & Sons, New York. 504 p.

Baily, J. B. and J. E. Swift. 1968. Pesticide Information and Safety Manual. Univ. Calif. Agri. Extension Serv., Berkeley. 147 p.

Bjornson, B. F., H. D. Pratt and K. S. Littig. 1972. Control of Domestic Rats and Mice. DHEW Publication No. (HSM) 72-8141. U. S. Dept. of Health, Ed., and Welfare, Health Services and Mental Health Admin., Center for Disease Control, Atlanta, GA 30333.

Borror, D. J., D. M. DeLong and C. A. Triplehorn. 1976. An Introduction to The Study of Insects. 4th Ed. Holt, Rinehart and Winston, New York. 852 p.

Brett, Charles H., K. A. Sorenson, and H. E. Scott. 1975. Vegetable Insect Control Using Less Insecticide. Circular 596. The North Carolina Agricultural Extension Service, North Carolina State University, Raleigh.

Burgerjon, A., and D. Martowut. 1971. Determination and significance of the host spectrum of *Bacillus thuringiensis in* Microbial Control of Insects and Mites. Edited by H. D. Burges and N. W. Hussey. Academic Press, New York, N.Y. 861 p. Ref. p. 305-325.

Burges, H. D. and N. W. Hussey, (Eds.) 1971. Microbial Control of Insects and Mites. Academic Press, New York. 861 p.

Carruth, L. A. 1975. Vegetable Garden Pests. Bull. A-81. Cooperative Extension Service, The University of Arizona, Tucson, AZ 85721.

Carruth, L. A. and G. S. Olton. 1976. Household Pests. Bull. A-72. Cooperative Extension Service, The University of Arizona, Tucson, AZ 85721.

Corbett, J. R. 1974. The Biochemical Mode of Action of Pesticides. Academic Press, New York. 330 p.

Davies, John E. 1976. Pesticide Protection. A Training Manual for Health Personnel. Univ. of Miami School of Medicine. Miami, FL. 71 p.

Erwin, Donald C. 1973. Systemic Fungicides: Disease control, translocation, and mode of action. Ann. Rev. Phytopath. 11: 389-422.

Falcon, L. A. 1971. Microbial control as a tool in integrated control programs. P. 346-364 in Biological Control, (Ed.) C. B. Huffaker. Plenum Press, New York, London.

Fichter, George S. 1966. Insect Pests. A Golden Guide. Golden Press. New York, N.Y. 160 p.

Fowler, D. Lee and John N. Mahan. 1980. The Pesticide Review 1978. U.S. Dept. Agr., Agr. Stabilization and Conserv. Serv., Wash., D.C. 42 p.

Goldstein, Jerome. 1978. The Least is Best Pesticide Strategy. JG Press, Emmaus, Pa. 205 p.

Good, Heidi B. and Dan M. Johnson. 1978. Nonlethal Blackbird Roost Control. Pest Control. 46(9):14-18.

Gray, F. A. and C. M. Sacamano. 1976. Diseases of Landscape Trees in Southern Arizona. Q-23. Cooperative Extension Service, The University of Arizona, Tucson, AZ 85721.

Hadden, Charles. 1972. Turf Diseases and Their Control. EC-808. Agricultural Extension Service, University of Tennessee, Knoxville 37901.

Hayes, Wayland J., Jr. 1975. Toxicology of Pesticides. Williams and Wilkins Co., Baltimore. 580 p.

Heathman, S. E. and K. C. Hamilton. 1978. Control weeds in urban areas. Publication Q-364. Coop. Extension Service, University of Arizona, Tucson, AZ. 5 p.

Heathman, S. E. and K. C. Hamilton. 1978. The care and weeding of desert landscape and rock-covered areas. Publication Q-349. Coop. Extension Service, University of Arizona, Tucson, AZ. 4 p.

Heimpel, A. M. 1967. A Critical Review of *Bacillus thuringiensis* and other crystalliferous bacteria. Annual Review of Entomology 12:287-322.

Hine, R. B., A. W. Johnson and L. F. True. 1975. Nematode Control in the Home Yard. Q-169. Cooperative Extension Service, The University of Arizona, Tucson, AZ 85721.

Janson, B. F. and R. L. Miller. 1977. Backyard Fruit Sprays for Insects and Diseases. Bull. L-1. Cooperative Extension Service, The Ohio State University, Columbus.

Johnson, W. T., J. E. Dewey, and H. J. Kastl. 1972. A Guide To Safe Pest Control Around The Home. Cornell Miscellaneous Bull. 74. Cooperative Extension Service, Cornell University, Ithaca.

Juska, F. V. and J. J. Murray. 1973. Lawn Diseases—How to Control Them. U.S. Dept. Agriculture, Home and Garden Bulletin No. 61. Superintendent of Documents, U.S. Government Printing Office, Washington, D.C. 20402.

Kenaga, E. E. and R. W. Morgan. 1978. Commercial and Experimental Organic Insecticides. Special Pub. 78-1, Entomol. Soc. Amer. 79 p.

Koval, C. F., G. C. Klingbeil and E. K. Wade. 1975. Grape Pest Control for Home Gardeners. Urban Phytonarian Series, A2129. University of Wisconsin-Extension. Madison.

Koval, C. F., G. C. Klingbeil and E. K. Wade. 1975. Strawberry Pest Control for Home Gardeners. Urban Phytonarian Series, A2127. University of Wisconsin-Extension. Madison.

Koval, C. F., G. C. Klingbeil and E. K. Wade. 1977. Plum, Cherry, and Peach Pest Control for Home Gardeners. Urban Phytonarian Series, A2130. Unicersity of Wisconsin-Extension. Madison.

Koval, C. F. and E. K. Wade. 1975. Apple Pest Control for Home Gardeners. University of Wisconsin-Extension. Madison.

Levi, Herbert W. and Lorna R. Levi. 1968. A Guide to Spiders and Their Kin. A Golden Guide. Golden Press, New York, N.Y. 160 p.

Lewis, K. R. and H. A. Turney. 1976. Insect Controls for Organic Gardeners. MP-1284. Texas Agricultural Extension Service, The Texas A&M University System, College Station.

Libby, J. L. 1977. Insect Control in the Home Vegetable Garden. Urban Phytonarian Series, A2088, University of Wisconsin-Extension. Madison.

Luckens, R. J. 1971. Chemistry of Fungicidal Action. Molecular Biology, Biochemistry and Biophysics Series, No. 10. Springer-Verlag. New York. 136 p.

Lyon, W. F. 1978. Safe Pesticides for Pets. Bull. 586. Cooperative Extension Service, The Ohio State University, Columbus.

Lyon, W. F. 1979. Pesticides for Household Pests. Control of Household Insect, Tick and Mite Pests. Bull. 512. Cooperative Extension Service, The Ohio State University, Columbus.

Lyon, W. F. 1979. Pesticides for Livestock and Farm Buildings. Bull. 473. Cooperative Extension Service, The Ohio State University, Columbus.

Marsh, R. E. and W. E. Howard. 1976. House Mouse Control Manual. (in 4 parts). Pest Control Magazine, August through November.

Martin, D. P., R. L. Miller, and J. D. Farley (Eds.). 1977. Control of Turfgrass Pests. Bull. L-187. Cooperative Extension Service, The Ohio State University, Columbus.

McCain, A. H. 1970. Chemicals for Plant Disease Control (Fungicides, Nematicides, Bactericides). Agri. Extension, Univ. Calif., Berkeley, 213 p.

Meehan, A. P. 1975. The Rodenticidal Activity of Some Cardiac Glycosides. International Pest Control, May/June.

Meister, R. T. (Ed.) 1980. Farm Chemicals Handbook. Meister Publishing Co., Willoughby, OH.

Miller, R. L. (Ed.). 1979. Home Vegetable Garden Insect Control. Bull. 498. Cooperative Extension Service, The Ohio State University, Columbus.

Miller, R. L. (Ed.). 1979. Insect and Mite Control on Ornamentals. Bull. 504. Cooperative Extension Service, The Ohio State University, Columbus.

Nutting, W. L. 1971. Termite Control for Homeowners. Intermountain Regional Publication 7. Cooperative Extension Service, The University of Arizona, Tucson, AZ 85721.

O'Brien, R. D. 1967. Insecticides Action and Metabolism. Academic Press, New York. 332 p.

Orlob, G. B. 1973. Ancient and Medieval Plant Pathology. Pflanzenschutz-Nachrichten. 26(2). Farbenfabriken Bayer GMBH, Leverkusen, Germany.

Pelczar, M. J., Jr. and Roger D. Reid. 1972. Microbiology. 3rd Ed. McGraw Hill Book Co., New York. 948 p.

Pratt, H. D. and K. S. Littig. 1974. Insecticides for the Control of Insects of Public Health Importance. DHEW Publication No. (CDC) 74-8229. U.S. Dept. of Health, Education and Welfare. Public Health Service. Center for Disease Control, Atlanta, GA. 30333.

Pratt, H. D. and K. S. Littig. 1974. Insecticide Application Equipment for the Control of Insects of Public Health Importance. DHEW Publication No. (CDC) 74-8273. U.S. Dept. of Health, Education and Welfare, Public Health Service, Center for Disease Control, Atlanta, GA 30333.

Pratt, H. D. and K. S. Littig. 1974. Insecticides for the Control of Insects of Public Health Importance. DHEW Publication No. (CDC) 74-8229. U.S. Dept. of Health, Education and Welfare. Public Health Service. Center for Disease Control, Atlanta, GA. 30333.

Russell, T. E. 1973. Diseases of Arizona Turf. Q-98. Cooperative Extension Service, The University of Arizona, Tucson, AZ 85721.

Schwartz, P. H., Jr. 1975. Control of Insects on Deciduous Fruits and Tree Nuts in the Home Orchard — Without Insecticides. U.S. Dept. of Agriculture Home and Garden Bulletin No. 211. Superintendent of Documents, U.S. Government Printing Office, Washington, D.C. 20402.

Scott, H. G. 1972. Household and Stored-Food Insects of Public Health Importance and Their Control. DHEW Publication No. (HSM) 72-8122. U.S. Department of Health, Education and Welfare, Public Health Service, Center for Disease Control, Atlanta GA 30333.

Slife, F. W., K. P. Buchholtz, and Thor Kommedahl (Eds.). 1960. Weeds of the North Central States. North Central Regional Publication No. 36. Circular 718 of the University of Illinois Agricultural Experiment Station, Urbana.

Spencer, E. Y. 1973. Guide to the Chemicals Used in Crop Protection. Pub. 1093, 6th ed., Information Canada, Ottawa, Ontario, Canada. 542 p.

Swan, D. G. 1978. Principles of Weed Control. Publication EB-698. Coop. Extension Service, Washington State University, Pullman. 16 p.

Thomson, W. T. 1979. Agricultural Chemicals. Book II, Herbicides. Thomson Publications, Fresno, CA 254 p.

Thomson, W. T. 1980 Agricultural Chemicals. Book III, Fumigants, Growth Regulators, Repellents, and Rodenticides. Thomson Publications, Fresno, CA 180 p.

Torgeson, DeWayne, C. (Ed.) 1967. Fungicides, an Advanced Treatise. Vol. 1. Agricultural and Industrial Applications, Environmental Interactions. Academic Press, New York. 697 p.

Torgeson, E. Y. (Ed.) 1969. Fungicides, an Advanced Treatise. Vol. 2. Chemistry and Physiology. Academic Press, New York. 742 p.

Univ. of California. 1972. Insects, Mites and Other Invertebrates and Their Control in California, Study Guide for Agri. Pest Control Advisers Agri. Publ., Univ. Calif., Berkeley. 138 p.

Univ. of California. 1972. Nematodes and Nematicides, Study Guide for Agri. Pest Control Advisers. Agri. Publ., Univ. Calif., Berkeley. 53 p.

Univ. of California. 1972. Plant Diseases, Study Guide for Agri. Pest Control Advisers. Agri. Publ., Univ. Calif. Berkeley. 232 p.

Univ. of California. 1972. Vertabrate Pests, Study Guide for Agri. Pest Control Advisers. Agri. Publ., Univ. Calif., Berkeley. 125 p.

Univ. of California. 1972. Weed Control, Study Guide for Agri. Pest Control Advisers. Agri. Publ., Univ. Calif., Berkeley. 64 p.

U.S. Dept. of Agriculture. 1976. Controlling Chiggers. Home and Garden Bulletin No. 137. Superintendent of Documents, U.S. Government Printing Office, Washington, D.C. 20402.

U.S. Dept. of Agriculture. 1976. Controlling Household Pests. Home and Garden Bulletin No. 96. Superintendent of Documents, U.S. Government Printing Office, Washington, D.C. 20402.

U.S. Environmental Protection Agency. 1975. EPA Compendium of Registered Pesticides; Vol. 1. Herbicides and Plant Regulators; 2. Fungicides and Nematicides; 3. Insecticides, Acaricides, Molluscicides and Anti-fouling Compounds; 4. Rodenticides and Mammal, Bird and Fish Toxicants; 5. Disinfectants. Tech. Serv. Div., Office of Pesticide Programs. Washington, D.C.

Ware, George W. Sr. and J. P. McCollum. 1980. Producing Vegetable Crops. 3rd ed. Interstate Printers and Publishers, Inc. Danville, IL. 607 p.

Ware, George W. 1975. Pesticides — An Auto-Tutorial Approach. W. H. Freeman and Company, San Francisco. 205 p.

Ware, George W. 1978. The Pesticide Book. W. H. Freeman and Company, San Francisco. 196 p.

Watson, T. F., Leon Moore, and G. W. Ware. 1976. Practical Insect Pest Management. W. H. Freeman and Company, San Francisco. 209 p.

Webb, Ralph. 1977. Insects and Related Pests of House Plants. U.S. Dept. Agri. Home and Garden Bulletin No. 67. Sup't. of Documents, U.S. Gov't. Printing Office, Washington, D.C. 20402. 14 p.

Weed Science Society of America. 1979. Herbicide Handbook, 4th Ed. Champaign, Illinois. 480 p.

Weidhaas, D. E. and G. S. Burden. 1978. Ants in The Home and Garden: How to Control Them. U.S. Dept. Agri. Home and Garden Bulletin No. 28. Sup't. of Documents, U.S. Gov't. Printing Office, Washington, D.C. 20402. 10 p.

Westreich, George A. and Max D. Lechtman. 1973. Microbiology and Human Disease. Glencoe Press, New York. 814 p.

Wilkinson, C. F. 1976. Insecticide Biochemistry and Physiology. Plenum Press, New York, London. 768 p.

Wiswesser, W. J. (Ed.) 1976. Pesticide Index, 5th Edition. Entomological Society of America. College Park, MD. 328 p.

ORGANIC GARDENING AND NATURAL PEST CONTROL

Alth, M. 1977. How to Farm Your Backyard the Mulch-Organic Way. McGraw-Hill Book Co., New York, NY 10010. 272 p.

Edinger, Philip, (Ed.) 1971. Sunset Guide to Organic Gardening. Lane Books. Menlo Park, CA. 72 p.

Hunter, Beatrice T. 1971. Gardening Without Poisons. 2nd ed. Houghton Mifflin Co., Boston, MA. 318 p.

Kramer, Jack. 1972. The Natural Way to Pest-Free Gardening. Charles Scribner's Sons. New York, NY. 118 p.

Moran, Clara S. and David O. Percy. 1978. A Companion Planting Dictionary. National Colonial Farm, Accokeek Foundation, Inc., Accokeek, MD 20607. 51 p.

Olkowski, Helga. 1971. Common Sense Pest Control. Consumers Cooperative of Berkeley, Inc. 4805 Central Ave., Richmond, CA 94804.

Philbrick, Helen, and Richard B. Gregg. 1966. Companion Plants and How to Use Them. Devin-Adair Co., 23 E. 26th St., New York, NY 10010. 111 p.

Philbrick, Helen and John Philbrick. 1974. The Bug Book — Harmless Insect Controls. Garden Way Publishing, Charlotte, VT. 126 p.

Riotte, Louise. 1975. Secrets of Companion Planting for Successful Gardening. Garden Way Publishing, Charlotte, VT 05445. 226 p.

The Staff of Organic Gardening and Farming. 1973. Getting the Bugs Out of Organic Gardening. Rodale Press Books, Inc., Emmaus, PA 18049.

The Staff of Organic Gardening and Farming. 1976. The Organic Way to Plant Protection. Rodale Press Books, Inc., Emmaus, PA 18049.

The Staff of Organic Gardening and Farming. 1977. Organoculture. Rodale Press Books, Inc., Emmaus, PA 18049.

Tyler, Hamilton A. 1970. Organic Gardening Without Poisons. Van Nostrand Reinhold, New York, NY. 111 p.

U.S. Dept. of Agriculture. 1971. Mulches for Your Garden. Home and Garden Bulletin No. 185. Superintendent of Documents, U.S. Government Printing Office, Washington, D.C. 20402.

Westcott, Cynthia. 1971. Plant Disease Handbook. 3rd ed. Van Nostrand Reinhold. New York, NY. 843 p.

Westcott, Cynthia. 1973. The Gardener's Bug Book. 4th ed. Doubleday and Co., Inc., Garden City, NY. 689 p.

Yepsen, Roger B., Jr. 1976. Organic Plant Protection. Rodale Press, Inc. Emmaus, PA 18049. 688 p.

Zim, Herbert S. and Clarence Cottam. 1956. Insects. Golden Nature Guide. Simon and Schuster, New York, NY. 160 p.

Index